EU Climate Change Policy

NEW HORIZONS IN ENVIRONMENTAL LAW

Series Editors: Kurt Deketelaere, *Professor of Law and Director, Institute of Environmental and Energy Law, University of Leuven, Belgium* and Zen Makuch, *Department of Environmental Science and Technology, Imperial College, London, UK.*

Environmental law is an increasingly important area of legal research. Given the increasingly interdependent web of global society and the significant steps being made towards environmental democracy in decision-making processes, there are few people that are untouched by environmental law-making processes.

At the same time, environmental law is at a crossroads. The command and control methodology that evolved in the 1960s and 1970s for air, land and water protection may have reached the limit of its environmental protection achievements. New life needs to be injected into our environmental protection regimes. This new series seeks to press forward the boundaries of environmental law through innovative research into environmental protection standards, procedures, alternative instruments and case law. Adopting a wide interpretation of environmental law, it will include contributions from both leading and emerging European and international scholars.

Titles in the series include:

Whaling Diplomacy
Defining Issues in International Environmental Law
Alexander Gillespie

EU Climate Change Policy
The Challenge of New Regulatory Initiatives
Edited by Marjan Peeters and Kurt Deketelaere

EU Climate Change Policy

The Challenge of New Regulatory Initiatives

Edited by

Marjan Peeters

*Institute for Transnational Legal Research – METRO,
University of Maastricht, The Netherlands*

and

Kurt Deketelaere

*Professor of Law and Director,
Institute of Environmental and Energy Law,
University of Leuven, Belgium*

NEW HORIZONS IN ENVIRONMENTAL LAW

Edward Elgar
Cheltenham, UK • Northampton, MA, USA

Published by
Edward Elgar Publishing Limited
Glensanda House
Montpellier Parade
Cheltenham
Glos GL50 1UA
UK

Edward Elgar Publishing, Inc.
136 West Street
Suite 202
Northampton
Massachusetts 01060
USA

A catalogue record for this book is available from the British Library

Library of Congress Cataloguing in Publication Data
EU climate change policy : the challenge of new regulatory initiatives / edited by Marjan Peeters and Kurt Deketelaere.
     p. cm. — (New horizons in environmental law series)
  Includes bibliographical references and index.
  1. Environmental law—European Union countries. 2. Global warming—Law and legislation. I. Peeters, Marjan. II. Deketelaere, K. (Kurt) III. Series.
  KJE6242.E9 2006
  344.2404'6—dc22

                             2006040018

ISBN-13: 978 1 84542 605 7
ISBN-10: 1 84542 605 3

Typeset by Manton Typesetters, Louth, Lincolnshire, UK
Printed and bound in Great Britain by MPG Books Ltd, Bodmin, Cornwall

Contents

v

Contributors

Véronique Bruggeman
Faculty of Law, Maastricht University, The Netherlands

Javier de Cendra de Larragán
Maastricht University, The Netherlands

Kurt Deketelaere
University of Leuven, Belgium

Bram Delvaux
University of Leuven, Belgium

Claudia Dias Soares
Portuguese Catholic University (Oporto), Portugal and London School of Economics, UK

Wybe Th. Douma
TMC Asser Institute, The Netherlands

Mar Campins Eritja
University of Barcelona, Spain

Joyeeta Gupta
UNESCO-IHE Institute Delft and Free University of Amsterdam, The Netherlands

Ludwig Krämer
University of Bremen, Germany

Karen MacDonald
Imperial College London, UK

Zen Makuch
Imperial College London, UK

Birgitte Egelund Olsen
Aarhus School of Business, Denmark

Marc Pallemaerts
Vrije Universiteit Brussel and Université Libre de Bruxelles, Belgium

Marjan Peeters
Maastricht University, The Netherlands

George (Rock) Pring
University of Denver, USA

Manfred Rosenstock
European Commission, Belgium

Bettina Schmitt-Rady
Heiermann, Franke, Knipp, Germany

Geert van Calster
University of Leuven, Belgium

Rhiannon Williams
Institute for European Studies, Vrije Universiteit Brussel, Belgium

PART I

Introduction

1. Key challenges of EU climate change policy: Competences, measures and compliance

Kurt Deketelaere and Marjan Peeters[1]

1. INTRODUCTION

This book explores and comments on, from a legal perspective, the regulatory framework the European Union (EU) has established to meet the international greenhouse gas emission reduction obligation it has taken up in Kyoto and to anticipate further emission reductions after 2012.[2]

We start from the perspective that legally binding obligations, which can be imposed through several forms of regulatory instruments, are needed in order to steer society towards low-carbon behaviour. The market, as such, does not provide for a reduction of greenhouse gas emissions, and the liability regimes in place, thus far, do not include important incentives to avoid excessive greenhouse gas emissions. Governmental action by meaningful policies is inevitably needed. Specifically for the climate change problem, economic instruments are expected to be the most suitable ones, although complementary approaches are also needed.[3] Moreover, perfect regulatory solutions probably do not exist for the complex climate change problem, and theory still develops different ideas for the best design and the best mix of instruments for climate change policies.[4] Furthermore, the regulatory approach which is developed in the real world is of course highly influenced by the political process, which often leads to a flawed and/or confusing approach.

As a consequence of a number of circumstances, among which is the refusal of the USA to ratify the Kyoto Protocol, the EU has become a global green leader in climate change policies.[5] However, are its current internal measures (such as the EU emissions trading scheme) a meaningful and sound approach? If a (and for the moment very unlikely to happen) 'trading up' effect might occur towards the EU's trading partners – stimulated through a serious climate change approach by the EU – it is important that the current measures are real and are not to be qualified as a kind of window dressing.[6] Within this respect, it is of utmost importance that the adopted regulatory arrangements within the EU are and will be implemented, executed and complied with at best in practice. Adopting an approach is only a first step, but ensuring that the effects take place in practice is the second and even more difficult step to address.

For the short term, the main challenge for EU climate change policy is to comply with the international obligation to reduce emissions during the period 2008–12 as this is concluded in the Kyoto Protocol. The European Community (EC) itself has an emission reduction commitment of –8% compared to the reference years 1990 and 1995. This emission reduction commitment is related to the group of 15 Member States who joined the EU before 1 May 2004 (the so-called 'EU-15'). Within the so-called EC Burden Sharing Agreement, the emission reduction commitments for these Member States, as stipulated in the Kyoto Protocol, are reallocated; each Member State has got a new emission reduction commitment which differs from the one in the Kyoto Protocol. For the Member States who joined in May 2004, the emission reduction commitments apply as agreed upon in the Kyoto Protocol. For Malta and Cyprus, the Kyoto Protocol does not mention any emission reduction commitment.

The EU has developed a mix of regulatory approaches in order to contribute to those targets, and a lot of discretion to set up climate change policies is left to the Member States. This already makes it hard to assess the real content of the climate change policies within the EU: we do have in fact one global problem, but 26 climate change policy approaches, ie one at the EU-level, and 25 Member State policies. However, as climate change is a transboundary environmental problem, with high consequences for the functioning of the internal market, there is a strong reason to develop a meaningful EU climate change policy.

This book argues that the current approach like emissions trading and energy taxation is not yet meaningful enough because of the fact that they, thus far, do not lead to impressive reductions. Nevertheless, these seeds, when carefully and seriously nurtured, may lead to a fruitful outcome.

For the longer term, after 2012, it is still very uncertain how or what kind of international agreements regarding climate change will be concluded. Of course there is the United Nations Framework Convention on Climate Change (UNFCCC), but experience shows that there are severe difficulties for the Treaty Parties to reach agreement on emissions reductions to be realized after 2012. The USA/Asia-Pacific Agreement, which was concluded in 2005, leads even to the assumption that a common global approach under the umbrella of the UNFCCC is almost impossible.

However, within the EU, some exploration of the necessary post-2012 policy options, including emission reductions, has been taking place. The 2002 Sixth Environmental Action Programme indicates climate change as a priority of the Community environmental policy.[7] This means that the temperature rise worldwide should not exceed 2°C compared with the pre-industrial era, and that the

CO_2 concentration should stay below 550ppm.[8] In the longer term, this means a reduction of 70% compared with the emissions of 1990.[9,10] Whether within the EU the ambitiously sounding targets of the Sixth Environmental Action Programme will be pursued in the future and whether there will be a readiness to take further measures than other UNFCCC Parties is hard to predict. In the spring of 2005, the meeting of the Council of the EU made it clear that the EU does not want to take responsibility alone. The meeting resulted in some, but not too promising, pronouncements about the future development of climate change policies, among which was the formulation of a specific target to be pursued: developed countries – meaning not the EU Member States alone – would have to set themselves a target of an emission reduction of 15–30% in 2020, compared with 1990. Just before this Council meeting, the Environmental Council indicated further that the developed countries would have to take the road to an emission reduction of 60–80% in 2050. The Council did not give its opinion about so far ahead into the future, and did not go any further for the time being than indicating emission reduction percentages for the year 2020.

It is uncertain whether the political process of the EU will indeed stay adhered to the forecasted emission reduction targets when other big world players would not be prepared to take effective measures.[11] The global context will influence the content and seriousness of the internal regulatory measures taken within the EU – and it is not clear whether the accidentally short-term European 'green leadership' will be prolonged into a long-term leadership.

However, let us turn now to the legal instruments for climate change policy which currently have been adopted on the basis of the EC Treaty. This book discusses secondary EC-legislation covering greenhouse gas emissions and energy use. Remarkably, the approach towards the climate change problem rests mainly on a regulatory concept that is really new for the EU, as it is using emissions trading and taxation to induce society to decrease greenhouse gas emissions. The main European climate change policy instrument, the Greenhouse Gas Emissions Trading Directive (the ET Directive) has led to an EU-wide market for greenhouse gas allowances. This innovative instrument – which simultaneously raises a lot of legal questions – is expected to contribute towards a large part of the emissions reductions for which the EU Member States and the EC are legally responsible. However, there is more: the ET Directive is also accompanied by directives on energy taxation, energy efficiency, the promotion of electricity produced from renewable energy sources, by voluntary agreements with the car-industry, and by a proposal for a regulation for certain fluorinated greenhouse gases. In fact, a broad and complex package of regulatory measures has now been established at the EU level, representing a mix of instruments, among which soft law approaches (like labelling, indicative non-binding targets, and voluntary agreements), market-based instruments (taxation and emissions trading) and traditional command and control regulation (the permit approach and the prescription of best available techniques through the Integrated Pollution Prevention and Control (IPPC) Directive). However, the policy regarding climate change within the EU is dependent on the interplay between the EU institutions and the Member States. Some of the EC measures, like the ET directive, leave a rather broad, maybe too broad discretion to the Member States to implement them, as is the case with the allocation of the greenhouse gas allowances. This multi-level approach makes the EU Climate Change Policy dossier, which is already hard to understand because of the diversity of measures undertaken at EC level, even more complex. The whole body of regulatory measures at EC level and at national level is of course meant to ensure that the internationally binding emission reduction targets will be complied with. From this perspective, the monitoring of the progress within the 25 Member States, and the enforcement actions by the Commission when measures and action are lacking, must be the ultimate part of the European climate change policies as well.

In this book, a group of distinguished authors discuss and comment upon the policy package as established at EC level, addressing from a legal point of view both theoretical and practical aspects. The overall finding is that the EC policy on climate change is still very 'green' and thus needs to be developed further in a comprehensive and meaningful way. We are indeed in a phase in which we need to learn about the usefulness of the instruments as adopted now, and try to improve them. With 'improvement' one should not only focus on another possible design of a specific instrument, but also on the way of implementing the regulatory framework in practice.

2.   OVERVIEW OF THE BOOK

Part I: Introduction

After this chapter, which introduces the book, the exploration of the real meaning of the regulatory measures within EU climate change policy starts with an extensive discussion of the international and European legal framework concerning the climate change problem. In Chapter 2, Marc Pallemaerts and Rhiannon Williams provide us with an historical overview of the development of the framework thus far. Key elements of the international framework are, of course, the UNFCCC, the Kyoto Protocol and the Bonn and Marrakesh Accords. The EU climate change policy can, of course, only be understood in the context of these legal agreements. Pallemaerts and Williams pay attention to the concept of 'differentiated responsibilities', and explain the specific responsibilities and commitments of the developed countries. Within this context, the provisions in the Kyoto Protocol regarding burden-sharing between parties are particularly interesting for the EC as such, and for the EU Member States. The Kyoto Protocol (art. 4) allows the EC and the Member States to agree internally on the reallocation of the emission reduction targets as concluded in the Kyoto Protocol. Consequently, the EC Burden Sharing Agreement includes different national targets, ranging from −21% (Germany and Luxembourg) to +27% (Portugal).

The historical overview shows that the EC and its Member States ratified, in 1993, the UNFCCC without having yet a concrete internal policy. Main initiatives were left to the Member States, to elaborate national emissions reduction programmes. Pallemaerts and Williams explain that, until 2001, the climate change policy within the EC was merely a collection of soft law measures. From 2001, some more serious measures were adopted, like the ET Directive, which presently can be seen as the main instrument, and, among others, Directive 2002/91/EC on the energy performance of buildings. Pallemaerts and Williams also show us in which policy context the legal measures are to be developed, like the European Council policy announcements in the Spring of 2005, following the 2004-presented Communication of the Commission with the title 'Winning the Battle against Climate Change'. They show that the European Parliament takes a more ambitious approach compared to the European Council, which in its Spring 2005 meeting refused to formulate a long-term target with regard to the reduction of greenhouse gases.

Chapter 3 provides us again with some historical insight, but now with regard to the difficult, sensitive and very important relationship between the EU and Russia regarding the ratification of the Kyoto Protocol. Building upon and improving the EU-Russia economic relationship, which is formalized in the 'Partnership and Co-operation Agreement' from 1994, was of crucial importance for the ratification of the Kyoto Protocol (KP) by Russia. The main idea is to develop a common European economic and sustainable space. However, Russia took rather a long time before it eventually decided to ratify the KP. It is well-known that the EU took a very active role in convincing Russia to ratify the Protocol, but that ratification was for Russia less attractive since the USA – the biggest big buyer – would not participate in the international emissions trading facilities. For Russia, the enlargement process of the EU was of bigger concern, as this would possibly lead to economic disadvantages (trade barriers for Russia to the new EU Member States), and as a consequence it demanded the relaxation of EU trade-related measures. The EU indeed dealt with the many trade-related concerns put forward by Russia, and extended the Partnership and Cooperation Agreement (PCA). After all, the author, Wybe Douma, argues that there indeed was a link between the agreements on PCA extension and the negotiations of the World Trade Organization (WTO) accession of Russia, and – finally – the Kyoto ratification, although this relationship is officially denied by the Russian leader. Douma ends with an outlook on the future cooperation between Russia and the EU towards climate change, discussing the possibility of establishing Joint Implementation (JI) projects, for which many hurdles are still on the way. Russia's monitoring systems and inventory procedures for greenhouse gases still need, for example, to be improved, being a condition for conducting international emissions trading. In fact, the EU is supporting Russia financially to provide those necessary institutional provisions. Another concern is the so-called 'hot air trading' that could occur between Russia and the EU. However, this remarkable possibility was taken into account when agreement took place on the Kyoto Protocol.[12] Hot air trading is, in this respect, just part of the current international climate change deal, and could be used by the EU, although it is a remarkable and contestable part of international emissions trading.

Part II: Greenhouse Gas Emissions Trading Within the EU

Within Part II, the focus is turned towards the parade horse of the EC climate policy package, which is Directive 2003/87/EC of the European Parliament and of the Council of 13 October 2003 establishing a scheme for greenhouse gas emission allowance trading within the Community. In Chapter 4, Mar Campins Eritja reviews the challenging task faced by Member States in implementing this Directive, and discusses from this perspective issues of Member State liability. The timetable for implementation as included in the ET Directive was remarkably short: the Member States had to prepare their national legislation in order to implement the new regulatory instrument within only a couple of months after the adoption of the directive. It is clear that this was hardly achievable. Moreover, one of the core and sensitive elements of the emissions trading scheme, the development of National Allocation Plans (NAPs), had to be executed by Member States within a very tight schedule.

   Mar Campins Eritja explains more specifically the difficulties the Member States met in implementing the ET Directive and drawing up the NAPS. They had to do that under strong pressure from domestic industry, which caused most Member States allocated more emission allowances than would be desirable in order to reach the Kyoto emissions reduction target. Due to the short timetable and the complexities of the instrument, it is questionable whether all Member States were really ready in the year 2005 for the emissions trading scheme. This may even result in a distorted internal market. The Commission, however, lacks an efficient and effective instrument to stimulate and enforce compliance behaviour by Member States. Nevertheless, it is not excluded that infringement procedures will ultimately lead to high fines imposed by the European Court of Justice, although there is limited jurisprudence on the issue of penalizing Member States thus far. Reference is made to the last judgement of the Court of 12 July 2005, in which the Court explicitly refers to the fact that penalty payments and/or a lump sum are meant to place Member States under economic pressure in order to induce it to put an end to its non-compliance behaviour.

Another liability issue is the possibility for European citizens and companies to hold Member States responsible for damages following from incorrect implementation of the ET Directive. More specifically, the position of the industry covered by the ET Directive is discussed. When a Member State fails to establish the necessary provisions for emissions trading, interested industries would be denied the right and full access to the EU allowance market and would suffer economic damage. The specific liability regime and its specific criteria are discussed, which shows that such a liability is not out of expectation. However, the assessment of the specific conditions to vest liability will depend on a rather detailed case-by-case analysis, and can thus not be generally predicted. Besides the internal EU liability concerns, attention is paid shortly to the external front, where Mar Campins Eritja explains that the shared nature of powers conferred to the EU and its Member States determines the issues related to the rules of standing within the international climate change framework.

Chapter 5 extensively assesses the allocation process within the greenhouse gas emissions trading system. The author, Bettina Schmitt-Rady, starts from the point of view that in an EU emissions trading scheme a level playing field between the covered industries should be the case. She gives, from a practitioner's perspective, a view on the chosen method of allocation as included in the ET Directive, and from this angle she concentrates on the allocation of allowances as done in Germany for the first trading phase 2005–07. Many concerns are addressed, and above all, the author states that the allocation as conducted for the first phase does not contribute to a level playing field between covered industries. The ET Directive is too vague, and the Commission has given Member States considerable lee-way in designing the specific allocation criteria. Even administrative over-regulation has taken place, especially in Germany, where many different allocation rules are established. Schmitt-Rady pleads for an improvement of the definitions stipulating the coverage of the directive, points at the variations between NAPs with regard to the treatment of new entrants, reveals the protectionist approach taken by the government with regard to the cessation of activities, and, among other things, argues against the possibility of ex-post allocation.

Chapter 5 makes clear that the present allocation framework needs to be critically reviewed, in order to make it simpler and ensuring as much as possible equal treatment of the industries covered. Of course, there is somehow a trade-off between on the one hand using simple formula, and on the other, treating industries fairly.[13] Both goals can probably not be reached to the same extent. Equal treatment means that firms in equal positions need to be treated equally – but when their positions are not equal, differences should in principle be taken into account. Maybe it is too complicated to ensure this notion of 'equal treatment' of the European industry within a system of free allocation of allowances, and the option of auctioning may be better. This option should at least be considered for the long term, especially now that emissions trading is seen as the main instrument for climate change policies in the EU, and because auctioning is the main approach recommended by economic literature.[14]

More or less a year after its adoption the ET Directive was already amended by Directive 2004/101/EC linking the KP's project mechanisms to the EU emissions trading scheme. In Chapter 6, Javier de Cendra de Larragán discusses these complicated concepts. The EU's position towards the project mechanisms evolved from a sceptical position to, ultimately, an adoption of the above mentioned directive by which the project mechanism will become an option for European industry, as far as indeed the Member States will allow them to make use of this. The linking directive also gives much discretion to the Member States. Again it can be said that the EU has run very fast, maybe too fast, with establishing the linking directive, as many questions regarding the practical application of the concepts of Clean Development Mechanism (CDM) and JI still need to be answered. Here again the EU follows a 'learning by doing' exercise in implementing a regulatory instrument. But already some shortcomings of the directive can be identified, like the possibility for industry to undermine the additional qualitative requirements stipulated by a Member State for the engagement in CDM or JI projects. A more uniform approach on the admissibility of projects would, in this light, be recommendable. The conclusion is that both from a perspective of environmental effectiveness and competitiveness many concerns can still be raised, and that more harmonization would be necessary. Moreover, with regard to the considerations on the EU level or on the Member State level to allow for CDM or JI projects, the precautionary principle should be taken into full account.

Emissions trading is recommended for its market-based approach, through which – at least, in theory – a cost-effective environmental policy can be reached. However, how does this new instrument fit into the fundamental rights for the public with regard to environmental policy? This basic and important question is extensively addressed by Karen McDonald and Zen Makuch in Chapter 7. They examine the extent to which some central elements of the EU emissions trading scheme (EU ETS) comply with the objectives of the 1998 Aarhus Convention on Access to Information, Public Participation in Decision-making and Access to Justice in Environmental Matters; thereby asking themselves whether 'environmental democracy' is ensured within the greenhouse gas emissions trading regime. It is important to note that 'Aarhus rights and obligations' are conventionally inscribed in traditional command and control legislation. McDonald and Makuch investigate how the rights of the public as ensured by the Aarhus Convention are to be applied within an emissions trading scheme. They point to the feature of emissions trading, which is that governmental and public interest is concerned with the total amount of pollution emitted in a certain area and not with individual decisions related to the permitting of installations, as is the case in traditional command and control approaches. However, with emissions trading, the public should have a right to the relevant information, and more specifically to be able to participate in decision-making procedures regarding the issuing of a permit and the allocation of emissions allowances. The public should be informed, involved and included in the process, on the basis that a stable 'atmosphere' is a common concern. The ET Directive does attempt to provide for both access to information and public participation in the emissions trading scheme, but these provisions are not as broad in scope as that of the Aarhus Convention. For instance, in the absence of explicit provisions in the ET Directive on the procedure for public consultation on the development of a National Allocation Plan or the allocation of allowances, the Aarhus Convention requirements can come into play. They also point at the 'wealth of other legal instruments' that exist in relation to public participation and access to environmental information. These instruments will contribute to enhanced access to information and public participation in emissions trading procedures. In particular, by 2009 the obligation to establish pollutant release and transfer registers will lead to greater consistency and uniformity in emissions data. However, legal provisions towards the Aarhus pillars as such would not be enough: a public information campaign should be developed in order to stimulate stakeholder involvement in greenhouse gas emissions trading in what is otherwise a technical and inaccessible field.

With the ET Directive, not only the trade in greenhouse gas allowances has been introduced, but at the same time a new permit scheme has been established within EC environmental law: installations that want to emit greenhouse gases can only do that when a so-called greenhouse gas permit has been issued. In practice, the permit-obligation is thus far only established for CO_2 emissions. In Chapter 8 Birgitte Egelund Olsen discusses the relationship between this greenhouse gas permit and the so-called IPPC permit.[15] She holds the view that carbon dioxide is clearly covered by the IPPC Directive, and that the other greenhouse gas emissions are as well. She demonstrates how the IPPC permit scheme is linked to the greenhouse permit scheme, and, from this perspective, she investigates in what way greenhouse gas emissions that are not covered by the Emissions Trading Directive are regulated.

Before going into the more procedural and detailed provisions, Egelund Olsen analyzes the main differences between the regulatory approach as included in the IPPC Directive and in the ET Directive. Just because of the rather fundamental differences (best available techniques (BAT) versus freedom of decision to emit), one might worry how the interplay between both instruments might work. These permits are parallel but mutually exclusive schemes, each reflecting a very different approach to regulation. The procedures between these permits need to be co-ordinated, but Member States are not required to establish one integrated permit scheme, covering the IPPC permit and the greenhouse gas permit (indeed, also within the EC-secondary legislation these permits have not been integrated). However, the co-ordination of the procedures is difficult to execute, because the procedures for both permits differ. There is even a large potential for conflict, especially when one considers the fact that the introduction of the emission trading regime may be qualified as a taking, which leads to the question of compensation. However, as long as historical emissions are taken into account when allocating the greenhouse gas allowances, this problem might only be a theoretical one. On the other hand, granting allowances equal to the historical emissions could be in conflict with the polluter pays principle.[16]

In conclusion, the ET Directive is seen by Egelund Olsen as reflecting a new regulatory tendency in Community environmental law, in which the regulatory framework is less flexible and the culture, legal traditions and diversity of the Member States are not taken into account to the same degree. The functioning of an emissions trading scheme implies that more uniform criteria will have to be used both for setting the standards for permits and for monitoring emissions and the verification and certification of emission reports. However, with respect to the allocation of the allowances – which was not the focus of Egelund Olsen's contribution – it can, on the other hand, be seen that this is currently a highly decentralized topic, as was already explained by Bettina Schmitz-Rady, but she indeed pleads for more harmonization of the allocation. When such harmonization or even centralization would be put through, the role of the Member States would then indeed become – in line with the view of Egelund Olsen – not that of a regulator, but more that of an administrator and enforcer.

In Chapter 9, the enforcement of the EU ETS scheme is discussed by Marjan Peeters. With the notion 'enforcement' she refers to the whole package of provisions meant to ensure compliance by industry with the obligations included in the legal framework for emissions trading. She emphasizes the need to set up an effective enforcement framework, because within an emissions trading scheme it might be financially attractive for firms to behave illegally, thus avoiding buying (expensive) allowances. From this perspective, Peeters discusses the current provisions as included within the ET Directive, where the greenhouse gas permit forms a prominent enforcement tool. The monitoring and reporting provisions, which need to be included within the greenhouse gas permit, are already harmonized to a large extent through the monitoring and reporting guidelines. In this respect, legal disputes may be expected with regard to the specific formulation of the monitoring provisions within the permits.

Quite unusually for EC environmental law, the directive prescribes the Member States to use some specific sanctions, like naming and shaming and a fixed and automatic penalty. Thus, a quite large harmonization has taken place. The question arises whether these provisions are the best choice that could have been made, as several comments can be raised against the way in which they have now been presented within the directive. Moreover, it is very important that Member States do not rely exclusively on the sanctions as prescribed by the directive, but consider what additional enforcement competences/instruments will be needed, eg the competence to close down an installation when acting without a greenhouse gas permit. Of course, the proportionality thereof should be considered on a case-by-case basis.

Ultimately, much will depend on the practical control strategies to be developed and executed by the Member States. It should be of no surprise when it appears that Member States fill in the given framework for monitoring, reporting and verification with different levels of ambition. A basic question in this respect is whether the Member States will provide for enough financial capacity for executing the necessary control tasks, such as making regular on-site inspections by competent civil servants. The most important enforcement task is, thus, probably the task of the Commission to control the enforcement duties of the Member States, and to start an infringement procedure when necessary. Ultimately, it could be considered to provide the Commission itself with some inspection tasks.

Part II also entails a comparative analysis. George (Rock) Pring unfolds in Chapter 10 the successful experiences with the acid rain allowance trading programme that already runs in the USA since 1995. Although the existence and successful performance of the USA acid rain allowance trading programme influenced the embracement of the instrument by the EU, there are some striking differences to be recognized between the acid rain trading programme and the EU ETS. The specific characteristics of the EU legal system and of the climate change problem as such have influenced the design of the EU ETS, and whether the approach adopted within the EU will be as successful as the USA acid rain allowance trading programme, remains to be seen. The biggest operational dissimilarity is that, within the USA, the scheme is governed under one unitary decision-making system, operated by the central government, with one set of rules. The individual states included in the programme have no role in determining caps, distributing allowances, record-keeping rules, enforcement strategies, etc. This centralistic feature makes the US emissions trading programme for acid rain a much simpler system compared with the EU ETS, which takes a decentralized approach towards the allocation of allowances through NAPs, and regarding the execution of enforcement tasks. The decentralized enforcement means a risk for the use of instruments within the EU, as it can be expected that enforcement would vary considerably under a decentralized system. Moreover, Pring points out that regulating CO_2 through an emissions trading scheme is more complicated than regulating SO_2, because CO_2 is produced by a very large spectrum of very different sources (energy, minerals and metals, pulp, transportation, households, etc.), giving rise to objections of unequal treatment if more than half of the CO_2 emitters are exempted from the programme. Moreover, CO_2 is more difficult to control compared with SO_2.

Pring does not state that emissions trading should not be used for greenhouse gas trading, but argues that the striking differences of any emissions trading scheme compared with the acid rain trading programme should be identified and be taken into account in setting up and executing such a scheme. In fact, there is currently a rich learning phase with several greenhouse gas trading experiments: also within the US a variety of CO_2 emissions trading programmes at the state and local government level are being developed. In addition, there are also voluntary programmes in the private sector. Ultimately, emissions trading is seen as a promising instrument, and it has indeed been successful in practice in the context of acid rain policies. However, Pring has shown that within that scheme the central government set strong long-term caps, a feature which still is lacking within the EU ETS.

It is hard to say how strong the European politicians will be in establishing long-term pathways, thereby it is difficult to prove that strong leadership would eventually lead to a trading up effect.[17] However, it can easily be assumed that the EU institutions are not ready to bring their economy into a too disadvantageous position (if this effect would ever happen) by imposing climate change measures, where other big world players would not be prepared to do so. At least European industry will ask their political leaders to prevent such a situation – although the negative effects of climate change measures on the European economy are, in fact, not so clear or not too impressive.

It is well-known that imposing trade barriers is a delicate and complicated topic, as discussed in Chapter 11 by Geert van Calster. His specific focus lies on climate change levies or taxes. However, before analyzing the possibilities of this potential instrument, he first relates the EU ETS to international trade rules. For instance, he points out that a too restricted access for new entrants from abroad would lead to an international trade concern. This is a point that needs to be taken into account in bringing forward the emissions trading regime.

Van Calster discusses several forms of carbon levies or taxes, eg imposing energy taxation upon imported products. In principle, environmental considerations might justify the use of a border tax adjustment on imports. However, in the author's view there is no room for so-called 'Kyoto' or 'climate change' levies, which simply levy a charge on products originating in States which do not join the Kyoto Protocol. This levy is even unworkable, as too many questions need to be answered for making it compatible under WTO law. For instance, one would have to put into place a regime which would not impose the climate levy on States which, even without adhering to the Protocol, put into place measures which with equal effect as the Kyoto Protocol, tackle climate change. Within this context, the effectiveness of the measures taken by the Kyoto Protocol Parties must be assessed, in what is already a sensitive point. When a climate tax is directed at a number of emerging economies, a lot of question marks and legal arguments will be raised: emerging economies, for instance, may even cite the principle of common but differentiated responsibilities, so as not to have to be subjected to a similar tax as the USA. Nevertheless, the imposition of border tax adjustments is believed to be one possible 'stick' to get the USA on the road towards meaningful action, as 'carrots' would not seduce this powerful state. The precise possibilities of border tax adjustments through import taxes or even export subsidies should, therefore, be investigated further, and subsequently some well prepared experiment on a limited scale, e.g. for a specific product, should be undertaken in practice.[18] With this, the green leadership role would indeed be reinforced and strengthened.

Part III: Energy and Climate Change Measures

The implementation of the UNFCCC and the Kyoto Protocol has a fundamental influence on EU energy policy and legislation. An overview of this impact in its different aspects is given in Chapter 12 by Véronique Bruggeman and Bram Delvaux, where they present an extensive discussion of the current EU regulatory measures. The 'EU energy-package' is diverse, using different regulatory approaches with regard to, on the one hand, the supply and, on the other hand, the demand of energy. The content of the measures is more than once 'soft' in its nature, like the 'indicative targets' the Member States have to set according to the Directive on the promotion of electricity from renewable energy sources. Also, the Directive on the promotion of the use of biofuels entails this instrument of indicative targets. The legal nature, and therefore the possibility of enforcing these targets, is questionable, and therefore not promising. Soft approaches are also part of the energy-efficiency policy: at present the political process of the proposal of the Commission for a Directive on energy end-use efficiency and energy services has evolved into a watered-down approach with a non-binding indicative target to increase energy-efficiency by 6% over 8 years, with no separate public sector target, allowing the use of voluntary agreements as instruments for achieving the indicative targets. Besides these indicative targets, labelling is a method used in EU energy measures, e.g. Directive 92/75/EC on the labelling of household appliances. As a continuation of this Household Labelling Directive, various daughter Directives for specific product groups have been drawn up, e.g. for household electric refrigerators, freezers and their combinations, washing machines, ovens, household lamps, and so on. The provision of accurate, relevant and comparable information on the specific energy consumption of household appliances may, in the authors' opinion, influence the public's choice in favour of those appliances which consume less energy, thus indirectly encouraging the efficient use of these appliances. However, although the fact that this measure will play a more important role as the energy prices will keep on increasing, one can doubt if this information will give sufficient stimulus to potential buyers. The authors plead for more action to be taken, especially by the Member States, in order to accelerate growth in the use of renewable energy. Moreover, in the field of energy-efficiency a lot of greenhouse gas emissions could be reduced as well, when the politicians are ready to take immediate and effective action.

It is already said that within climate change policies the market-based approach is prominent. Within the EU an energy taxation directive has been adopted in 2004. This directive is discussed in two contributions: the first one starts from an economic analysis, while the other gives a deep and critical legal analysis of the directive, showing that improvements could be made.

The Chapter 13 contribution, written by Manfred Rosenstock, takes the economic perspective in analyzing the 2003 Energy Taxation Directive. He shows that the criticisms that were originally raised against the proposal, namely (1) that it would have a negative impact on the competitiveness of European industry, in particular the energy intensive part of it, and (2) that it would raise consumer prices, were not accepted by the Commission. However, given the concerns about competitiveness and also inflationary impacts, some Member States obtained transitional periods to adjust to higher tax rates for specific products. Also, the new Member States got transitional periods for specific products through the adoption of a separate directive. Furthermore, to address the concerns that higher tax rates could harm the international competitiveness of energy-intensive industries, the directive allows for tax reductions for energy-intensive firms.

Rosenstock welcomes the introduction of minimum rates on energy products from both an economic and an environmental point of view: on the one hand, this development reduces distortions between different energy products as these products are substitutes for each other, and on the other, it leads to progress towards the internalization of external costs – and thus most likely to a more efficient energy use and a reduction in emissions. Rosenstock is positive about the way in which the environmental state aid guidelines combined with the relevant provisions of the energy taxation directive have ensured that any derogations from taxes are designed in such a way as to minimize both distortions of competition and a loss of incentives for environmental protection.

In Chapter 14, Claudia Dias Soares conducts a critical legal assessment of the present EC-regime for energy taxation. She explains that within the EU the insertion of environmental concerns within energy policy still lies very much in the hands of the Member States. Furthermore, as an explicit energy chapter in the EU Treaty is still not established, there is a lack of explicit authority for EC standard setting in the energy field.

Nevertheless, Directive 2003/96/EC concerning the taxation of energy products and electricity represents, in her opinion, an important step towards a level playing field for renewable energy sources. However, it is still possible to improve the present legal framework. Some aspects of the legal framework set by Directive 2003/96/EC fall short of full compliance with the polluter pays principle, more specifically when there is an uneven distribution of emissions control costs. The Directive might allow some distortion between industries due to an asymmetric tax treatment of different energy sources not supported by internal market reasons, neither competitiveness arguments nor environmental concerns. On the one hand, the aluminium industry is favoured over the steel industry. On the other hand, nuclear power plants are more advantageous than power plants using either traditional energy sources or renewable energy sources.

The delicate issue of taxing kerosene is also analysed and it is explained that the aviation sector got a win-win situation by accepting to be submitted to the EU ETS (in order to avoid taxation), and thereby gaining at least for the moment an advantageous position compared with other greenhouse gas emitters.

Part IV: Good Governance for Climate Change: Reflections and Perspectives

Part IV begins with reflections on the current mix of EU instruments by Ludwig Krämer. He examines whether the Commission's announced fight against climate change has been followed up with serious measures. Krämer argues that the biggest problem is probably the fact that the EU lacks an adequate and coherent framework of competences to address the multifaceted climate change problem that is caused, for instance, by energy issues, transport issues, and environmental issues. The wording 'climate change' or any similar notion does not exist in the EC Treaty. Moreover, as stated before, the EU does not have explicit competences in the energy field, which means that the Member States are primarily responsible for energy policy questions. Krämer shows that there is in fact no EU-wide plan or programme on climate change – we only have a chapter within the 6th environmental action programme.

The political commitment of other EU administrative departments, other than the environmental one, is low. The different Commission administrations and the Council in its different formations had, and are probably still having their own agenda towards climate change. A meaningful European Climate Change Programme should be elaborated through art. 175(3), thus involving the European Parliament, but this has as such not yet occurred. Krämer also recommends the initiation of a European Treaty on alternative energies. Moreover, a strategic, visionary approach should be launched within the EU.

Despite all this, it can be concluded that a lot of instruments, representing different regulatory approaches, have been adopted. The EU managed largely to overcome the constitutional problem not having a general competence for an EU energy policy. However it is not yet certain that the measures will be effective enough; that will depend on their real application in practice. According to Krämer, the instrument mix is rather complete. Nevertheless, a lot of institutional recommendations can still be made, such as the establishment of a scientific advisory body on climate change, the appointment of a commissioner for climate change, and providing more budgets to stimulate alternative and clean energy. Moreover, a clear EU leadership on the global scene would even ask for a more serious approach through meaningful instruments within the EU. The measures undertaken are not promising enough when one realizes what necessary and dramatic changes in behaviour are needed in the next 50 years.

The contribution of Joyeeta Gupta, in Chapter 15, reflects on the dominant position of the rich countries – among EU Member States – within the international climate change negotiations. She argues that they do not take into account in a sufficient way the needs of the developing countries. Gupta starts from the point of view that a durable, global regime is needed that addresses both current as well as future aspects. She discusses the gap between the Europeans and the Americans – and explains that the developing countries are, meanwhile, worst hit. However, not these victims, but the USA and the EU are both very much in a position to protect their own interests – among which to present 'science' in a way that suits them best.

Furthermore, the climate change negotiations are so complex that it is very difficult for developing countries to represent themselves at best. Also, a large number of proposals on distribution of the climate change burden are very complex, and the implications for developing countries are not transparent. Gupta discusses problems of technology transfer to developing countries. Ultimately, she presents four criteria and two behavioural changes that need to be explored in developing a durable global regime on climate change, which respects the needs of the poorest countries. In conclusion, she argues for retrospection for the North, and proactivity for the South, or in other words, the developed and the developing world.

3.   CONCLUSION

The EU and its Member States have, in the short term, a very simple emissions reduction commitment, but it is for them a big challenge to establish and to execute a regulatory package that is sufficient for complying with the committed emissions reduction. Only with this outcome can a convincing role on the global scale be vested by the EU. Famous scholars like Stewart and Wiener have, in fact, stipulated the assumption or fear in the USA that international greenhouse gas limitations obligations would be more effectively implemented and enforced within the USA than would be the case in other countries or blocs like the EU.[19] It is up to the EU and its Member States to show that compliance with the international emissions reduction obligations is indeed not problematic. However, the EU is in fact disabled: there is no coherent framework of competences to address the complex climate change problem. Despite this, a complex regulatory package has been adopted, containing a lot of measures. Due to its complexity and diversity it is hard for citizens to understand what it really means, and whether it is equitable and effective. And, after all, the possibility for the Commission to enforce compliance by Member States is not a very effective one.

With respect to compliance, within the EU an important role is provided for by Decision 280/2004/EC of 11 February 2004, through which a mechanism is installed to acquire information about the (results of the) climate change policies and the level of compliance of each Member State with its emissions reduction target.[20] This decision replaces a decision of 1993, which set up a control system for the emission of CO_2 and other greenhouse gases in the Community.[21] An important target of that decision was to gain insight into the joint emissions and the possibilities to reach at least a stabilization of the greenhouse gas emissions. There was a need for a new decision because of the fact that the Kyoto Protocol and, in particular, the Marrakech agreements with its elaborated rules needed implementation. On the basis of the experiences gained with the former decision, Decision 280/2004 realizes a further harmonization of the reporting commitments. Through the new decision, moreover, additional information on the emission projections on Member State and Community level will be pursued.[22]

The monitoring system as included in Decision 280/2004/EC has become a large and complex administrative technical control structure.[23] The monitoring of the greenhouse gas emissions in the EU is especially complex because of the diversity of the regulatory interventions, the shared responsibility between the EU and its Member States for environmental policy issues, and the fact that the promulgated measures on Community level often imply a lot of discretion for the Member States.

The ultimate goal of the monitoring decision is to map the anthropogenic emissions in the Member States per source and removals per sink (e.g. through forestations) of greenhouse gases, so that it can be determined whether the emissions and the emission removals (sinks) correspond with the committed maximum emission budget of a Member State, and that of the EC and the EU-15 jointly. Member States shall report their emissions to the Commission by 15 January of each year on the next-to-last year (the report on 2005 therefore needs to be submitted by 15 January 2007 at the latest); by 15 March of each year at the latest, a complete national inventory report needs to be submitted.[24] The Commission shall annually report on the Community greenhouse gas inventory and submit it to the UNFCCC Secretariat by 15 April of each year.[25] Before doing so, a draft has to be submitted to the Member States by 28 February. The European Environmental Agency (EEA) has an important supportive role in this process.

One of the latest reports of the EEA, however, concluded that the EU is getting even further away from compliance with the emissions reduction commitments. The report shows that with the existing policies and measures, the EC and the EU-15 will not fulfil their commitments by the end of the first commitment period, and that the use of flexible instruments will be needed.[26]

This book argues that there is a case to improve the internal measures and institutions in order to address the climate change problem. There are a lot, and in fact too many, soft law approaches within the EU climate change package. The most promising instruments, which are emissions trading and taxation, are not yet applied in such a way that they lead to an impressive reduction of greenhouse gas emissions. However, the positive thing is that these market-based instruments are there, and although a lot of recommendations can be made for improving the design of these instruments, they certainly can contribute to a significant reduction of emissions when they would be applied in a meaningful way in practice.

NOTES

1. Kurt Deketelaere is Full Professor of Law, University of Leuven, Faculty of Law, Institute for Environmental and Energy Law. Marjan Peeters is Associate Professor of Transboundary Environmental Law, Maastricht University, Faculty of Law, Metro Institute.
2. The contributions to this book were finalized in September 2005.
3. Richardson, B.J. (2004), 'Climate law and economic policy instruments: a new field of environmental law', *Environmental Liability*, **12**(1), 19–32.
4. There is an abundant amount of literature on climate change instruments. The idea that no perfect regulatory approach exists is argued by Daniel H. Cole (2002), *Pollution & Property*, Cambridge, UK: Cambridge University Press.
5. Gupta, J., Grubb, M. (eds) (2000), *Climate Change and European Leadership: A Sustainable Role for Europe*, Dordrecht: Kluwer Academic Publishers.
6. Vogel, D. (1995), *Trading Up: Consumer and Environmental Regulation in a Global Economy*, Cambridge MA: Harvard University Press.
7. OJ L 242/1–15, 10.9.2002, Decision No. 1600/2002/EC of the European Parliament and the Council of 22 July 2002, art. 1(4)). This Sixth Environmental Action Programme runs until 22 July 2012.
8. Or, due to recent scientific insight, should even go further: European Environment Agency, *Climate Change and a European Low-carbon Energy System*, EEA Report no. 1/2005, p. 22.
9. Sixth Environmental Action Programme, article 2(2).
10. It is probably not exactly determinable whether this is the most correct interpretation of the UNFCCC stabilization target of art. 2. For instance, in Germany, a report was published which proposes 450ppm: EEA Report, *Impacts of Europe's Changing Climate*, Report no. 2/2004, p. 15.
11. However, a European leadership and measures at EU-level are, for instance, supported by the House of Lords of the UK: House of Lords, European Union Committee, 30th report of Session 2003–04, *The EU and Climate Change*, published on 10 November 2003, p. 19.
12. Woerdman, E. (2005), 'Hot Air Trading Under the Kyoto Protocol: An Environmental Problem or Not?', *European Environmental Law Review*, **14**(3), pp. 71–7.
13. Already recognized by the Commission in its Green Paper on greenhouse gas emissions trading within the European Union, COM(2000) 87 Final.

14. Already described by Dales, J.H. (1968; reprinted in 2002), *Pollution, Property & Prices*, Cheltenham UK/Northampton, MA, USA: Edward Elgar Publishing Ltd.

15. IPPC stands for Integrated Pollution and Prevention Control, and has been introduced by Directive 96/61/EC, PbL 302, 26 November 1996, p. 28.

16. Peeters, M. (1993), 'Towards a European System of Tradable Pollution Permits?', *Tilburg Foreign Law Review*, **2**(2), pp. 117–34; Peeters, M. (2003), 'Emissions trading as a new dimension to European Environmental Law: The Political Agreement of the European Council on Greenhouse Gas Allowance Trading', *European Environmental Law Review*, **12**(3), pp. 82–92.

17. David Vogel has argued that the trading up effect presumes a strong leadership: Vogel, D. (1995), *Trading Up: Consumer and Environmental Regulation in a Global Economy*, Cambridge MA: Harvard University Press.

18. This corresponds to one of the conclusions at the symposium 'Moving beyond 2012: International perspectives on Future Climate Change Policy Options', Institute for Environmental Studies, Free University of Amsterdam, 21 October 2005.

19. Stewart, R.B. and Wiener, J.B. (2003), *Reconstructing Climate Policy: Beyond Kyoto*, Washington DC, US: The AEI Press, p. 86.

20. Decision of 11 February 2004 PbL 049, 19 February 2004, pp. 1–8. The legal basis is art. 175 EC Treaty.

21. Decision no. 93/389/EEG, PbL 167, 9 July 1993 , pp. 31–3 Amended by decision no. 99/296 of 26 April 1999, PbL 117, 5 May 1999, pp. 35–8.

22. Proposal of the Commission of 5 February 2003, COM(2003) 51 final – OJ.

23. This image is confirmed by the subsequent decision of the Commission of 10 February 2005. PbL 55, 1 March 2005, pp. 57–91.

24. Decision no. 280/2004/EU, art. 3(1): see the elaborate enumeration of aspects which require reporting.

25. Decision no. 280/2004/EU, art. 4(1).

26. 'Greenhouse Gas Emission Trends and Projections in Europe 2004. Progress by the EU and its Member States towards achieving their Kyoto Protocol targets', EEA Report, No. 5/2004.

2. Climate change: The international and European policy framework

Marc Pallemaerts[1] and Rhiannon Williams[2]

1. THE PROCESS OF REGIME DEVELOPMENT

1.1 The Early Scientific and Political Debate in the 1970s and 1980s

The first scientific hypotheses suggesting that the concentration of carbon dioxide (CO_2) in the atmosphere might lead to global warming date from the end of the nineteenth century. The first systematic measurements of the average atmospheric concentration of carbon dioxide, confirming its increase, were made from the late 1950s. Meeting in Stockholm in June 1972, the United Nations (UN) Conference on the Human Environment adopted an action plan recommending, inter alia, the establishment of a network of stations to measure atmospheric pollution with a view to studying, over the long term, atmospheric variations likely to cause climate change.[3] In Geneva, in 1979, the first World Climate Conference, was organized by the World Meteorological Organization (WMO), which led to the launch, one year later, of the World Climate Research Programme (WCRP), co-ordinated by the WMO and the International Council of Scientific Unions (ICSU). The United Nations Environment Programme (UNEP) became involved with the work of the WCRP as a result of its work on environmental assessment. Nevertheless, in the early 1980s, the attention of UNEP and the policy makers was focused on the question of the protection of the ozone layer, climate change being seen as a more hypothetical problem, on which there was no scientific consensus. It was only from 1985 that the issue started to become a more mainstream political concern, as some measure of scientific consensus began to emerge.

In October 1985, the WMO, UNEP and ICSU held an international scientific meeting at Villach in Austria. The meeting concluded that the extent and speed of future climate change could be influenced by government policies. It therefore recommended consideration of the elaboration of a world treaty aimed at countering climate change.[4] Following this meeting, the WMO and UNEP began to collaborate on promoting the need for *intergovernmental* negotiations on climate change, moving the debate out of the purely scientific arena. Towards the end of the 1980s, the issues became increasingly politicized, with meetings moving through a politico-scientific phase towards full-fledged multilateral diplomacy.

1.2 From Science to Diplomacy: Towards Intergovernmental Negotiations

The Intergovernmental Panel on Climate Change (IPCC) was created in 1988 by resolution of the Executive Council of the WMO.[5] It was placed under the joint authority of the political organs of the WMO and UNEP, its members to be chosen by the participating governments and to be of 'as high a level as possible'. Its purpose was to develop international consensus on the science of climate change, its causes, repercussions and possible response strategies.

At the same time, the Canadian Government, encouraged by the recent success of the signing of the Montreal Protocol for the protection of the ozone layer in 1987, organized in Toronto, in June 1988, a World Conference on the Changing Atmosphere.[6] This focused equally on stratospheric ozone and climate change. Its final declaration, which did not commit governments, recognized the urgency of climate change and called for action to limit the emission of greenhouse gases. For the first time a specific (global) target was proposed: to reduce world emissions of carbon dioxide by 20% of 1988 levels by 2005. It also launched the idea of a framework convention and the creation of a 'world atmosphere fund' financed by a tax on fossil fuels in order to fund measures against climate change.

In the UN a few months later, a suggestion by Malta that the conservation of climate should be dealt with as part of the common heritage of mankind was summarily rejected. The General Assembly limited itself to acknowledging that climate change should be recognized as a 'common concern of mankind' necessitating action 'within a global framework'.[7] Its Resolution invited UNEP and the WMO to suggest, through the IPCC, strategies to 'delay, limit or mitigate' climate change and 'elements for inclusion in a possible future international convention on climate'.

The problem of climate change was pushed to the forefront of the international political agenda at The Hague summit on the protection of the atmosphere, the first North-South environmental summit, held in March 1989, at the instigation of President Mitterrand of France, and the Dutch and Norwegian Prime Ministers Lubbers and Brundtland. Their aim was to gather support from countries perceived as environmentally 'progressive' in order to put pressure on the rest of the international community. At The Hague, over 20 heads of state and government signed a political Declaration which stated that, faced with a problem which was 'vital, urgent and global, we are in a situation that calls not only for implementation of existing principles but also for a new approach, through the development of new principles of international law including new and more effective decision-making and enforcement mechanisms'. The signatories agreed to promote within the framework of the UN a 'new institutional authority ... responsible for combating any further global warming of the atmosphere' which (somewhat audaciously), 'shall involve such decision-making procedures as may be effective *even if, on occasion, unanimous agreement has not been achieved*'.[8]

A few weeks later, when discussing the further multilateral process to reach agreement on a convention on climate change, UNEP's Governing Council merely took note of the Hague Declaration. It invited the WMO and UNEP to begin preparations for negotiations on a framework convention on climate, and recommended that these should begin as soon as possible after the adoption of the IPCC's interim report.[9] One would have supposed that these negotiations would have taken place under the joint aegis of the WMO and UNEP, but they were placed, for political reasons, under the direct authority of the UN General Assembly.

The next political stage was the Noordwijk Ministerial Conference, organized by the Dutch Government in November 1989. Despite this taking place outside any established institutional framework, it was more formal and of greater diplomatic weight than the 1988 Toronto meeting. It was the first high-level intergovernmental meeting specifically devoted to climate change, with 66 countries formally represented, together with all directly-concerned international organizations – not just the WMO and UNEP, but also the World Bank, the UN Development Programme (UNDP), the Organization for Economic Cooperation and Development (OECD) and the International Energy Agency (IEA). This conference reinforced the politicization of the problem and reflected the positions of the main interest groups and emerging political alliances. For the first time, because of the participation by the developing countries, the stakes in the struggle against climate change were explicitly seen from the point of view of North-South relations.

The Noordwijk Declaration, adopted at the end of the conference, was fiercely negotiated and ultimately laid down the political bases of a number of principles which would later be concretized in the framework convention elaborated under the auspices of the UN. In particular, it first established what would later become known as the principle of 'common but differentiated responsibilities'. The Declaration called on all countries to take action to control, limit or reduce the emission of greenhouse gases, though the industrialized countries were, in addition, called upon to set an example in doing so, and to provide financial assistance to those countries for which adaptation to climate change would be excessively burdensome.

In the long term, the common action envisaged would be designed to lead to the stabilization of greenhouse gases in the atmosphere, an objective described as 'imperative'. No specific target level was identified. This was to await the initial conclusions of the IPCC, which was invited to study the different options in its report. The Declaration mentioned a number of industrialized countries who believed that such a stabilization should take place by 2000, but there was no consensus: the US resolutely opposed the adoption of any quantified objective, invoking scientific uncertainty as to global warming. Certain European Community (EC) countries were also reluctant to commit to any stabilization target prior to achieving further growth.

The developing countries made their mark clearly on the text of the Noordwijk Declaration, not only because of the emphasis on differentiated responsibility, but also because of the numerous references to their needs with regard to technical and financial assistance and the transfer of clean technologies. These demands would be concretized in the negotiations on the framework convention.

1.3 Setting the Stage for Treaty Negotiations: the UN General Assembly and the 2nd World Climate Conference

In December 1989, the General Assembly adopted its second resolution on climate change, entitled 'Protection of global climate for present and future generations of mankind'.[10] Comparison of this with the first one, which had followed Malta's proposal, shows the accrued politicization of the debate and the raising of the stakes within it. The urgency of the matter is stressed, and the need to prepare for future intergovernmental negotiations. It invited active participation in the IPCC by all interested governments, intergovernmental and non-governmental organizations, expressed concern at the limited participation by the developing countries and requests immediate measures to increase this. The General Assembly made clear its intention to take control of the negotiation process at the end of the preparatory scientific-technical phase within the IPCC, whose first report was expected in 1990.

The object of the anticipated negotiations was the preparation of 'a framework convention on climate and associated protocols containing concrete commitments'.[11] The resolution encouraged the development of international funding mechanisms, 'bearing in mind the need to provide new and additional financial resources to support developing countries', and decided that 'the concept of assured access for developing countries to environmentally sound technologies and assured transfer of those technologies to developing countries on favourable terms' should be explored 'in the context of the elaboration of a framework convention on climate', two themes that would become fundamental issues during the subsequent negotiations.

Shortly after the publication of the IPCC's first summary report in August 1990, the second World Climate Conference took place in Geneva, at the behest of the WMO, in collaboration with UNEP, UNESCO, the Food and Agriculture Organisation of the UN (FAO) and ICSU. While the first such conference in 1989 was of a pre-eminently scientific character (the co-ordination of research), the second had a more political objective: drawing conclusions from the evaluation of the state of scientific knowledge by the IPCC. Observers saw it as a dress rehearsal for the intergovernmental negotiations on the framework convention. Scientific and political discussions were held in tandem.

The differences between the respective conclusions drawn by scientists and policy makers were the points at which the parties' positions crystallized, constituting the main issues at stake in the negotiations to come. The experts agreed that the conclusions of the IPCC's first report – that if no measures were taken to reduce greenhouse gas (GHG) emissions, there would be a rise in the planet's average temperature of 2–5°C over the next century – reflected the international consensus on the state of scientific knowledge relating to climate change.[12] However, the Ministerial Declaration was more ambiguous, citing economic and scientific uncertainties, while calling for world action based on the best available knowledge, of which the IPCC evaluation was cited as but one example.[13] The ministers nevertheless recommended that response measures should be adopted without delay, despite significant scientific uncertainties. They declared the importance of conserving forests as carbon sinks, in contrast to the scientists, who had commented that increasing forest cover could not be considered as 'the major cure for the problem'. Following this, the respective importance to be accorded in the struggle against climate change to emission reduction measures and to measures to enhance carbon sinks would be one of the major issues at stake in the negotiations on the framework convention and its protocol.

Different North-South, North-North and South-South fault lines began to appear. The Ministerial Declaration established the bases of the principle of common but differentiated responsibilities. It set out, as a principle, that the developed countries should lead by example because most GHG emissions were imputable to them, while the developing countries should only be required to take appropriate measures, to the extent possible. This commitment was subject to the transfer of sufficient additional financial resources and environmentally sound technologies. Allowance was made for developed countries which consumed relatively little energy to establish objectives taking account of national economic growth imperatives. There was profound dissent between OECD members, and even between EC Member States as to the level and nature of commitment foreseen. The EC, Australia, Canada, the Nordic countries, Japan and Switzerland were prepared to do what was necessary to stabilize, by 2000, their GHG emissions at a level generally that of 1990. The US refused, judging such a commitment to be unrealistic. The reference to developed countries whose energy consumption was still relatively low and who were to be allowed to increase their 1990 GHG emissions, was judged necessary to maintain the

EC's common front, as it was negotiating on the basis of stabilization of emissions by the EC *as a whole*. The South were far from united with regard to the importance for them of the struggle against global warming: the Ministerial Declaration recognized that it could threaten the very existence of certain countries, while still stressing the importance of the export of fossil fuels for other developing countries, i.e., the interests of small island states and those suffering desertification conflicted with those of the petroleum-exporting countries. The Declaration concluded by calling for the signing of a framework convention at the UN Conference on Environment and Development to be held in Rio de Janeiro in 1992, if possible.

1.4   United Nations Mandate for the Negotiation of a Framework Convention

In December 1990, the UN General Assembly formally launched the negotiation of a framework convention.[14] An *ad hoc* forum for these negotiations was created, under the name of the 'Intergovernmental Negotiating Committee for a Framework Convention on Climate Change', thus reducing the WMO and UNEP to a purely technical role. Their approach was believed to be too technical, and insufficiently sensitive to the interests of the developing countries, which had participated little in the IPCC.

With regard to the aim of the negotiations, the General Assembly reduced its level of ambition from that of the previous year, no longer foreseeing the entering into of concrete commitments, but simply stating that the framework convention should comprise 'appropriate' commitments. The negotiations began in Washington in February 1991, and were concluded *in extremis* in New York in May 1992, after seven meetings of the Intergovernmental Negotiating Committee.

2.   THE UN FRAMEWORK CONVENTION ON CLIMATE CHANGE

Like all other 'products' of Rio, the text of the UN Framework Convention on Climate Change (UNFCCC)[15] reflects a subtle balancing exercise between national sovereignty and collective state responsibility for the protection of the global environment, and between economic development and ecological constraints.[16]

Thus, the Convention's Preamble, whilst acknowledging 'that change in the Earth's climate and its adverse effects are a common concern of humankind' and that 'the global nature of climate change calls for the widest possible cooperation by all countries and their participation in an effective and appropriate

international response', equally 'reaffirms' the 'principle of sovereignty of
States in international cooperation to address climate change'. In the same spirit,
it recognizes in addition that, while 'various actions to address climate change
can be justified economically in their own right and can also help in solving
other environmental problems', all such responses to climate change must nevertheless 'be co-ordinated with social and economic development in an
integrated manner with a view to avoiding adverse impacts on the latter, taking
into full account the legitimate priority needs of developing countries for the
achievement of sustained economic growth and the eradication of poverty'. The
preambular provisions underline that 'the largest share of historical and current
global emissions of greenhouse gases has originated in developed countries,
that per capita emissions in developing countries are still relatively low and that
the share of global emissions originating in developing countries will grow to
meet their social and development needs'. It was therefore recognized that there
was 'the need for developed countries to take immediate action ... as a first step
towards comprehensive response strategies at the global, national and, where
agreed, regional levels', while, in order for developing countries to progress
towards the goal of sustainable social and economic development, 'their energy
consumption will need to grow taking into account the possibilities for achieving
greater energy efficiency and for controlling greenhouse gas emissions in general, including through the application of new technologies'. It is on these terms
that the Convention's Preamble introduces the principle of 'common but differentiated responsibilities', which is not only explicitly written into the body
of the treaty itself, but underlies its entire structure and constitutes one of the
cornerstones of the international climate change regime, as will be seen
below.

2.1   Ultimate Objective

The objective of the regime instituted by the Convention is formulated as follows
in its art. 2:

> The *ultimate* objective of this Convention ... is to achieve, in accordance with the
> relevant provisions of the Convention, *stabilization* of greenhouse gas concentrations
> in the atmosphere at a level that would prevent *dangerous* anthropogenic interference
> with the climate system. Such a level should be achieved within a time-frame suffi-
> cient to allow ecosystems *to adapt naturally to climate change*, to ensure that food
> production is not threatened and *to enable economic development to proceed in a
> sustainable manner*. (emphasis added)

It is notable that stabilizing GHG concentrations is described as an 'ultimate',
long-term objective. The *level* at which and the *timeframe* within which such
stabilization must be realized are set out in terms which are particularly vague

and subject to interpretation. The Parties obviously had very divergent opinions on what constitutes 'dangerous' interference with the climate system. The wording of art. 2 comprises, moreover, the implicit acceptance of climate change as, to a certain degree, inevitable. In addition, the realization of the objective, or at least the timeframe within which it could be realized, is explicitly subject to the condition that economic development should be able to 'proceed', albeit 'in a sustainable manner'.

2.2 Principles

Article 3 sets out a series of 'principles' which must 'guide' the implementation of the Convention. These principles are, in fact, variations on the general principles of the Rio Declaration on Environment and Development. By being so incorporated into the body of a multilateral convention, a legally binding international instrument, these principles acquired a legal nature, even if, as for every legal principle, their content remains abstract, often ambiguous and subject to divergent interpretations.

The first principle set out in art. 3 is that of 'common but differentiated responsibilities', which is formulated in the following manner:

> The Parties should protect the climate system for the benefit of present and future generations of humankind, on the basis of equity and in accordance with their common but differentiated responsibilities and respective capabilities. Accordingly, the developed country Parties should take the lead in combating climate change and the adverse effects thereof.

Echoing the provisions of the Preamble analyzed above, this part of art. 3 initially invokes the common responsibility of all the Parties to contribute to the preservation of the global climate system, an obligation that is incumbent on them 'for the benefit of present and future generations of humankind'. It leads on to the theme of the differentiated responsibilities of the developed countries on the one hand and the developing countries on the other hand, a differentiation based on the principle of equity and the differing capacities of the two categories of country.

Equity has both an ecological and economic dimension here. In fact, as apparent from the Preamble, the primary responsibility which falls to the industrialized countries derives as much from the scale of their historic contribution to the increase in GHG concentrations in the atmosphere, as from their level of economic development and their greater scientific, technological and financial capacity. It results from this state of affairs that these countries must be the first to act and therefore lead by example in the international response to climate change. It is by subscribing to this principle upfront in the Framework Convention itself that the industrialized countries accepted the asymmetry of obligations

between North and South and, as it were, renounced immediate reciprocity between the respective commitments of the contracting Parties, a classic feature
of conventional international law. The fact that the Kyoto Protocol, adopted
later, does not impose supplementary obligations on the developing countries
in addition to those resulting from the Framework Convention and fixes GHG
emission reduction or limitation objectives only in respect of the industrialized
countries is the logical consequence of this concession made in 1992. The reasons invoked by the US and Australia to justify their refusal to ratify the Protocol
therefore seem contrary to the fundamental principle of good faith which must
direct the implementation of their obligations under the Framework Convention
that they ratified *in tempore non suspecto* and for which they continue, moreover, to affirm their support.

The second paragraph of art. 3 confirms the necessity to give full consideration to 'the specific needs and special circumstances of developing country
Parties, especially those that are particularly vulnerable to the adverse effects
of climate change, and of those Parties, especially developing country Parties,
that would have to bear *a disproportionate or abnormal burden* under the Convention' (art. 3(2) – emphasis added). This last notion implicitly echoes the
equity principle set forth in the preceding paragraph.

The very controversial precautionary principle is the next to be enumerated
in art. 3. Its formulation in the context of climate change is similar to its more
general exposition in Principle 15 of the Rio Declaration. Just as there, the word
'principle' is not part of the provision's wording, only of its title or introductory
clause. The provision talks of 'precautionary *measures* to anticipate, prevent or
minimize the causes of climate change and mitigate its adverse effects', specifying that '[w]here there are threats of serious or irreversible damage, lack of full
scientific certainty should not be used as a reason for postponing such measures'. This precautionary duty is, nevertheless, qualified by the affirmation that
'policies and measures to deal with climate change should be *cost-effective so
as to ensure global benefits at the lowest possible cost*' (art. 3(3) – emphasis
added).

One could question the possibility of being able concretely to evaluate cost-
effectiveness and the benefits of measures to deal with climate change in the
absence of scientific certitude. Is this a surreptitious manner of re-introducing
into the equation the pretext of the absence of scientific certitude which the
precautionary principle intended to remove? Moreover, the general and abstract
manner which the authors of this provision chose to evoke the cost of the policies and measures is clearly not accidental: does it concern public or private
costs? Which costs are to be borne by which actors in applying these
principles?

In the name of the abstract pursuit of cost-effectiveness, art. 3(3) provides in
addition that the measures to be taken by the Parties should 'cover all relevant

sources, sinks and reservoirs of greenhouse gases ... and comprise all economic sectors'. The regime set up by the Framework Convention does not set out to be prescriptive with regard to the choice of measures and refrains from targeting any one sector in particular. The accent is placed as much on preventative measures reducing emissions at source as on curative measures envisaging either the elimination of GHGs from the atmosphere by sequestering them in sinks or in natural or artificial carbon reservoirs, or adaptation to the effects of climate change where these happen to be less costly.

The two last paragraphs of art. 3 identify the principles supposed to govern the balancing of economic and environmental interests. Article 3(4) affirms that the right to development should be conceived as a right, and even a duty of states: 'Parties have a right to, and should, promote sustainable development'. The economic connotation of the right to development is only a little attenuated by the qualification 'sustainable'. In the same provision, the integration principle is again evoked, intimately bound to the concept of sustainable development itself. The policies and measures for the Convention's implementation should 'be integrated with national development programmes, taking into account that economic development is essential for adopting measures to address climate change'. The studied ambiguity of this formula should be noted, as it may be interpreted in two ways: is it the economy which must determine the environmental policies, or vice versa?

An equal ambiguity characterizes the following paragraph, which elevates to the status of a basic principle of the global climate change regime the promotion of an 'international economic system that would lead to sustainable economic growth and development in all Parties, *particularly developing country Parties, thus enabling them better to address the problems of climate change*' (art. 3(5) – emphasis added). In addition, art. 3(5), whose wording is inspired by Principle 12 of the Rio Declaration, stipulates that '[m]easures taken to combat climate change, including unilateral ones, should not constitute a means of arbitrary or unjustifiable discrimination or a disguised restriction on international trade'. Thus, paradoxically, the Framework Convention explicitly sanctions the primacy of economic and trade imperatives and the rules of the multilateral trade system over the implementation of its own provisions relating to the struggle against climate change.

2.3 The Parties' Concrete Commitments

The principle of 'common but differentiated responsibilities' is very concretely expressed in art. 4 of the Convention entitled 'Commitments', which differentiates between the obligations of the Parties according to whether they fall into specific categories or sub-categories. With regard to commitments, four main categories of Party can be distinguished:

1. The developed country Parties and other Parties (in this case the EC, Party to the Convention as a regional economic integration organization) included in Annex I to the Convention;
2. The developed country Parties and other developed Parties (including the EC) included in Annex II to the Convention (a sub-category of the Parties in Annex I, consisting of those countries who were OECD members when the Convention was adopted in 1992);
3. Parties included in Annex I undergoing the process of transition to a market economy (Parties included in Annex I, but not in Annex II); and
4. Developing country Parties (all Parties not included in Annex I).

Furthermore, the provisions of art. 4 recognize the existence of other categories of Party which have a particular status insofar as their 'special' situation must be taken into account in the implementation of the Convention. These Parties benefit from preferential rights by virtue of their membership of one or more of these categories.

2.4 Common Responsibilities

The principal common obligations of all Parties are the formulation, implementation and regular updating of 'national and, where appropriate, regional programmes containing measures to mitigate climate change' (art. 4(1)(b)) as well as the development and periodic updating of national inventories of anthropogenic emissions by sources and removals by sinks of all greenhouse gases (art. 4(1)(a)), and the communication of information on national inventories and implementing measures to the Conference of the Parties (art. 12(1)(a) and (b)). Moreover, art. 4(1) of the Convention sets out further obligations which, in principle, are incumbent on all Parties, but the wording of the pertinent provisions is so vague that their implementation is, in practice, left to each Party's discretion. They concern commitments to 'promote' and 'cooperate in' the implementation of various measures such as the application and diffusion of clean technologies, the conservation and enhancement of carbon sinks, scientific and technological research, preparing for adaptation to the impacts of climate change, or further exchange of scientific and technical information.

2.5 Differentiated Responsibilities

In addition to these general obligations, some of the developed countries have, by virtue of the principle of 'common but differentiated responsibilities', certain more onerous, supplementary obligations.

First of all, the Annex I Parties undertake to take measures to *limit* their 'anthropogenic emissions of greenhouse gases' and to protect and enhance their greenhouse gas sinks and reservoirs whilst 'recognizing that the return by the end of the present decade to earlier levels of anthropogenic emissions ... would contribute to' a modification 'of longer term trends ... consistent with the objective of the Convention' taking into account, in particular, 'the need to maintain strong and sustainable economic growth' (art. 4(2)(a)). This provision, the result of a compromise between the American and European positions, does not explicitly impose a *reduction* in emissions, just a *limitation* at a non-specified level. The tautological sentence with its non-prescriptive formulation evoking the 'return by the end of the present decade to earlier levels of anthropogenic emissions' permitted the EC to save face, as it had been negotiating for the stabilization of carbon dioxide emissions at their 1990 level by 2000, while the US had been opposed to the setting of any specific target figure.

There is, however, an explicit reference to 1990 emission levels in the following paragraph, which commits each Annex I Party to provide detailed information on its policies and measures 'as well as on its resulting projected anthropogenic emissions by sources and removals by sinks of greenhouse gases ... *with the aim of returning individually or jointly to their 1990 levels* these anthropogenic emissions of carbon dioxide and other greenhouse gases' (art. 4(2)(b) – emphasis added). The provisions of art. 4(2) are often interpreted as implying the obligation for the developed countries to stabilize by 2000 their emissions at 1990 levels, but that is a moot point legally as no such obligation is made explicit in either the first or second paragraph. The main aim of the provisions is the adoption of programmes and the taking of measures and the communication of information and projections. In both, any reference to stabilization is made in a non-prescriptive sub-clause. In addition, the formulation used does not apply to each Party individually.

Article 4 concerns the financial commitments by the wealthier Annex II countries to the developing countries. The developed countries commit themselves to providing 'new and additional financial resources' to cover certain costs arising out of the implementation of the Convention's provisions by the developing countries (art. 4(3)), as well as financial assistance to the developing country Parties that are particularly vulnerable to the adverse effects of climate change in meeting the costs of adaptation to those adverse effects (art. 4(4)).

Another obligation specific to the Annex II Parties is 'to promote, facilitate and finance, as appropriate, the transfer of, or access to, environmentally sound technologies and know-how to other Parties ... to enable them to implement the provisions of the Convention' (art. 4(5)). Technology transfer may therefore benefit not only the developing countries, but also those in transition.

The Annex I Parties are also bound, under art. 12, to more onerous obligations than the other Parties with regard to the communication of information. They must communicate a 'detailed description of the policies and measures' that they have adopted to implement their commitment under arts 4(2)(a) and (2)(b) and 'a specific estimate of the effects' of those policies and measures on anthropogenic emissions (art. 12(2)). The Annex II Parties must also provide 'details of measures taken' in accordance with their obligations of assistance to the other Parties (art. 12(3)).

2.6   Preferential Status of Certain Parties

Certain provisions of the Convention accord rights or preferential facilities to certain categories of contracting Parties by dint of recognizing their special status. The beneficiaries are mainly developing countries, particularly certain sub-categories of them, who benefit from differential treatment, but certain categories of 'developed' countries have equally obtained preferential status under the Convention. The three main groups in question are: particularly environmentally vulnerable developing countries (arts 4(4) and (8)(a–g)); particularly economically vulnerable developing countries (art. 4(8–10)); and particularly economically vulnerable Annex I Parties (arts 4(6) and (10)).

The most important provision of the Framework Convention according preferential status to a category of Parties is the conditionality clause which explicitly links '[t]he extent to which developing country Parties will effectively implement their commitments under the Convention' to 'effective implementation by developed country Parties of their commitments under the Convention related to financial resources and transfer of technology' as well as to the fact that 'economic and social development and poverty eradication are the first and overriding priorities of the developing country Parties' (art. 4(7)). Thus, the developing country Parties may, in appropriate circumstances, cite insufficiency of financial and technological assistance as well as their economic and social priorities to justify a delay in the performance of their obligations.

The countries in transition to a market economy also benefit from a preferential status as they can, under art. 4(6), claim that the Conference of Parties grant them 'a certain latitude' in the performance of their art. 4(2) commitments. In contrast to art. 4(7), however, this is not a self-executing provision, but one dependent upon the making of a specific prior decision by the Conference of the Parties.

The Least Developed Countries are also guaranteed preferential status by art. 4(9) of the Convention which requires the contracting Parties 'to take full account of the specific needs and special situations of the least developed countries in their actions with regard to funding and transfer of technology'. By its nature, this is an obligation for the developed country Parties, particularly those listed in Annex II, who have specific obligations in those two domains.

Article 4(10) originated in the concerns of the petroleum and coal-exporting countries, above all the members of the Organization of the Petroleum Exporting Countries (OPEC), about the potential consequences of concerted international action against climate change on their commercial and economic interests. This provision demands that:

> [t]he Parties shall ... take into consideration in the implementation of the commitments of the Convention the situation of Parties, particularly [but not exclusively] developing country Parties, with economies that are vulnerable to the adverse effects of the implementation of measures to respond to climate change. This applies notably to Parties with economies that are highly dependent on income generated from the production, processing and export, and/or consumption of fossil fuels and associated energy-intensive products and/or the use of fossil fuels for which such Parties have serious difficulties in switching to alternatives.

The fossil energy exporters are also specifically covered by art. 4(8)(h).

Article 4(8) provides a detailed list of Parties benefiting from a particular status. It requires the Parties to 'give full consideration to what actions are necessary under the Convention, including actions related to funding, insurance and the transfer of technology, to meet the *specific needs and concerns of developing country Parties* arising from the adverse effects of climate change and/or the impact of the implementation of response measures' (emphasis added). Certain such Parties are specifically cited by dint of their environmental or economic vulnerability. The obligation is not particularly onerous because it is only to 'give full consideration' – there is no pre-determined result. The Conference of the Parties may, however, 'take actions, as appropriate, with respect to this paragraph'. Unsurprisingly, the developing countries collectively put pressure on the Conference of Parties from its first meeting to adopt concrete measures implementing art. 4(8), which it did in the global framework of the Marrakesh Accords at its seventh session in November 2001.

2.7 Institutional and Procedural Framework for Regime Development

Analysis of the provisions of the Framework Convention shows that they do not directly impose precise material obligations on the contracting Parties with regard to the reduction of their greenhouse gas (GHG) emissions. They do create a normative and institutional framework for the pursuit of further negotiations on emissions reduction and on other aspects of a multilateral regime against climate change and its effects. The normative elements (principles and general commitments) were analyzed above. The institutional architecture, which has served as a framework for the further development of the regime, will now be sketched.

The 'supreme' political organ created by the Convention is the Conference of the Parties (COP), which has met annually since 1995. It must 'keep under regular review the implementation of the Convention and any related legal instruments that the Conference of the Parties may adopt, and shall make, within its mandate, the decisions necessary to promote the effective implementation of the Convention' (art. 7(2)). The same Article goes on to enumerate a range of specific tasks, but this list is not exhaustive, as it provides as its final provision that the COP may '[e]xercise such other functions as are required for the achievement of the objective of the Convention as well as all other functions assigned to it under the Convention'.

So as to allow the Conference effectively to play its role of ongoing negotiating forum and to carry out its multiple tasks, two permanent subsidiary bodies were established by the Convention. The Subsidiary Body for Scientific and Technological Advice (SBSTA) was 'established to provide the Conference of the Parties ... with timely information and advice on scientific and technological matters relating to the Convention', 'drawing upon existing competent international bodies', an implicit reference to the IPCC. The Subsidiary Body for Implementation (SBI) is charged with assisting the Conference of the Parties 'in the assessment and review of the effective implementation of the Convention' and 'in the preparation and implementation of its decisions'. It is primarily responsible for regularly examining the information communicated by the Parties under art. 12 and formulating opinions on this for the COP.

With a few exceptions, such as the adoption of amendments to the Convention and its annexes, the COP's decision-making procedures are not fixed in the Framework Convention. Article 7(3) charges the Conference to adopt its own internal rules of procedure, for example with regard to the taking of decisions and the necessary majorities where these are not already specified in the Convention. The only decision which *must* be made by consensus, according to the Convention itself, is that relating to the adoption of the rules of procedure, in accordance with art. 7(2)(k). As the Parties have never been able to agree that any decision can be made other than by consensus, no rules of procedure have ever been formally adopted by the COP. They have been applied 'provisionally' since 1985. In practice, therefore, unless the Convention otherwise states, all COP decisions must be taken by consensus.

Only Parties to the Framework Convention can vote in the COP, though many observers may participate in its work, for example, the United Nations (UN), its specialized agencies and the International Atomic Energy Agency, as well as any State member of those organizations not Party to the Convention, and any body or agency, whether national or international, governmental or nongovernmental, which is 'qualified in matters covered by the Convention'. A large number of organizations have seized the opportunity afforded by this provision.

The Convention is silent on the legal nature of COP decisions. The references made in the Convention to the nature of its tasks seem to imply that its decisions have a force greater than that of simple recommendations, whilst not legally binding. Their normative force is, in practice, significant.

Should the COP wish to adopt legally binding measures, two instruments lie at its disposal. It can adopt amendments to the Framework Convention. If no consensus can be reached on these, 'amendment[s] shall as a last resort be adopted by a three-fourths majority vote of the Parties present and voting at the meeting' (art. 15(3)). Such an amendment would only bind those Parties which voted in its favour. The COP can also adopt additional Protocols to the Convention (art. 17). As there is no other provision in the rules of procedure, such a decision must be taken by consensus. Under the general rules of the law of treaties, only ratifying states can be bound by such protocols. Neither method, therefore, can bind a Party without its consent. It was on the basis of art. 17 that, from 1995, what would become the Kyoto Protocol was elaborated.

3. THE KYOTO PROTOCOL

The Kyoto Protocol completed the UNFCCC with supplementary obligations for the industrialized or 'Annex I' countries. The first COP, at Berlin in 1995, found that the vague commitments of the Annex I Parties set out in art. 4(2) of the Framework Convention were inadequate to obtain its objective, and called for the elaboration of a complementary legal instrument. This was to define concrete policies and measures to be brought into play by the industrialized countries, and would fix quantified objectives for the limitation and reduction of GHG emissions beyond 2000. It was specified in the 'Berlin Mandate' that no new obligations further to those in the Framework Convention would be imposed on the developing countries by the new instrument, in conformity with the principle of 'common but differentiated responsibilities'.

The Protocol, adopted at Kyoto on 11 December 1997, was the end-result of laborious negotiations. It fixed national emission quotas for the period 2008–12 for each Annex I Party that had ratified the Convention (including a quota for the EC as a whole, as an Annex I contracting Party), and defined how the quotas should be calculated and how contracting Parties could demonstrate that they had respected their obligation not to exceed their respective quotas. It authorizes the use of a series of 'flexible mechanisms' which allow the transfer between Annex I Parties of parts of quotas and even the acquisition, by those Parties bound by a quota, of 'certified emission reductions' which can be used to comply with their obligations through the implementation of emission-reduction projects in the developing countries not bound by a quota. The Protocol leaves the contracting Parties absolute discretion as to how they fulfil their obligations. In this chapter, we will examine only the provisions concerning the quotas or 'quantified emission limitation and reduction commitments' which are at the heart of the Kyoto Protocol.

3.1   Main Provisions

The main obligation of the Annex I Parties is to ensure that their anthropogenic emissions of six GHGs (listed in Annex A to the Protocol) do not exceed a certain quota, described as an 'assigned amount' during the first 'commitment period' from 2008 to 2012 (art. 3(1)). The quota includes all six gases, but is expressed in terms of 'carbon dioxide equivalent emissions'. It is the 'aggregate emissions' of these gases as a whole over the 5-year period which is taken into account. The conversion formula for the different gases is based on their relative 'global warming potential'. A 5-year period was chosen to allow for the neutralization of the effect on emissions of natural, random weather variations.

Each Party's assigned amount results from a calculation based on parameters fixed in the Protocol. The main parameters are the 'quantified emission limitation or reduction commitments' subscribed to by each of the Parties in Annex B to the Protocol and their respective emissions during the base year, which is, in principle, 1990. However, the Parties may choose 1995 as the base year for the emission of three of the six GHGs (the fluorinated gases), in respect of which the systematic recording of emission data began later in most countries (art. 3(8)).

Annex B to the Protocol sets out the quantified commitments of the Parties as a percentage of emissions in the base year. These percentages range from 92% to 110%, which correspond to an 8% reduction and 10% increase respectively in relation to annual emissions in the base year. Most of the Parties are committed to emitting less than 100% of their 1990 emission level at an average annual rate over the 2008–12 period. Those few parties with a commitment of 100% or higher do not have to reduce their annual emission in absolute terms, but must ensure that they do not exceed the fixed level, ie they have a limitation rather than a reduction commitment.

After having calculated the assigned amount, the figure is corrected to reflect the variations in carbon stocks on the Parties' territories which result from certain human activities relating to land-use change and to forestry, which may constitute either a source of emissions or a net carbon sink (arts 3(3), (4) and (7)). The accounting rules for these sinks were the subject of arduous negotiations after the adoption of the Protocol and were only established in 2001 in the Marrakesh Accords.

In order to prove compliance, the Annex I Parties must set up national systems for the estimation of anthropogenic emissions by sources and removals by sinks of all greenhouse gases not controlled by the Montreal Protocol in accordance with internationally agreed methodologies (art. 5) and must communicate detailed annual inventories and other pertinent information to the UNFCCC Secretariat for the purpose of verification (art. 6). Compliance depends not only on activities in a Party's own territory, because of the three 'Kyoto' or 'flexible' mechanisms: emissions trading (ET); joint implementation (JI); and the clean development mechanism (CDM).

Under art. 17, '[t]he Parties included in Annex B may participate in emissions trading for the purposes of fulfilling their commitments under Article 3'. These Parties may agree upon the transfer of 'any part of an assigned amount' to be added to the assigned amount for an acquiring Party and subtracted from the assigned amount for a transferring Party (arts 3(10) and (11)). In addition, the Annex I Parties may also participate in JI projects under art. 6 of the Protocol. Under this provision, a Party may finance a project on the territory of another Annex I Party with a view to reducing emissions by sources or enhancing removal by sinks. The 'emission reduction units' resulting from these projects may be transferred, subject to certain conditions, between the Parties concerned. Finally, art. 12 of the Protocol introduces the CDM, a mechanism under which similar projects may be financed by the Annex I Parties on the territory of non-Annex I Parties, that is, developing countries not bound by an assigned amount under the Protocol. The 'certified emission reductions' resulting from such projects may be added to the assigned amount of the Annex I Party participating in the project, which will see its emissions quota rise proportionately (art. 3(12)). In contrast to JI, the CDM is not neutral from an environmental point of view as there is no corresponding reduction of the amount assigned to another Party.

3.2 Provisions for Burden-sharing between Parties

Certain provisions of the Protocol are particularly interesting for the EC and its Member States. First, like the Framework Convention, the Kyoto Protocol contains the now classic provision allowing 'regional economic integration organisations' to become contracting Parties jointly with their Member States, on condition that they are already Parties to the Convention. In such cases, the organization and its members must 'decide on their respective responsibilities for the performance of their obligations under th[e] Protocol' and are not 'entitled to exercise rights under th[e] Protocol concurrently'. In their 'instruments of ratification, acceptance, approval or accession' regional economic integration organizations must 'declare the extent of their competence with respect to the matters governed by th[e] Protocol' (arts 24(2) and (3)). These provisions permitted the EC and its Member States to become joint contracting Parties under the Community law 'mixed agreement' regime.

More unusual are the art. 4 provisions, which are addressed, formally, not only to regional economic integration organizations, but were included in the Protocol at the express request of the EC and its Member States to allow them to agree internally on the thorny question of 'burden-sharing'. An examination of the Annex B table in the Kyoto Protocol of quantified commitments shows that the same figure (92%) is given for the EC and each of its 15 Member States which were Parties to the UNFCCC in 1997. However, it has been known since the October 1990 Council Conclusions that not all the Member States intended to reduce their GHG emissions by the same proportion and that some were, in fact, planning to let them increase in relation to 1990 levels (see above). This differentiation in national objectives was confirmed by the June 1998 Council Conclusions which agreed a span ranging from –21% (Germany and Luxembourg) to +27% (Portugal).

This Luxembourg agreement on the determination of the contributions by the individual Member States to the Community's agreed 8% reduction was rendered possible by art. 4 of the Protocol, which allows Parties to agree to fulfil their commitments jointly. Those who wish to do so must reach a formal agreement of which the terms must be notified to the UNFCCC Secretariat when the Protocol is ratified and shall remain in force until 2012.

In cases of art. 4 agreements, should the group of Parties pursuing a pooled target default on their joint obligations, each country would be individually responsible for the emission level attributed to it in the agreement (art. 4(5)). If the Parties concerned are acting in the framework of a regional economic integration organization, which is itself a Party to the Protocol, then each of them would again be individually responsible for the emission level notified under the agreement, '*together with* the regional economic integration organization' (art. 4(6) – emphasis added).

The authors of the Protocol took into account the possibility of further EU enlargement, providing that:

[i]f Parties acting jointly do so in the framework of … a regional economic integration organization, any alteration in the composition of the organization after adoption of this Protocol shall not affect existing commitments under this Protocol. Any alteration in the composition of the organization shall only apply for the purposes of those commitments under Article 3 that are adopted subsequent to that alteration. (art. 4(4))

The purpose behind this was to ensure that, should the EU enlarge, the Community could not more easily acquit itself of its obligations by 'absorbing' new Member States whose GHG emissions fell strongly after 1990 as a result of their economic transition. The accession of 10 new Member States in May 2004, therefore, had no effect on the 1998 burden-sharing agreement between the then 15 Member States. The new Member States remain outside the European 'bubble' and are still bound by their individually calculated commitments as stipulated in Annex B to the Protocol until 2012. Two of the new Member States, Malta and Cyprus, do not even have such commitments as they are not yet Annex I Parties to the Convention.

3.3 Entry into Force

To conclude this overview of the principal provisions of the Kyoto Protocol, the clause governing its entry into force will be examined. The Protocol comes into effect 'on the ninetieth day after the date on which not less than 55 Parties to the Convention, incorporating Parties included in Annex I which accounted in total for at least 55 per cent of the total carbon dioxide emissions for 1990 of the Parties included in Annex I, have deposited their instruments of ratification, acceptance, approval or accession' (art. 25(1)). This double threshold, unusual in multilateral environmental agreements, explains why the Protocol took so long to enter into force. The first threshold of 55 Parties was relatively easy to reach because of all the non-Annex I Parties which did not incur new responsibilities in ratifying the Protocol. It was the second threshold, of 55% of emissions, which posed the problem because those Annex I Parties that ratified the Protocol for a long while represented only 44.2% of the total base year CO_2 emissions of the Annex I Parties. The 55% condition had been inserted at the behest of some industrialized countries which did not want to find themselves bound by their commitments if there was no reciprocal obligation for a sufficiently representative number of other industrialized countries. The American rejection of the Protocol, unanticipated when it was adopted, made it particularly difficult to reach the critical threshold, the US being responsible for 36.1% of the total base year emissions. The entry into force of the Kyoto Protocol then became dependent on ratification by the Russian Federation.

Russia confirmed its intention to ratify the Protocol at the World Summit on Sustainable Development at Johannesburg in September 2002, as well as in the Joint Declaration of the EU–Russia Summit which took place at St Petersburg on 31 May 2003. These political declarations still left some doubt as to Russia's commitment, but these doubts were dissipated on 30 September 2004, when the Russian Government officially announced its decision to submit the Protocol to the Duma for ratification. After years of procrastination, the political process rapidly accelerated in Russia. The Duma gave its assent to ratification on 22 October 2004, which was followed several days later by a favourable vote in the second parliamentary chamber, the Federation Council. The ratification instrument was signed by President Putin on 5 November and officially transmitted to the UN Secretary-General on 18 November 2004. The Kyoto Protocol entered into force 90 days later, on 16 February 2005 – allowing the first COP serving as the meeting of the Parties to the Protocol (COP/MOP) to take place in late 2005 in Montreal. According to the terms of the Protocol, 2005 was also the year in which the Annex I Parties had to have made 'demonstrable progress' in implementing their commitments under the Protocol (art. 3(2)) and when negotiations were to begin on the Parties' commitments for the period beyond 2012 (art. 3(9)).

4.   THE BONN AND MARRAKESH AGREEMENTS

Due to the lack of sufficient negotiating time, indeed of political will, a number of crucial questions concerning the implementation modalities of the Kyoto Protocol could not be resolved before the adoption of the instrument by COP3. So as not to delay that adoption, those questions were put off for later negotiation in either the COP or the COP/MOP by different enabling provisions of the Protocol.

Meeting at COP4 in Buenos Aires a year later, the negotiators realized that, because of the extreme complexity and technicality of the questions and the divergent interests of the Parties, it would be impossible to decide on the modalities within the timeframe originally anticipated at COP3. The failure of COP4 was sweetened by the adoption of the 'Buenos Aires Plan of Action',[17] a menu of options and agenda for further negotiations. The sixth COP meeting in The Hague in November 2000 was unable to reach the hoped-for global agreement because of the irreconcilable positions of the EU and the US and their respective allies in the G77, despite the efforts of Jan Pronk, the Dutch Minister for the Environment, who chaired the conference. He did, however, manage to secure agreement to convene a 'COP6bis' in Bonn 6 months later. The situation became even more difficult when President Bush announced, in March 2001, that he would not be submitting the Kyoto Protocol to the US Senate for ratification. This made COP6bis in July 2001 a crucial test for the survival of a multilateral approach to solving global environmental problems within the framework of the UN.

The long-sought-after global political agreement based on the Buenos Aires Plan of Action was finally reached in Bonn in July 2001 after laborious ministerial-level negotiations. Translating that agreement into formal decisions at the same meeting proved impossible, however, and the task was put off again to COP7, which was to meet in Marrakesh in November 2001.

The outcome of COP7 was the 'Marrakesh Accords'. These were a set of 23 decisions, comprising 220 pages of detailed rules for the implementation of the Kyoto Protocol and certain provisions of the UNFCCC. They covered subjects such as capacity-building in developing countries and countries in transition, technology transfer, funding under the Convention, the particular needs of the developing countries in the face of climate change, and very detailed provisions relating to the operation of the three Kyoto mechanisms. Within the space available for this chapter, it is not possible to address the substance of these wide-ranging provisions.

For institutional reasons, most of the COP7 decisions were drawn up in the form of *draft* decisions of the COP *serving as meeting of the Parties to the Kyoto Protocol* (COP/MOP), and intended to be formally confirmed after the entry into force of the Protocol. All the draft decisions that were part of the Marrakesh

Accords were endorsed as expected by the first meeting of COP/MOP, which was held in November/December 2005 in Montreal, in conjunction with COP11.

5.   EUROPEAN UNION POLICY ON CLIMATE CHANGE

The EU's policy on climate change developed in parallel, and in close interaction, with the multilateral regime-building process described above. The year 1989 saw the beginnings of Community climate change policy during the preparatory process for the Rio summit. In October 1990, a meeting of the energy and environment ministers concluded that, assuming similar commitments from other countries, the EC would be prepared to take measures to stabilize CO_2 emissions at their 1990 levels *in the Community as a whole*. These conclusions allowed the EC to position itself as a key player and leading actor in the international negotiations. They also took the place of internal policy for a long time as they were not immediately followed by any measures to achieve the objective. On the eve of the Rio summit, the Commission did, however, submit to the Council a very controversial proposal for a directive introducing a tax on carbon dioxide emissions and energy,[18] which would have enabled the EC to significantly reduce its emissions.

Initially the EC and its Member States ratified the Framework Convention[19] without having a concrete internal policy. The proposed CO_2/energy tax was never adopted, and the Community left it to the Member States to elaborate national emission reduction programmes, while establishing an exchange mechanism for information on national policies allowing the Commission to play a co-ordinating role. The first Community measures, adopted on its conclusion of the UNFCCC, were largely symbolic.

Directive 93/76/EEC (SAVE) required the Member States to establish 'programmes' to limit CO_2 emissions through improvements in energy efficiency, particularly in areas such as energy certification and thermal insulation of buildings, regular inspection of boilers, and energy audits of undertakings with high energy consumption.[20] No quantitative objective was set, and the content of the programmes was left to the Member States' discretion. At the same time, the Council adopted the 'Altener' programme, a Community programme for the promotion of renewable energy sources.[21]

Earlier that year, the Council had addressed a Decision to the Member States establishing a monitoring mechanism of Community CO_2 and other GHG emissions.[22] This transposed into Community law, with no objective of harmonizing national legislation, some general obligations arising out of the Framework Convention. Under this Decision, the Member States had to establish and communicate to the Commission 'national programmes for limiting their anthropogenic emissions of CO_2 in order to contribute to: ... the stabilization of CO_2 emissions by 2000 at 1990 levels *in the Community as a whole*' (art. 2 – emphasis added). They were also bound to communicate to the Commission information as to the levels of their GHG emissions. On the basis of the information received, the Commission was mandated to 'evaluate the national programmes, in order to assess whether progress *in the Community as a whole* is sufficient' to ensure fulfilment of the Community's international obligations (art. 5(3) – emphasis added). *Community* compliance is referred to, but not any Member State's individual commitments. The issue of 'burden sharing' was carefully avoided until 1997, when the Kyoto Protocol was adopted (see above).

From the entry into force of the UNFCCC in 1994, the attention of Community politicians was concentrated on the next phase of international negotiations, to the detriment of internal implementation measures. The first COP, in 1995, launched negotiations to strengthen the industrialized countries' commitments by a protocol or other such measure fixing quantified limitation and reduction targets. It was only a few months before the end of the negotiations on the Kyoto Protocol (in order to establish credibility for its negotiating position on the quantified objectives to be agreed in the Protocol), that the Council decided for the first time, in its Conclusions of 3 March 1997, on a provisional table setting out the emission limitation and reduction targets for each Member State, and a menu of various possible 'common and co-ordinated policies and measures'. The table, in fact, showed that the Member States would achieve only two-thirds of the Community's proposed target in the Kyoto Protocol negotiations of 15% GHG emission reductions for industrialized counties by 2010.

The Protocol eventually adopted the much less ambitious target of –8% of 1990 emission levels to be reached in 2008–12. Paradoxically, the impact of the Protocol on Community policy was further to distract attention from the domestic front, while the negotiations which would lead to the Bonn and Marrakesh Accords seemed to go on interminably. The political agreement reached by the Council in June 1998 on the sharing between the Member States of the EC's 8% reduction commitment under the Protocol was more difficult to reach than the March 1997 agreement on the sharing of the more ambitious (but conditional) emission reduction target of 10%. This Luxembourg 'burden sharing' agreement of 16 June 1998 was only formalized 4 years later in the Council's Decision of 25 April 2002 on the conclusion of the Kyoto Protocol.[23]

In October 1998, the Council gave the political green light to the conclusion of a voluntary agreement between the Commission and the association of European car manufacturers on CO_2 reductions for passenger cars. This agreement was formalized by a Commission Recommendation of February 1999,[24] and flanked by two measures aiming to inform consumers and decision-makers about the evolution of the CO_2 emission levels of vehicles on the Community market.[25]

Furthermore, in April 1999, the Council decided to amend Decision 93/389/ EEC in order to strengthen the monitoring mechanism of Community GHG emissions and the exchange of information on national emission reduction programmes.[26] The amendment increased the range of information to be provided by the Member States, and envisaged an annual evaluation by the Commission of progress achieved. Bizarrely, the Decision did not set out the Member States' national targets and contained no explicit reference to the Council's important conclusions of June 1998 on burden-sharing.

Finally, in September 2001, on the eve of the Marrakesh COP7, Directive 2001/77/EC was issued.[27] This bound Member States to take appropriate measures to *promote* the growth of electricity produced from renewable energy sources in accordance with *indicative* national objectives.

Despite having recourse to legally binding acts, Community climate change policy was, until 2001, merely a collection of incentive, informative and co-ordinating measures, without any real constraining effect. It has only been since the Bonn and Marrakesh Accords, opening the way to ratification of the Kyoto Protocol, that internal EU policy has moved into the 'fast lane' and that legislative measures have lived up to their nomenclature.

It was not until October 2001 that the Commission presented to the Council, simultaneously with the draft decision to approve the Kyoto Protocol and formalize the burden-sharing between the Member States,[28] its draft directive establishing a scheme for GHG emission allowance trading within the Community.[29] At the same time, it presented a Communication to the European Parliament and Council on the implementation of the first phase of the 'European Climate Change Programme' (ECCP), announcing various supplementary measures.[30] The ECCP had been instituted by the Commission in 2000 so as to prepare, in collaboration with Member State experts, business and non-governmental organizations (NGOs), common and co-ordinated policies and measures against climate change,[31] as had been demanded by the Council repeatedly since 1997. Working groups were set up on subjects such as flexible mechanisms, the production and consumption of energy and transport. Their recommendations would later be used as the bases for the drawing up of various proposals by the Commission.

In the stride of its April 2002 Decision on the ratification of the Kyoto Protocol, the Council intensified its work on internal measures contributing to the reduction of GHG emissions. A growing number of measures have been adopted since the end of 2002 and others, resulting from the ECCP exercise, are under examination. Despite some important advances, the ambitions of the Commission and, above all, the Council with regard to common and co-ordinated policies and measures have clearly gradually been reviewed downwards since the end of the 1990s.[32]

Directive 2002/91/EC on the energy performance of buildings[33] and Directive 2003/87/EC establishing a scheme for GHG emission allowance trading within the Community[34] constitute the first harmonized measures with significant legal and economic impact taken by the EU institutions in the framework of their climate change policy. As has been shown, the main Community climate change measures adopted at the beginning of the 1990s were extremely timid with regard to the use of fiscal instruments and, in the energy production and consumption and transport sectors, measures were, in the main, of an incentive rather than binding nature. In these circumstances, the scheme for GHG emission allowance trading, set up by Directive 2003/87/EC, effectively became the main instrument of Community policy and is likely to remain so in the years to come. As this measure is analyzed in detail by other contributors in the rest of this volume, it will not be further examined here.

Since the ratification of the Kyoto Protocol, a number of other measures have, however, also been adopted.[35] One of those recent measures is Directive 2005/32/EC establishing a framework for the setting of eco-design requirements for energy-using products.[36] It does not actually set any harmonized standards, but enables the Commission to do so via a comitology procedure 'where market forces fail to evolve in the right direction or at an acceptable speed'. The Directive aims to improve the design from an environmental point of view of goods such as computers, fridges and office equipment. It remains to be seen to what extent its coming into force will really contribute to the reduction of energy consumption and GHG emissions, rather than hinder action by the most progressive Member States with regard to energy efficiency standards.

As a consequence of the ratification of the Kyoto Protocol by the EC, the Council has also completely revised the GHG emissions monitoring mechanism set up by Decision 93/389/EEC and first revised in 1999. The new Decision 280/2004/EC establishes a 'mechanism' which is intended to monitor not only GHG emissions within the Community, but also to implement the Kyoto Protocol.[37] It transposes certain of the Marrakesh Accord provisions into Community law, notably with regard to a Community inventory of emissions and removals, inventories and national registers and the communication of information. It formalizes to an extent the existence of the ECCP, but without binding the Council and Parliament, as any measures conceived by the Commission within the framework of the Programme will still need the approval of the other institutions in order to take effect.

Other legislative proposals from the Commission resulting from the ECCP Programme remain under consideration by the Parliament and Council. Among these, one of the most advanced is a proposed regulation on certain fluorinated GHGs (used in, for example, refrigeration and air conditioning systems) on which the Council adopted its common position on 20 June 2005.[38] It decided to split the proposed legislation in two:[39] a directive to phase out a fluorinated gas from vehicle air conditioning, and a regulation for other stationary applications. The final adoption of this legislation on which political agreement was reached between Parliament and Council in the Conciliation Committee in March 2006 is expected later this year.

The Transport, Telecommunications and Energy Council is considering a proposed directive on energy end-use efficiency and energy services,[40] which aims to increase savings on the sale of energy to end-users. The energy in question would range from electricity, to fuel for heating and petrol for cars. In June 2005, MEPs voted to amend the draft by setting higher energy-saving targets. However, those (binding) targets were all replaced by indicative targets on 28 June 2005 by the Energy Ministers, with the sop of a remaining requirement that Member States would be obliged to *try* to achieve a 6% reduction in energy consumption over a non-specified 6-year period. The ministerial agreement needs to be formalized by the Council, and will then pass back to the Parliament for second reading.

In March 2004, the European Council announced that in its Spring 2005 meeting it would consider 'medium and longer term emission reduction strategies, including targets'. In response to the European Council's request that it analyze the benefits and costs of action against climate change, taking account of environmental and competitiveness concerns, the Commission adopted a Communication on 'Winning the Battle against Climate Change' on 9 February 2005. The Communication focuses on the Community's strategy for 2012 onwards, and argues for greater participation by countries and sectors not yet committed to emissions reduction, the development of low-carbon technologies, greater utilization of market mechanisms and the need to adapt to the effects of climate change.

The Environment Council welcomed the Communication in the Conclusions to its meeting of 10 March 2005[41] and invited the Commission to continue its cost-benefit analysis, focusing on the costs of adaptation to climate change, the benefits of climate change policies, the costs of inaction, and an economic evaluation of the damages caused by climate change. With the initial focus on costs of adaptation, there might have been some concern that EU policy would move away from trying to keep climate change to a minimum to coping with its effects (which the North is far better equipped to manage than the developing countries). However, when considering the ongoing multilateral negotiations, the Environment Council summarized recent research, then stated that the EU believes that 'reduction pathways by the group of developed countries in the order of 15–30% by 2020 and 60–80% by 2050 compared to the baseline envisaged in the Kyoto Protocol should be considered'. Unfortunately, the European Council, meeting on 23 March 2005, limited itself to approving the first, but ignoring the second recommended reduction target, referring only to a possibility of going beyond the 15–30% target 'in the spirit of the conclusions of the Environment Council'.[42]

The European Parliament, in a resolution of 12 May 2005, regretted this attitude of the European Council, and suggested the reintroduction of the second target, as formulated by the Environment Council.[43] The Parliament's resolution generally calls for a more ambitious approach to the challenges ahead. In this, it 'stresses that a three-prong approach will be necessary to underpin the EU's climate policy at home: yearly reductions in the energy intensity of the EU economy in the range of 2–2.5%, a significant increase in the share of alternative energy in the overall energy mix and substantial increases in the support for R&D in sustainable energy'. The resolution also 'emphasizes that the transportation sector presents a major challenge, and that new and innovative policies are needed to curb emissions from road transport, aviation and shipping'. The Commission has already announced its intention to publish a communication on measures to reduce the climate change impacts of aviation, one of which is expected to include the aviation sector in the EU's emissions trading system. It has also recently issued a proposal for restructuring and harmonizing the system of taxation of passenger cars in the internal market,[44] as part of a three-pronged overall strategy for reducing CO_2 emissions from cars, which also includes the voluntary agreement with the automobile manufacturers and consumer information measures discussed above.

The debate on the further evolution of the EU's policy on climate change is closely linked with that on the future development of the multilateral regime under the UNFCCC. This debate will start in 2006, as a result of the decisions taken at COP11 and COP/MOP1 in Montreal to launch negotiations on future commitments for Kyoto Protocol Parties beyond 2012, in parallel with a broader 'dialogue' involving all UNFCCC Parties on long-term cooperative action to address climate change and promote enhanced implementation of the Convention. Thus, climate change is likely to remain high on the agenda of EU environmental policy for years to come.

NOTES

1. Professor of international environmental law, Université Libre de Bruxelles and Vrije Universiteit Brussel, and Senior Fellow, Institute for European Environmental Policy, Brussels/London.
2. Doctoral Research Fellow, Institute for European Studies, Vrije Universiteit Brussel.
3. Action Plan for the Human Environment, adopted by the UN Conference on the Human Environment, Stockholm, 16 June 1972, Recommendation 79.
4. UNEP/ICSU/WMO, *Report of the International Conference on the Assessment of the Role of Carbon Dioxide and of Other Greenhouse Gases in Climate Variations and Associated Impacts* (Villach, Austria, 9–15 October 1985), WMO Report No. 661, World Meteorological Organisation, Geneva, 1986.

5. Resolution 4 (EC-XL) of the Executive Council of the WMO.
6. The UNEP/WMO Conference on the Changing Atmosphere: Implications for Global Security, Toronto, June 1988.
7. UN General Assembly Resolution 43/53, 6 December 1988.
8. Declaration of The Hague, 11 March 1989, UN Doc. A/44/340 (emphasis added).
9. UNEP Governing Council Decision 15/36 , 25 May 1989, UN Doc. A/44/25, pp. 177–81.
10. UN General Assembly Resolution 44/207, 22 December 1989.
11. *Ibid*, para. 12.
12. Second World Climate Conference, Geneva, 29 October–7 November 1990, Conference Statement – Scientific/Technical Sessions, Part I, Main Conclusions and Recommendations, para. A.1.
13. Ministerial Declaration, Second World Climate Conference, Geneva, 7 November 1990, para. 5.
14. UN General Assembly Resolution 45/212, 21 December 1990.
15. UN Framework Convention on Climate Change, New York, 9 May 1992, available at http://unfccc.int/essential_background/convention/background/items/1349.php.
16. See generally Pallemaerts, Marc (1992), 'International Environmental Law from Stockholm to Rio: Back to the Future?', *Review of European Community and International Environmental Law*, **1**, 254–66.
17. Decision 1/CP.4, *The Buenos Aires Action Plan*, Doc. FCCC/CP/1998/16/Add.1, 25 January 1999, p. 4.
18. COM(1992) 226 final, 2 June 1992.
19. Council Decision 94/69/EC of 15 December 1993 concerning the conclusion of the United Nations Framework Convention on Climate Change, OJ L 33, 07.02.1994, p. 11.
20. Directive 93/76/EEC of 13 September 1993, to limit carbon dioxide emissions by improving energy efficiency (SAVE), OJ L 237, 22.09.1993, p. 28.
21. Council Decision 93/500/EEC of 13 September 1993, concerning the promotion of renewable energy sources in the Community (Altener programme), OJ L 235, 18.09.1993, p. 41.
22. Council Decision 93/389/EEC of 24 June 1993, establishing a monitoring mechanism of Community CO_2 and other greenhouse gas emissions, OJ L 167, 09.07.1993, p. 31.
23. Council Decision 2002/358/EC of 25 April 2002 concerning the approval, on behalf of the European Community, of the Kyoto Protocol to the United Nations Framework Convention on Climate Change and the joint fulfilment of commitments thereunder, OJ L 130, 15.05.2002, p. 1.
24. Commission Recommendation 1999/94/EC of 5 February 1999 on the reduction of CO_2 emissions from passenger cars, OJ L 040, 13.02.1999, p. 49. For a more detailed analysis and critique of this voluntary agreement, see Pallemaerts, Marc (1999), 'The Decline of Law as an Instrument of Community Environmental Policy', *Law & European Affairs*, **9**, 338–54.
25. Directive 1999/94/EC of the European Parliament and of the Council of 13 December 1999, relating to the availability of consumer information on fuel economy and CO_2 emissions in respect of the marketing of new passenger cars, OJ L 012, 18.01.2000, p. 16, and Decision 1753/2000/EC of the European Parliament and of the Council of 22 June 2000 establishing a scheme to monitor the average specific emissions of CO_2 from new passenger cars, OJ L 202, 10.08.2000, p. 1.
26. Council Decision 1999/296/EC of 26 April 1999, amending Decision 93/389/EEC for a Monitoring Mechanism of Community CO_2 and Other Greenhouse Gas Emissions, OJ L 117, 05.05.1999, p. 35.
27. Directive 2001/77/EC of the European Parliament and of the Council of 27 September 2001 on the promotion of electricity produced from renewable energy sources in the internal electricity market, OJ L 283, 27.10.2001, p. 33.
28. COM(2001) 579 final, 23 October 2001.
29. COM(2001) 581 final, 23 October 2001.
30. COM(2001) 580 final, 23 October 2001.
31. COM(2000) 88 final, 8 March 2000.
32. For a systematic comparison of planned and adopted measures, see Table 4 in Pallemaerts, Marc (2004), 'Le cadre international et européen des politiques de lutte contre les changements

climatiques', *Courrier hebdomadaire du Centre de recherche et d'information socio-politiques*, (1858–59), 58–60.

33. OJ L 1, 4.1.2003, p. 65.
34. OJ L 275, 25.10.2003, p. 32.
35. Eg Directive 2003/30/EC of 8 May 2003 on the promotion of the use of biofuels or other renewable fuels for transport, OJ L 123, 17.05.2003, p. 42; Directive 2004/8/EC of 11 February 2004 on the promotion of cogeneration based on a useful heat demand in the internal energy market, OJ L 52, 21.02.2004, p. 50; Directive 2003/96/EC of 27 October 2003 restructuring the Community framework for the taxation of energy products and electricity, OJ L 283, 31.10.2003, p. 51.
36. Directive 2005/32/EC of 6 July 20005 establishing a framework for the setting of ecodesign requirements for energy-using products and amending Council Directive 92/42/EEC and Directives 96/57/EC and 2000/55/EC of the European Parliament and the Council, OJ L 191, 22.07.2005, p. 29.
37. OJ L 49, 19.02.2004, p. 1.
38. Austria and Denmark voted against and Belgium, Portugal and Sweden abstained.
39. See Commission Press Release IP/04/1231.
40. COM(2003) 739 final.
41. 6693/05(Presse 40).
42. Presidency Conclusions, Brussels European Council, 22–23 March 2005, 7619/1/05 Rev 1.
43. European Parliament Resolution on the Seminar of Governmental Experts on Climate Change, 12 May 2005, P6_TA-PROV(2005)0177, B6-0278/2005.
44. Proposal for a Council Directive on passenger car related taxes, COM(2005) 261 final, 5 July 2005.

3. The European Union, Russia and the Kyoto Protocol

Wybe Th. Douma[1]

1. INTRODUCTION

The entry into force of the Kyoto Protocol depended on Russia's ratification for a long time. The ratification process took place against the background of the European Union's (EU) efforts to reshape its relationship with our largest neighbour since the 1 May 2004 enlargement. This relationship is formalized in a Partnership and Cooperation Agreement (PCA) from 1994.[2] In May 2003, it was agreed that the long-term goals of EU-Russia cooperation would be the development of four 'common spaces' in Europe: a common economic space (with specific reference to environment and energy); a common space of freedom, security and justice; a space of cooperation in the field of external security; and one of research and education, including cultural aspects.[3] The EU-Russia relationship constantly proves to be a demanding one for both sides. This became especially clear in the wake of the EU enlargement, as major differences of opinion existed on issues like the extension of the PCA to the new Member States, the EU conditions for Russia's World Trade Organization (WTO) accession and Russia's ratification of the Kyoto Protocol. In this chapter, the struggle between the EU and Russia about the ratification of Kyoto will be examined against the background of the development of the EU-Russia relationship. Attention will also be paid to the alleged links between Russia's ratification and the EU's green light for Russia's WTO accession, and to developments in Russia's environmental protection system. In the last paragraph, a brief look at the prospects for EU-Russia cooperation under the Kyoto Protocol will be presented.

2. HISTORICAL BACKGROUND

2.1 The 1997 PCA

A PCA between the Russian Federation and the European Communities and their 15 Member States was signed in Corfu on 24 June 1994. It entered into

force on 1 December 1997.[4] The delay was caused by a number of circumstances, and was even temporarily suspended by the EU, due to the Russian
forces' military hostilities in Chechnya. With the start of peaceful negotiations
in Chechnya, the ratification process was resumed. In October–November 1996,
the PCA was ratified by the State *Duma* and the Federation Council, and in
October 1997 its ratification was completed by the EU Member States.

The PCA forms an extensive and comprehensive agreement. It deals first and
foremost with strengthening economic, political and trade relations between the
parties, but protection of the environment forms an important issue throughout
the PCA. Already in the preamble, it is mentioned that the parties are desirous
of establishing close cooperation in the area of environmental protection taking
into account the interdependence existing between the parties in this field. In
art. 55, the parties recognize that an important condition for strengthening the
economic links between Russia and the Community is the approximation of
legislation. Russia promises that it shall 'endeavour to ensure'[5] that its legislation will be gradually made compatible with that of the Community, *inter alia*,
in the areas of the environment and the protection of health and life of humans,
animals and plants. In art. 57 on industrial cooperation, the promotion of public
and private sector efforts to restructure and modernize industry under conditions
ensuring environmental protection and sustainable development are mentioned.
Article 69 deals with environmental cooperation and stresses, among other
things, the need to improve laws to Community standards.[6]

2.2   The 1999 Common Strategy

From the side of the EU, a so-called Common Strategy[7] on Russia was adopted
soon afterwards, namely at the Cologne European Council on 4 June 1999.[8] The
document aims at strengthening the relationship with Russia on the basis of the
PCA through 'binding orientations' and increased coherence of EU and Member
States' action. The main objectives of the Common Strategy are consolidation
of democracy, the rule of law and public institutions in Russia, integration of
Russia into a common European economic and social space, cooperation to
strengthen stability and security in Europe and beyond, and addressing common
challenges on the European continent.

One of these common challenges is the environment, described as the common property of the people of Russia and the EU. The sustainable use of natural
resources, management of nuclear waste and the fight against air and water
pollution, particularly across frontiers, are described as priorities in this area.
To this end, the EU pledges to support the integration of Russia into a wider
area of economic cooperation in Europe, *inter alia*, by promoting progressive
approximation of environmental legislation. The document also sets out that the
EU is to cooperate with Russia to improve energy efficiency and energy con-

servation, by supporting the integration of environmental considerations in economic reform and the efforts to strengthen the enforcement of national environmental legislation, and finally by working with Russia, especially in areas adjacent to the enlarging Union, to reduce water and air pollution and to improve environmental protection.

2.3   Environmental Aspects of the 2001 Summits

The Joint Statements issued at the 2001 summits between Russia and the EU confirmed that the environment remained a 'major priority' in their relations. In Moscow, on 17 May 2001, the leaders of Russia and the EU even referred to cooperation on this issue as a 'key priority'. Particular attention was drawn to energy efficiency as a way to limit the growth in demand and enhance environmental protection.[9] At the Brussels Summit of 3 October 2001, both sides welcomed the agreements on climate change reached by Ministers in Bonn on 23 July 2001 during the meeting of the Conference of Parties to the UN Framework Convention on Climate Change (UNFCCC) and promised to 'work together for its full implementation with a view to early ratification and entry into force of the Kyoto protocol, particularly at the 7th meeting of the Conference of parties to the UNFCCC in Marrakech from 29 October to 9 November 2001'. It was also indicated that both sides 'believe that implementation of the Kyoto Protocol may help not only to reduce greenhouse gas emissions, but also to increase European investment in the energy sector in Russia in order to improve its energy efficiency and economic performance'. The EU expressed that it was very favourably disposed towards the Russian proposal to host a world conference on climate change in 2003, which could be both the annual Conference of Parties to the Framework Convention on Climate Change and the first meeting of the Conference of Parties to the Kyoto Convention.[10]

2.4   Commission Communication on EU–Russia Environmental Cooperation

In December 2001, a Communication on EU–Russia environmental cooperation[11] was released in which the Commission set out its ideas and proposals in order to work towards the objectives sketched at the two summits. *Inter alia,* the Commission argued that the time was ripe to develop a closer and more coordinated bilateral dialogue on environmental issues through the PCA, with a common strategic agenda and reinforced procedures, and stressed the importance of linking environmental, economic and social objectives in constructing a common European economic space that would be sustainable.

In the view of the Commission, Russia would gain a lot from such cooperation. The economy would become more efficient and productivity would be raised, for instance by applying cleaner and more modern technology and management techniques and by reducing the wasteful use of energy, water and raw materials. Exports would be facilitated and the climate for foreign investment in Russia would improve, as investors want a clear, stable and efficiently enforced regulatory framework so that they can assess with confidence what their environmental responsibilities are. Last but not least, the health of those living in densely populated industrial areas would benefit from reductions in pollution. President Putin himself admitted that '[u]p to 15 percent of Russia's regions are in critical or near-critical condition'.[12]

3.   WORDS AND DEEDS ON THE ENVIRONMENT

3.1   Kyoto, Johannesburg and Moscow

Russia's promises from October 2001 concerning an early ratification of the Kyoto Protocol were not kept. One year later, the Protocol was still not ratified. At the Johannesburg World Summit on Sustainable Development in September 2002, Russia's Prime Minister, Mikhail Kashyanov, did state that 'ratification will occur in the very near future'. At the same venue, Deputy Minister Mukhamed Tsikanov of the Economic Development (and Trade Ministry) indicated that there was a risk that Russia would not ratify because 'we don't have the economic stimulus, the economic interest in the Kyoto Protocol'. He did add that, for the moment, the plan in Moscow was still to ratify. By March 2003, this had still not occurred, prompting a visit to Moscow by EC Environment Commissioner Margot Wallström and the Greek and Italian Environment Ministers to encourage Russia to follow through on its pledges. 'The world is waiting for Russia to demonstrate that it is ready and willing to become a major player in the multilateral efforts to combat climate change', Ms Wallström stated ahead of her visit.[13]

At the same time, the protection of the environment in general was receiving less and less attention in Russia. In 1996, the part of the Ministry of Environmental and Natural Resources dealing with environmental protection had been transformed into a State Committee. This Committee was abolished in 2000, and its functions relegated to the Ministry for Natural Resources (MNR). The main function of the MNR is to provide the national economy with resources. On top of that, a steep reduction of staff in territorial bodies of the former environmental enforcement and permitting authority was undertaken. Inspectors became responsible for ever larger areas as their colleagues were dismissed.[14]

Authorities dealing with issuing permits and licences are part of the public administration in relation to which Russia itself recognizes that major improvements are necessary. Such improvements are to be achieved via administrative reform, as this is a key factor to combat widespread corruption. Such reforms, announced in 2000 as being one of President Putin's key policy priorities, would need to tackle the very root of the problem, i.e. the bureaucratic over-regulation of economic activity, which creates the opportunity for corruption.

Unfortunately, the efforts undertaken in the area of administrative reform have, so far, not lead to many concrete results.[15] The consolidation and strengthened position of President Putin after his re-election in May 2004 might mean that more progress can be made in the future. As for environmental legislation, a Federal Law on Environmental Protection was adopted in February 2002, but – due to a multitude of reasons – this law has failed to form a clear framework for achieving results and is often not applied in practice.[16] Stimulating Best Available Technologies (BAT) is possible under this law but does not take place in practice. Likewise, the State Environmental Expertise system (a type of environmental impact assessment) has severe shortcomings.[17] As for the actual standards that are to be observed, a large number of them still stem from the former Soviet Union and are outdated, impractical, inflexible or uneconomic in today's Russian Federation.

In many respects, the legacy of the former planned economy still needs to be overcome. Natural resources are still significantly under-priced, obsolete production technologies use energy and resources in quantities significantly exceeding those used in Western Europe, there is a lack of environmental management systems, and ecologically clean and safe technologies are not stimulated. Incentives to modernize are, as yet, virtually non-existent. Thus, there is a clear need for movement towards a compliance-oriented approach, based on equitable and environmentally sound criteria.

Since 1991, through its Technical Assistance to the Commonwealth of Independent States (TACIS) programme, the EU has tried to stimulate, *inter alia*, improvements in the field of environmental protection and energy efficiency. In the period from 1998 to 2003, numerous projects were supported.[18] One of these was a project entitled *Harmonisation of Environmental Standards – Russia*, which received €2 million in funding. It was aimed at supporting the State *Duma* in its efforts to align Russia's environmental legislation with the EU environmental *acquis*. The project resulted in Russian and international experts (including the author of this contribution) carrying out studies of the present Russian system and comparing it with the EU system, notably in the field of industrial pollution, and suggesting improvements. Initially, some of the Russian experts wondered how the EU planned to approximate EU environmental legislation to that of Russia. Reference to the text of the PCA and studies highlighting that only some aspects of Russian environmental legislation are better than that of the EU – on paper and certainly not in practice – convinced them that the focus should be on approximating Russia's environmental law to that of the EU. A more persistent issue concerned the initial costs of cleaner production methods and the way in which these would harm Russia's economic growth – an issue that also recurs where the non-ratification of the Kyoto Protocol is concerned.

Draft conclusions were discussed with numerous stakeholders, including permitting and enforcement authorities, NGOs and industry, and were finalized taking their comments into account. The recommendations were also investigated and tested through case studies in three pilot regions.[19] The main recommendation, thus developed, was that Russia should introduce a system of integrated permitting based on BAT for those branches of industry that are the main polluters. This system would stimulate resource efficiency, innovation and more effective protection of human health and the environment. The final report in which this recommendation and others were laid down was offered to and accepted by the State *Duma* Committee on Environment in November 2003. Through awareness-raising actions during the start of 2004, the project team ensured that the newly elected *Duma* and others were provided with information on these recommendations, in order to ensure that future legislative changes can build on the recommendations. A follow-up TACIS project has been announced in 2005. It might smooth the road towards actual implementation of the recommendations in practice, i.e. the actual adoption of amendments to the existing environmental legislation. As already indicated above, Russia is under no strict legal obligation actually to approximate its legislation to that of the EU. In this respect, the view expressed by Commissioner Chris Patten on the extension of key parts of the EU single market comes to mind: 'It is not a matter of imposing EU legislation word for word on Russia, but of offering a model for economic and legislative integration'.[20]

3.2   Four Common Spaces

At the St Petersburg Summit in May 2003, the EU and Russia agreed to reinforce their cooperation in the framework of the PCA. As evident from the Joint Statement issued at the end of the summit,[21] it was agreed that the goal of common economic and social spaces set out in the 1999 Common Strategy would be expanded to four common spaces that would need to be created in the long term: a common economic space (in a new, broader sense); a common space of freedom, security and justice; a space of cooperation in the field of external security; and one of research and education, including cultural aspects.

The environment was one of the areas touched upon extensively in connection with the common economic space. It was made known in the Joint Statement that the two sides had agreed 'to make every necessary effort to ensure that the Kyoto Protocol becomes a real tool for solving the problems of global warming and to this end we shall seek its entry into force as soon as possible'. Russia and the EU also confirmed their readiness to cooperate closely in preparing the

World Conference on Climate Change, which was to be held in Russia later that year. The EU asked Russia to submit the Kyoto Protocol to the *Duma* for ratification as soon as possible. At the Conference, President Putin explained how Russia interpreted 'as soon as possible'.

3.3 The World Conference on Climate Change and Further Russian Arguments

This Conference, convened between 29 September and 3 October 2003, was intended to be the first meeting of the parties to the Kyoto Protocol. Since Russia had not ratified, this was not possible. Instead, it was hoped that Russia would use this opportunity to announce when it would ratify. However, President Putin's opening speech dashed any hopes for a swift entry into force of Kyoto:

> Russia is being actively called on to ratify the Kyoto protocol as soon as possible. I am certain that these appeals will also be heard many times at your meeting. I want to say that the Government of the Russian Federation is carefully examining and studying this issue, studying the entire range of complex problems connected with this.

Putin also said 'A decision will be made after this work is finished. And, of course, it will be made in accordance with the national interests of the Russian Federation'. The President later added, during the forum:

> In Russia, you can often hear, either in joke or seriously, that for a northern country like Russia, it would be no big deal if it gets 2 or 3 degrees warmer. Maybe it would be even better – we would spend less money on fur coats and other warm things, and agriculture specialists say our grain production will increase, and thank God for that. All that is true. But we must also think about something else. We much think about the consequences of these possibly global climate changes.[22]

Joke Waller-Hunter, the executive secretary of the UN Framework Convention on Climate Change, which also oversees the Kyoto Protocol, was diplomatic in her praise of Russia for its commitment to climate change issues. At the same time, she was frank in expressing her disappointment: 'We had hoped that [President Putin] would have been somewhat more specific on the date when he would expect the Russian ratification to take place'. She also said: 'Last year in Johannesburg, the [Russian] prime minister announced Russia would ratify the Kyoto Protocol in the nearest future, and we had hoped the nearest future had come today and that we would have a clear signal'.[23]

The Conference itself definitely examined the issue of climate change and a very wide range of problems linked to it, but consensus on the benefits of Russia's ratification of the Kyoto Protocol was not reached. The opponents of Kyoto brought forward fierce arguments. Andrei Illarionov, economic adviser to President Putin, warned for instance that 'the Kyoto Protocol will stymie economic growth; it will doom Russia to poverty, weakness and backwardness', claiming that each percentage point of GDP growth is accompanied by a 2% growth in CO_2 emissions.[24] Kyoto would thus stand in the way of reaching President Putin's 2003 goal of doubling Russia's GDP by the year 2010. As Russia's GDP grew by 7.3% in 2003 and 7.1% in 2004, average annual growth during the 2005–10 period would need to be more than 9.6% in order to reach this goal – a highly unlikely scenario, with or without ratification of the Kyoto Protocol. Such arguments did meet with opposition,[25] showing some of the mistakes in Illarionov's reasoning (notably with regard to the link between reductions in CO_2 emissions and GDP growth in many transition economies), but these counterarguments were far less captivating than the original messages of doom and received little attention in the Russian media.

One opinion piece by Vyacheslav Nikonov, Director of the Politika Foundation, argued that, with the United States (US) out of the game, Russian gains from emissions trading would be much smaller than earlier projected. He also objected that the EU had created its emissions trading scheme without consulting Russia, claiming that the scheme would not provide Russia with the kind of market that it had hoped for. In his view, this fairly closed scheme would lead to trade being focused on the new EU Member States, which would benefit from the influx of money and new technology. Nikonov also pointed to concessions by the EU in the area of trade that Russia had hoped for, but which had not been granted, including concession with regard to WTO accession, visa-free entry and anti-dumping investigations. He warned that with a number of uncertainties surrounding the effects of the Protocol on Russia and the potential of the emissions trading market, Russia could not 'sell our future economic growth for an unspecified price'.[26]

Later on, Illarionov even went as far as stating that:

> The Kyoto Protocol is a death pact, however strange it may sound, because its main aim is to strangle economic growth and economic activity in countries that accept the protocol's requirements. At first we wanted to call this agreement a kind of international Gosplan, but then we realized Gosplan was much more humane and so we ought to call the Kyoto Protocol an international gulag.[27] In the gulag, though, you got the same ration daily and it didn't get smaller day by day. In the end we had to call the Kyoto Protocol an international Auschwitz.[28]

These verbal assaults seem to indicate that Russia was not aiming at a serious debate on these matters but rather that it was looking for concessions in other areas, like its WTO accession and the extension of the PCA. Before commenting on the likelihood of this hypothesis being true, some of the events on the eve of the enlargement will be scrutinized.

3.4 Italian Alleingänge

From the point of view of a European common foreign policy, a rather disastrous general EU-Russia summit took place in Rome in November 2003. At a joint news conference with President Vladimir Putin, Prime Minister Berlusconi endorsed Kremlin policies ranging from the arrest of oil magnate Khodorosvsky to the war in Chechnya. The Commission and most Member States dissociated themselves from Berlusconi's statements.

In the Joint Statement, Russia and the EU stated that they recognized their 'responsibility to tackle together and in the framework of relevant international organisations, instruments and fora, common environmental challenges and shared concerns regarding climate change', but did not refer to the Kyoto Protocol. While the two sides solemnly pledged to reinforce their strategic partnership, progress on the extension of the PCA, needed before 1 May 2004 because of the imminent enlargement of the EU, was also left unmentioned.[29]

In December 2003, Italian Environment Minister Altero Matteoli criticized the European Commission for wrongly using strong-arm tactics to try to get Russia to ratify Kyoto. In his opinion, softer methods would be more effective. He failed to spell out what was wrong with the Commission's tactics, and claimed that 'there has never been a declaration by Russia against Kyoto',[30] overlooking the numerous times at which Kyoto was condemned for standing in the way of economic growth by people like Andrei Illarionov.

3.5 The Controversies Surrounding Extension of the PCA

On 14 January 2004, Russia presented a list of 14 specific concerns with respect to the imminent EU enlargement. Among these concerns were issues such as Russian agricultural exports to the new Member States, which would need to comply with stricter hygiene and sanitary standards, quantitative restrictions regarding the export of Russian steel and grain, and higher import tariffs on goods. Furthermore, Russia demanded the non-application of EU barriers against Russian exports of nuclear fuel materials to the trade in such materials carried out under agreements and contracts signed between Russia and the relevant acceding countries before 1 May 2004, as well as guarantees regarding the non-application of restrictions under EC directives on the diversification of energy supply sources regarding quantitative or other ceilings for energy imports from Russia. Some Russians even put a price tag on all of these concerns, claiming that the enlargement process would cost Russia €150 million per year,[31] while Deputy Foreign Minster Vladimir Chizhov – in charge of relations with the EU – even estimated that it would cost Russia $375 million.[32] Consequently, Russia claimed that the enlargement of the EU called for substantial changes to the content of the 1994 PCA.

These demands resulted in a sharp reaction from the EU. In its conclusions of 23 February 2004, the External Relations Council confirmed that the 1994 PCA 'remains the essential cornerstone of the European Union's relations with Russia'. It emphasized that the PCA 'has to be applied to the EU–25 without pre-condition or distinction by 1 May 2004. To do so would avoid a serious impact on EU–Russia relations in general'. According to the Council, the timely extension of the PCA would allow Russia to benefit from the many advantages accruing to it from EU enlargement, including in the trade field. The EU was open to discussing any of Russia's legitimate concerns over the impact of enlargement, but insisted that this should remain entirely separate from the PCA extension.[33]

At the same meeting, an internal assessment of the Common Strategy for Russia was presented with recommendations to strengthen the partnership. The *Financial Times* revealed that the memo concluded that the EU failed to speak with one voice and that it contained the following passage: 'Russia has often made use of differences among the Member States and EU Institutions and used a policy of linking often unrelated issues'.[34]

This assessment obviously did not lead to immediate improvements within the EU. The following day, French President Jacques Chirac stated that the EU should show more respect for Russia's national interests as Moscow adjusts to EU enlargement, and that the EU had been too tough on Moscow in the past: 'Moscow is making an enormous effort to regain its rightful place in the world', Chirac said, 'We must convince Russia that we regard its efforts with friendship'.[35]

Hectic negotiations followed right until the end of April, i.e. just before the actual EU enlargement. Only on 27 April 2004 was agreement reached to extend the 1994 PCA as the EU had demanded. The Protocol in which this extension was laid down applies provisionally as of 1 May 2004, until all instruments of approval have been laid down.[36] A Joint Statement on EU Enlargement and EU–Russia Relations was adopted alongside the Protocol that affirmed the extension itself. The Statement deals with many of the concerns put forward by Russia. For instance, both sides acknowledged that the overall level of tariffs for the import of goods of Russian origin to the new Member States would decrease from an average of 9% to around 4%, due to the application by the enlarged EU of the Common Customs Tariff to imports from Russia, as of 1 May 2004, leading to improved conditions for trade. Furthermore, the Statement underlined that agreement had been reached on numerous other issues ranging from steel, anti-dumping and the import of fossil fuels and electricity to visa requirements and noisy aircraft.

3.6 WTO and Kyoto Progress: EU–Russia Summit of 21 May 2004

If a country wants to join the WTO, it first needs to conclude bilateral agreements with any WTO member demanding this. The EU, Russia's major trading partner, demanded, among other things, that Russia would increase its energy prices for the internal market, notably for gas. The discussions were long and hard. Prodi, thanking the negotiating team at the May 2004 summit where an agreement was finally struck, put it as follows: 'They worked like hell'.

At the same time, progress was booked on Kyoto. President Putin stated that 'We will speed up Russia's movement towards ratifying the Kyoto Protocol'.[37] He did stress, however, that ratification is not an issue for him but for the Parliament to decide. Considering the overwhelming support for Putin in the State *Duma* elected in December 2003, this might have been true on paper. In practice, however, the *Duma* would follow Putin on this matter. Fortunately, this time Putin must have urged Parliament to ratify, because on 22 October 2004 the *Duma* members voted 334 in favour of ratification, with 73 against and two abstentions. As a result, the Kyoto Protocol entered into force on 16 February 2005.

4.   A LINK BETWEEN PCA, WTO AND KYOTO?

Was there a link between the agreements on PCA extension, WTO accession and Kyoto ratification? According to Sergey Yastrzhembsky, envoy of the Kremlin to the EU, there was. Shortly before the May 2004 summit, he stated that the apparent breakthrough on Russian accession to the WTO could pave the way for Russia to ratify the Kyoto Protocol by the end of 2004, and that a deal could be done. 'We don't say no [to Kyoto]' he said, adding that Russia would look for certain measures taken in return: 'We would like to see that our interests are welcomed and satisfied in different spheres … for example, the WTO'.[38] At the summit itself, Prime Minister Bertie Ahern of Ireland also appeared to describe a link. He underlined that the WTO talks had brought new clarity and that this could simplify ratification of the Kyoto Protocol.[39] President Putin explained first of all that Russia did not link WTO accession and the Kyoto Protocol. He elaborated, saying that Russia had long since made its position clear: 'We are for the Kyoto process and we support it. We do have some concerns regarding the commitments we would have to take on. But one of the biggest concerns was that we had to face several issues all at once that imposed serious commitments on us and gave rise to a great number of risks: EU enlargement, Kyoto itself with its potential complications and consequences, and the process of our accession to the World Trade Organization. The European Union has made some concessions on some points during the negotiations on the WTO.

This will inevitably have an impact on our positive attitude to the Kyoto process. We will speed up Russia's movement towards ratifying the Kyoto Protocol'.[40]

These elaborations indicate that, in practice, a link did exist. They also show that Illarionov himself was distorting facts in the Kyoto debate. Asked by a journalist whether his fierce opposition to Kyoto reflected the Kremlin's position, Illarionov replied that 'Putin didn't say he supports the Kyoto Protocol, he said he supports the Kyoto process'.[41] The opposite was true,[42] fortunately.

5.   PROSPECTS FOR EU–RUSSIA COOPERATION UNDER THE KYOTO PROTOCOL

With the Kyoto Protocol having entered into effect, the EU and Russia seem two ideal partners to achieve the targets that this protocol sets for the two sides. By the years 2008–12, the EU Member States will need to cut emissions corresponding to their target as set in the Kyoto Protocol or the EC burden-sharing agreement. In order to reach this target, the measures to decrease emissions inside the individual Member States can be accompanied by measures in other countries like Russia via the Protocol's so-called flexible mechanisms, notably Emissions Trading (ET), under which industrialized nations with targets under Kyoto can meet these targets by trading emission allowances among themselves and gaining credits for emission-curbing projects abroad, and Joint Implementation (JI), i.e. emission-reducing projects in countries that have a surplus of emission credits in the period 2008–12.

Russia has a zero emissions reduction commitment, meaning that the yearly emissions in the commitment period 2008–12 should not exceed the 1990 level of emissions. According to the Russian Government's own estimates in its Energy Strategy until 2020, Russia is not likely to exceed 1990 emission levels before 2020. Even with a doubling of its GDP by the year 2010 (a scenario that in all likelihood will not become reality, as was explained above), Russia's annual total greenhouse gas (GHG) emissions in 2008–12 will not exceed the 1990 level. Thus, Russia – with its considerable surplus of emission credits – could benefit a lot from the Kyoto Protocol and its flexible mechanisms. The country could benefit by selling emissions credits through JI projects or directly to other industrialized countries.[43] Any JI could attract foreign investment from the EU in concrete emissions reductions projects in exchange for the credits being transferred from Russia. Thus, Russia certainly has a large potential to reduce emissions: its energy efficiency is notoriously low.

In practice, making the most out of the ET and JI opportunities will be possible only if considerable steps forward are taken in Russia. A national system for estimating emissions and removals of GHG within Russia's territory will need to be established. Moreover, a national registry must record and track the creation and movement of assigned amount units (AAUs), emission reduction units (ERUs) etc. Furthermore, Russia must annually report information on emissions and removals to the Kyoto Protocol secretariat.

A joint statement by two important Russian companies to Russia's Prime Minister Fradkov in September 2005 expressed that 'Russian business is worried about the rates of preparing the rules and procedures for such projects, especially for joint implementation projects'.[44] These concerns are understandable, as JI might be the only effective way to attract foreign activity under the Kyoto protocol. This possibility exists as Russia could end up becoming a Track II Party, i.e. a country not satisfying the Track I requirements and thus unable to sell any potential assigned amount surpluses to other Parties, and thus only able to sell credits via JI projects. The Track I status is reserved for countries that can demonstrate that their national inventories are sufficiently reliable, and enables them to assigned AAUs trading under art. 17 of the Kyoto Protocol.[45]

Several initiatives have been taken that could contribute to a successful Russian–European cooperation in the fight against climate change and for energy efficiency. In the Netherlands, a submission from Russia was selected under the ERUPT-5 tender.[46] Unified Energy System (UES) subsidiaries signed Russia's first Kyoto-linked contracts in June 2005. The projects involve the upgrading of the Mednogorsk Power Plant and fuel-switch at the Amursk Power Plant. The total investment in the two plants will amount to €20 million and resulting in a reduction of CO_2, of which DanishCarbon has purchased 1.2 million tons. The two projects were identified in DanishCarbon first tender round in September 2004, and DanishCarbon expects the two projects to pave the way for approval of further JI projects in Russia. RAO UES still has to obtain a 'Letter of Approval' from the Russian Government for the deal in order for the two contracts to enter into force. It is feared, however, that such approval will not be given in time: 'If, by the end of October this year, the contract is not approved by the Russian government, it will be severed by the Danish side, which will undermine foreign investors' trust in the Russian greenhouse gas market' the joint statement from September 2005 (discussed above) stated.

The lack of major investments in Russia so far was also discussed at the International Summer Academy, 'Energy and the Environment', organized in Irkutsk from 21–27 August 2005. The missing institutional and legal framework for JI projects, a lack of political will to focus on climate change issues and the difficult investment environment in Russia were mentioned there as the most pressing reasons for the lack of initiatives so far.[47]

The EU is aware of the hurdles that need to be overcome and has offered its support in the form of a €2 million project aimed to build the capacity of the Russian Federation to implement the Kyoto Protocol. Concretely, the project aims to recommend improvements to Russia's monitoring systems and inventory procedures for GHG emissions; to make proposals for the legal basis of the

Russian monitoring and reporting systems concerning GHG emissions, to draft proposals for a national GHG emissions registry, including its design, organization and legal basis, to transfer knowledge to key Russian stakeholders of 'best practice' in monitoring and reporting GHG emissions and setting up national registries. Furthermore, the project aims at making recommendations for a Russian system to carry out projects under the Kyoto Protocol, with national guidelines for their preparation and approval, together with the institutional organization and appropriate legal provisions of the system. The leader of the consortium carrying out the project indicated in June 2005 that: 'It is important to note that the consequences of Russia not being ready to participate in the carbon market could be severe for companies in Europe, North America, and Japan, which have emissions reduction obligations'.[48]

If Russia does succeed in establishing an effective Kyoto Protocol implementation system, one more worry exists – the 'hot air' trading. If Russia's emissions budgets under the Kyoto Protocol exceed its actual emissions, it will end up with unused AAUs. These AAUs could be transferred without having to take mitigation measures. Russia could thus sell this surplus to other countries. The latter can use this to cover emissions that would not have been allowed without the transfer of this surplus, making overall emissions higher than without such trading. An excellent analysis of the many aspects to 'hot air' trading indicates that 'hot air' trading might be considered unavoidable in a system such as the one established via the Kyoto Protocol.[49]

To conclude, it seems safe to say that the relationship between Russia and the EU could benefit from cooperation on Kyoto-related issues. Resourcefulness and ingenuity will be asked for to achieve concrete results and to overcome the many hurdles on the way. In the first place from the side of Russia,[50] but also from the side of the EU. There, concerted action will be needed more than ever before. Having persuaded Russia to ratify the Kyoto Protocol might yet turn out to have been simple compared to what lies ahead.

NOTES

1. Dr Wybe Th. Douma, Head of the Department of European Law, TMC Asser Institute, The Hague, e mail w.t.douma@asser.nl. This contribution forms a revised and updated version of parts of my article (2004) 'Russia and the enlarged European Union. Mending fences whilst shaping a Wider Europe's environment', in Jaap de Zwaan, Steven F. Blockmans and Frans A. Nelissen (eds), *The European Union, a Gradual Process of Integration (Liber Amoricum for Alfred E. Kellermann)*, The Hague, The Netherlands: Asser Press, pp. 319–36.
2. Agreement on Partnership and Cooperation between the European Communities and their Member States and the Russian Federation, OJ 1997 L 327/3.
3. See section 3.2 below.
4. Council and Commission Decision of 30 October 1997, OJ 1997 L 327/1–2.
5. The careful phrasing of this duty indicates that it constitutes an obligation to perform to the best of one's abilities rather than an obligation to ensure a certain result.

6. Other provisions in the PCA mentioning the environment are those on agriculture (art. 64), energy (art. 65), construction (art. 68), the Coal and Steel Contact Group (Protocol 1) and the Joint Declaration in relation to art. 6 of Protocol 2.

7. Common strategies set out overall policy guidelines for activities with individual countries. This instrument was introduced by the Amsterdam Treaty in 1999 (art. 13(2) EU). They are decided by the European Council, on a recommendation from the Council, in areas where the Member States have important interests. Each strategy specifies its objectives, its duration and the resources that will have to be provided by the EU and the Member States. The Council implements them by adopting joint actions and common positions.

8. OJ 1999 L 157/1. The Common Strategy was valid until 24 June 2003 and was extended without further amendment until 24 June 2004 (OJ 2003 L 157/68).

9. See www.eur.ru/en/p_239.htm.

10. See www.eur.ru/en/p_238.htm.

11. COM (2001) 772 final, Brussels, 17 December 2001.

12. 'Putin calls for overhaul of environmental policy', *Moscow Times*, 5 June 2003.

13. Press release IP/03/322, Brussels, 5 March 2003.

14. For a full overview of these developments, see Kolbasov, O. and Krasnova, I. (2003), 'Environmental Law of the Russian Federation', in *International Encyclopaedia of Law*, The Hague/London/New York: Kluwer Law International.

15. The overall responsibility for administrative reform lies with the Government Commission, headed by P.M. Kasyanov and set up in August 2001.

16. See M.M. Бринчук и О.Л. (2003), 'Федеральный закон 'Об охране окружающей среды': теория и практика', **3**, 30–41. For a detailed study in English and Russian of this law, see *Inception Report* of TACIS Project 'Harmonisation of Environmental Standards – Russia', Annex 1, Review of Legislative and Institutional Framework, and the final report of the same project.

17. See Cherp, A. and Golubeva, S. (2004), 'Environmental assessment in the Russian Federation: evolution through capacity building', in *Impact Assessment and Project Appraisal*, **22**(2), 121–30; von Ritter, Konrad and Vladimir Tsirkunov (2002), *How Well is Environmental Assessment Working in Russia: A Pilot Study to Assess the Capacity of Russia's EA System*, Moscow and Washington: World Bank.

18. These included, *inter alia*, assistance to the Ministry of Natural Resources on water management (€2.4 million); assistance to the Ministry for Emergencies to improve industrial accident prevention (€3 million); the inclusion of an environmental protection element in the conversion of the chemical weapon industry (€4 million); continued institutional support to the State Committee for Environmental Protection (SCEP) and to regional authorities in the area of waste management (€2.5 million) and environment monitoring systems (€2.5 million).

19. Moscow Oblast, Penza Oblast and Arkhangelsk Oblast.

20. Chris Patten, speech at the European Business Club Conference: 'Investing in Russia', Brussels, 2 October 2001.

21. See www.eur.ru/en/p_234.htm.

22. See www.kremlin.ru/eng/text/speeches/2003/09/29/1942_53028.shtml. See also Lambroschini, S. (2003), 'Putin remarks start off international climate conference on sour note', available at www.rferl.org/features/2003/09/29092003152837.asp.

23. See www.unfccc.int/press/stat2003/stat_290903.pdf.

24. See www.rusnet.nl/news/2003/10/07/print/commentary_01_5423.shtml.

25. For instance, by Bert Bolin (chair emeritus of IPCC) and others, who at the conference presented 'Answers to the questions raised by AN Illarionov during his talk 'Anthropogenic factors in global warming: some questions at the World Climate Conference 2003, prepared from material of the IPCC Third Assessment Report (TAR) by attending scientists'. Document on file with author. Igor Bahmakov, 'Russian GDP doubling, District heating and climate change', paper presented at UNFCCC Workshop on 19 June 2004 in Bonn, Germany – pointed out that in order to reach Kyoto targets, Russia needs to radically improve its energy efficiency. See www.cenef.ru/bulletin/Bonn.ppt.

26. Nikonov, V. (2003), 'Costs of Kyoto', *The Statesman*, 21 September 2003.

27. The Soviet labour camps where millions of people died.

28. 'Illarionov likens Kyoto to Auschwitz', *Moscow Times*, 15 April 2004.
29. See http://www.eur.ru/en/p_236.htm.
30. 'Italians assail treatment of Moscow over Kyoto', *Moscow Times*, 10 December 2003.
31. According to Konstantin Kosachyov, head of the State *Duma*'s international affairs committee, in an interview with *Interfax* on Monday, 2 February 2004.
32. *Financial Times*, 23 February 2004.
33. Conclusions of the Foreign Affairs Council, Brussels, 23 February 2004, 6294/04 (Presse 49).
34. Dempsey, J. (2004), 'EU report criticises relationship with Russia', *Financial Times*, 22 February 2004.
35. See http://www.eubusiness.com/archive/afp/040224141512.ytennu6a, 24 February 2004.
36. Protocol, art. 5. The texts of the Protocol and of the original PCA are available at www.eur.ru/en/p_243.htm.
37. Press Conference following the EU-Russia Summit, 21 May 2004. A transcription is available at www.eur.ru/en/news_584.htm.
38. Interview with the *EU Observer*, 18 May 2004, see www.euobserver.com.
39. Press Conference of 21 May 2004, see n. 37 above.
40. *Ibid*. He also stated: 'We expressed concern that EU enlargement, the complications involved in the WTO negotiating process and our full adherence to the Kyoto process could give rise to problems for the Russian economy in the long and medium term. ... This was our concern and it was listened to. I want to emphasise that we did not push our demands at all costs, but rather, we listened to each other and reached agreement on taking certain steps regarding EU enlargement'. President Putin further explained that the agreement on Russia's WTO accession 'reduces the risk for our economy in the medium term and it frees us to a certain extent to move faster to resolve the question of Russia joining the Kyoto process. I cannot tell you with 100 per cent certainty exactly how this will proceed because ratification is the responsibility not of the President but of the Russian parliament. But we will speed up this process, as I already said'.
41. 'Illarionov attacks Britain, vows to bury Kyoto', *Moscow Times*, 12 July 2004. In the same article, Illarionov is also quoted as accusing British Prime Minister Tony Blair's government of declaring 'all-out and total war on Russia' and using 'bribes, blackmail and murder threats' to force Russia to ratify the Kyoto Protocol.
42. For further details, see Douma, W.Th. (2004), Letter in response to 'Illarionov attacks Britain, Vows to Bury Kyoto', *Moscow Times*, 23 July 2004.
43. See http://www.delrus.cec.eu.int/en/p_231.htm.
44. Unified Energy System (UES) chairman Anatoly Chubais and Sistema head Vladimir Yevtushenkov.
45. See further on this issue Jepma, C.J. (2005), 'JI's tricky tracks', *Joint Implementation Quarterly*, **11**(3), 1.
46. KronoClimate Fuel Switch project, ERU05/20, contracted amount 1,653,812 ERUs and 597,569 AAUs. See www.carboncredits.nl.
47. See www.ecologic.de/modules.php?name=News&file=article&sid=1508.
48. The Consortium contracted by the European Commission to implement the project is led by ICF Consulting Ltd (UK), together with the Institute of Global Problems of Energy Efficiency and Ecology (Russia), Hogan and Hartson (UK). See www.icfconsulting.com. The major Russian partner is the Economic Development and Trade Ministry.
49. Woerdman, E. (2005), 'Hot air trading under the Kyoto Protocol: an environmental problem or not?', *European Environmental Law Review*, **14**(3), 71–77.
50. For further information on this topic, see Yulkin, M. (2005), 'Russia and the Kyoto Protocol: how to meet the challenges and not to miss the chances', www.eel.nl/index.asp?ssc_nr=202, 1 October 2005.

PART II

Greenhouse gas emissions trading within the EU

4. Reviewing the challenging task faced by Member States in implementing the Emissions Trading Directive: Issues of Member State liability

Mar Campins Eritja[1]

1. THE EU ETS DIRECTIVE

The Directive 2003/87/EC adopted in July 2003[2] establishes a legal framework for an emission trading scheme (ET) covering carbon dioxide releases. The deadline for implementing the Directive within domestic law was 31 December 2003. By 31 March 2004, national allocation plans for the period covering 2005–07 had to be submitted to the EC Commission for approval.[3] Drawing up these national allocation plans was the most critical challenge facing Member States as they prepared to implement the ET Directive. In doing so, they were bound by articles 9 and 10 of the ET Directive and, more specifically, by the 11 criteria of Annex III, which had been interpreted by the EC Commission in its Decision 2004/156/EC of 20 January 2004.[4] By 1 January 2005, Member States had to ensure that no facility undertook any activity listed in Annex I resulting in emissions unless its operator held a site-specific and non-transferable greenhouse gas (GHG) emissions permit, issued by the competent authorities. At the same time, Member States had to guarantee that emission allowances were properly allocated to relevant installations. Thus, a system for granting allowances also had to be available at the national level. By 1 April 2006, companies and corporations have to deliver for the first time a sufficient number of allowances to cover emissions during 2005.

From the point of view of the European Union (EU), the implementation of the ET Directive raises a number of important issues. First, Member States have responded to the ET Directive in different ways. Their reactions were shaped by their policy preferences and capacity of action, particularly in the area of regulatory policies. While some Member States have actively pushed for an ET scheme as a policy preference, and have undertaken specific commitments to reduce CO_2 emissions, other Member States have been strictly reactive since they feared such reductions would restrict their economic growth and that this would have significant political consequences. Furthermore, implementation should be based, as far as possible, on the existing administrative and legal structures. Within each Member State, different levels of power may interact, and here too the responses amongst Member States are different. In Spain, for instance, powers relating to climate change are shared by central, regional and local authorities. The Spanish legislation implementing Directive 2003/87/EC[5] makes specific mention of this issue and reserves certain aspects to the jurisdiction of central government (emissions permits, monitoring, reporting and verification) while others are covered by regional powers. At the other extreme, it may be that central government is powerless because these matters come entirely under the jurisdiction of the regions, as is the case in Belgium, where the Walloon region, the Flemish region and Brussels are implementing their own separate allocation methodologies and plans.[6] In addition, legal administrative traditions and existing national procedures differ between Member States, and different approaches to granting permits and allowances are used, as well as diverse systems for monitoring, validation, registry, reporting and verification of compliance at the national level, and as regards to which authorities enforce them.[7] Other collateral issues addressed in different ways by Member States include public education; transparency, accountability and participation in the scheme; acknowledgment of policy synergies between climate change and other policy areas; abolition of polluter subsidies; and the use of green technologies and the promotion of research to fight the effects of climate change.

2.   STATE LIABILITY BEFORE THE EU'S INSTITUTIONS

The biggest challenge facing Member States was how to implement the ET Directive within a short period of time. For its part, the EC Commission's task was to oversee the national allocation plans in order to ensure that they were consistent with the ET Directive and Decision 2004/156/EC, as well as with the EU's overall strategy for reaching its Kyoto Protocol target.

The main concern of market actors about the process before the EC Commission was, and remains, whether or not the EU would be able to challenge Member States and get them to reduce their GHG emissions in line with the Kyoto Protocol target. Under pressure from domestic industry,[8] most Member State governments allocated more than the baseline, though the EC Commission had urged them to allocate less. From the environmental perspective, it is true that most of the national allocation plans finally approved by the EC Commission indicate a modest reduction of emissions from a business-as-usual scenario (from 0.5–5%), while the countries' emissions trends and Kyoto targets dictated the opposite (excessive allocation and ex-post adjustments have been mentioned by the EC Commission as two elements that may jeopardize the achievement of the Kyoto Protocol target).

Thus, it is still doubtful whether all Member States are ready to enter fully the ET scheme. Although Member States have already passed national legislation, most of them have not yet faced its practical implementation. The overall result, in general terms, is a lack of a consistent regulatory environment, at the national level at least, which may end up distorting the internal market.

In short, it is not the system's lack of economic efficiency that is at stake, but whether or not the EU as a whole would be able to meet its Kyoto Protocol target. It is clear that the cost of meeting the target in the first Kyoto commitment period (2008–12) will finally depend not on the national allocation plans of each Member State, but on the EU's success, or lack of it, in limiting the amount of allowances for the whole of the EU. In addition, it should be emphasized that the presentation of credible domestic abatement programmes by the EU and its Member States will be essential in the international negotiations regarding the Kyoto Protocol's second commitment period (2013–17).

The current situation is as follows: in July 2004, the EC Commission accepted the first eight National Allocation Plans (NAPs).[9] Five plans – from Denmark, Ireland, the Netherlands, Slovenia and Sweden – were accepted unconditionally. Three others – from Austria, Germany and the United Kingdom[10] – were approved on the condition that technical changes be made. In October 2004,[11] the Commission accepted another eight NAPs. Six of them – from Belgium, Estonia, Latvia, Luxembourg, Portugal and the Slovak Republic – were accepted unconditionally.[12] Another two – from Finland and France – were approved, provided that some amendments were adopted.[13] In December 2004, the EC Commission accepted unconditionally the NAPs from Malta, Cyprus, Hungary and Lithuania.[14] The national plan from Spain[15] was also accepted, on condition that technical changes be made. In March and April 2005, the EC Commission approved the NAPs from Poland and the Czech Republic,[16] respectively. The Italian NAP was finally accepted in May 2005, subject to certain changes,[17] and the Greek NAP was approved without objections in June 2005.[18]

The EC Commission is responsible for ensuring that all Member States implement the ET Directive (and particularly for the implementation of measures related to the adoption of NAPs). Should the EC Commission consider that Member States are failing to take all the necessary measures to ensure the correct implementation of the Directive, it can start infraction proceedings under art. 226 EC Treaty.

However, as regards the implementation of the ET Directive, there are a number of obstacles that might question the feasibility and effectiveness[19] of such a proceeding: the EC Commission's lack of an effective structure or decentralized administration to monitor Member States' implementation; the formal process of infringement proceedings, which do not provide the most effective means for solving implementation problems at regional levels; and the average time between the start of proceedings by the EC Commission and the Court of Justice's judgment, which can take between 2–4 years. Another important point is the fact that implementation of the ET Directive can raise sensitive political questions relating to the consequences for domestic economies of its implementation, which may fall beyond the jurisdiction of the Court of Justice.

At the end of the procedure before the Court of Justice, should the latter find the Member State to be guilty, and further find that the said Member State does not comply with the Court's judgment, high fines may be imposed under EC law and upon the EC Commission's request. However, the case-law of the Court of Justice on this issue remains unclear. Its first judgment[20] was criticized for being limited and for causing imbalances between the EC Commission and the Court of Justice. Later judgments[21] reflected a rather narrow conception of both the execution procedure and the infringement procedure as a complementary instrument, and dealt only with its specificities incidentally and casuistically, particularly as regards the judgment execution term and the determination of the type and amount of the penalty. This issue has been especially addressed by the Court of Justice in its last judgment of 12 July 2005,[22] in which it establishes the imposition of a lump sum in addition of the penalty payment, in order to deter offenders from similar infringements. Against the objections raised by some Member States concerning legal certainty, transparency of penalties and the principle of equal treatment (since a lump sum was not envisaged in previous judgments), the Court of Justice decides that it has full discretion as to the financial penalties that can be imposed. As for the punitive nature of this procedure, the Court of Justice is clear when stating that:

> Article 228(2) EC is a special judicial procedure, peculiar to Community law, which cannot be equated with a civil procedure. The order imposing a penalty payment and/or a lump sum is not intended to compensate for damage caused by the Member State concerned, but to place it under economic pressure which induces it to put an end to the breach established. The financial penalties imposed must therefore be decided upon according to the degree of persuasion needed in order for the Member State in question to alter its conduct.

Also, where 'a breach of Community law persists, the rights of defence which are to be accorded to the defaulting Member State so far as concerns the financial penalties envisaged must take account of the objective pursued, namely securing and guaranteeing that the law is again complied with'.[23]

3.   STATE LIABILITY BEFORE EU CITIZENS

Beyond the EC Commission's infringement action, individuals, non-governmental organizations and companies can bring law suits against Member States for incorrect implementation of the ET Directive, provided that they may be able to establish the conditions set out by the Court of Justice case-law. According to the *Francovich* (1991) ruling:

> The full effectiveness of Community rules would be impaired and the protection of the rights which they grant would be weakened if individuals were unable to obtain redress when their rights are infringed by a breach of Community law for which a Member State can be held responsible ...
>
> It follows that the principle whereby a State must be liable for loss and damage caused to individuals as a result of breaches of Community law for which the State can be held responsible is inherent in the system of the Treaty ...
>
> It follows from all the foregoing that it is a principle of Community law that the Member States are obliged to make good loss and damage caused to individuals by breaches of Community law for which they can be held responsible.[24]

In the ET Directive, both the executive and the legislative power are involved in the implementation process. According to the *Brasserie du Pêcheur* (1996) ruling, a Member State may be held liable for the activity of any of its organs, whether legislative, administrative or judicial. The Court of Justice stated that:

> the principle holds good for any case in which a Member State breaches Community law, whatever be the organ of the State whose act or omission was responsible for the breach', and that 'the obligation to make good damage to individuals by breaches of Community law cannot depend on domestic rules as to the division of powers between constitutional authorities.[25]

In addition, and even if this is not the case here, it is worth mentioning that according to the *Klober* (2003) ruling, the principle of the State's liability also applies 'where the alleged infringement stems from a decision of a court adjudicating at last instance'. As for the Court of Justice:

> the principle of liability in question concerns not the personal liability of the judge but that of the State. The possibility that under certain conditions the State may be rendered liable for judicial decisions contrary to Community law does not appear to entail any particular risk that the independence of a court adjudicating at last instance will be called in question.[26]

Moreover, in countries with a federal structure, Member States may also be liable for the actions of decentralized institutions. The international dimensions of the climate change cannot be separated from the constitutional issue of the distribution of powers between different levels of government, since policies affecting climate change are typically cross-cutting. As for policies aimed at addressing climate change, decision-making as well as implementation processes need to be integrated both vertically (that is, internationally, nationally and sub-nationally) and horizontally (intra and inter-sectorally). Therefore, compliance with the commitments undertaken through the Kyoto Protocol and the ET Directive requires the identification, for each single case, of the roles to be played by the competent authorities at both State and sub-national level, as the actions to be implemented concern primarily regional or local powers.

Notwithstanding, and with respect to the particular question about liability within a federal, regional or autonomous State, the Court of Justice has not made it sufficiently clear whether these competent authorities may be liable for the improper implementation of Directives. According to its case-law, however, 'Member States cannot, therefore, escape that liability either by pleading the internal distribution of powers and responsibilities as between the bodies which exist within their national legal order or by claiming that the public authority responsible for the breach of Community law did not have the necessary powers, knowledge, means or resources'. With regard to federal Member States, the Court of Justice also stated that:

> if the procedural arrangements in the domestic system enable the rights which individuals derive from the Community legal system to be effectively protected and it is not more difficult to assert those rights than the rights which they derive from the domestic legal system, reparation for loss and damage caused to individuals by national measures taken in breach of Community law need not necessarily be provided by the federal State in order for the Community law obligations of the Member State concerned to be fulfilled.[27]

The question thus raised is how the case-law that derives from the *Francovich* judgment can be evaluated in terms of the ET Directive's implementation.

According to the Court of Justice's subsequent case-law, the right to reparation of an individual who has suffered damage is subject to the following conditions: 'the rule of law infringed must have been intended to confer rights on individuals, the breach must be sufficiently serious, and there must be a direct link between the breach of the obligation resting on the State and the damage sustained by the individuals.'[28]

3.1   Breach of Rights

First, the principle established in the *Francovich* ruling only applies to a breach of rights conferred by EC law. However, this gives rise to a real problem insofar as most of the EC environmental rules do not create subjective rights, but specially vest interests in individuals. Two different positions can be considered here. The first considers the protection of individual rights conferred by the EC law exclusively, understanding that such expression only includes subjective rights. The second, in contrast, conceives this requirement in a broader sense to include legitimate collective rights or interests within the concept of rights to be conferred by EC law.

It would be feasible to allow the possibility of claiming for the damages that non-implementation by Member States may cause to both the individual and collective rights that EC law intends to protect. This is the case when non-implementation affects certain rights conferred by environmental Directives or primary EC law, such as those related to public participation, access to information, or the protection of health.[29] In addition, an open interpretation has recently been accepted by the Court of Justice, for instance, in the *Wells* (2004) ruling concerning the Directive on the assessment of the environmental effects of certain projects. In such case-law, the Court of Justice ruled that 'mere adverse repercussions on the rights of third parties, even if the repercussions are certain, do not justify preventing an individual from invoking the provisions of a directive against the Member State concerned'.[30]

In the ET Directive, the situation is slightly different from what happens in most environmental cases. The ET regime may also include the protection of economic interests of companies or other actors involved in the EC emissions trading. As a consequence of the ET Directive's implementation, companies, for instance, can choose between investing in emission-reducing technologies or buying extra allowances in the market, and therefore avoid having to take more expensive measures. Thus, companies, as participants in the ET market, are qualified as beneficiaries of the ET Directive, and have right of access to national courts should they suffer economic damage (for example, damages caused by violation of the freedom of establishment, violation of the fundamental rights to property and the pursuit of an economic activity, or infringement of the principles of proportionality and of the principle of equality)[31] due to a Member State failing to implement the ET Directive on time or in full. In strict environmental interest terms, this situation may be extended to other natural or legal persons (for instance, non-governmental organizations which obviously cannot be allocated allowances but which may buy them). Thus, in principle, such individuals could be regarded as legally entitled and may claim against the Member State in order to defend their individual or collective rights or interests.

Therefore, individuals, companies and non-governmental organizations need Member States to implement the ET Directive correctly, and thus Member States must establish the proper legislative setting; ensure the issuance of allowances and permits; set the structure for monitoring, reporting, inspection and enforcement and define the competent authorities; provide for the registry of transfers of emission allowances; and supervise the procedure to acquire and account for

emission rights. Failure on the part of a Member State to comply with these requirements would mean that the interested actors would be denied the right and
full access to the EU allowance market, and therefore their guarantee of free
movement throughout the internal market would also be affected.

3.2   Serious Breach of EC Law

Secondly, a sufficiently serious breach of EC law is deemed to be a general
condition of State liability at EU level. However, the Court of Justice is far too
vague about what is considered a sufficiently serious breach of EC law, and apparently it is for the national courts to decide upon such an issue.

In the *Brasserie du Pêcheur* ruling, the Court of Justice stated that 'the decisive test for finding that a breach of Community law is sufficiently serious is
whether the Member State ... manifestly and gravely disregarded the limits on
its discretion'.[32] In *Dillenkofer* (1996) it also considered the breach to be sufficiently serious 'if, at the time when it committed the infringement, the Member
State in question was not called upon to make any legislative choices and had
only considerably reduced, or even no, discretion'.[33]

In the *Brasserie du Pêcheur* ruling, the Court of Justice listed several behav-
iours that may be considered as serious breaches of EC law; for instance, 'if
(the Member State) has persisted despite a judgment finding the infringement
in question to be established, or a preliminary ruling or settled case-law of the
Court on the matter from which it is clear that the conduct in question constituted an infringement'.[34] According to the *Dillenkofer* ruling, the Court of
Justice also stated that 'the mere infringement of Community law may be sufficient to establish the existence of a sufficiently serious breach'.[35] It also pointed
out in the *Klober* ruling that 'an infringement of Community law will be sufficiently serious where the decision concerned was made in manifest breach of
the case-law of the Court of Justice'.[36] However, the Court of Justice also stated
in *Dillenkofer* that:

> to make the reparation of loss or damage conditional upon the requirement that there
> must have been a prior finding by the Court of Justice of an infringement of Com-
> munity law attributable to a Member State would be contrary to the principle of
> effectiveness of Community law, since it would preclude any right to reparation ...
> The obligation for Member States to make good loss or damage caused to individuals
> by breaches of Community law attributable to the State cannot be limited to damage
> sustained after the delivery of a judgement of the Court finding the infringement in
> question.[37]

In *Brasserie du Pêcheur*, the Court of Justice also introduced several criteria
that should be taken into account by national courts when establishing the seriousness of the breach. They include:

the clarity and precision of the rule breached, the measure of discretion left by that rule to the national ... authorities, whether the infringement and the damage caused was intentional or involuntary, whether any error of law was excusable or inexcusable, the fact that the position taken by a Community institution may have contributed towards the omission, and the adoption or retention of national measures or practices contrary to the Community law.[38]

In most national legal systems, however, the State's liability derives from the objectively normal or abnormal functioning of public administrative authorities, now extended to the actions of any State organ. Therefore, it does not matter whether or not the fault or negligence caused the harm, nor whether the State bodies behaved in accordance with the rule.[39] This can be checked against the case-law of the Spanish Supreme Court, which overruled a claim brought by *Canal Satélite Digital* concerning property rights. The Spanish court broadened considerably the scope of the characterization given by the Court of Justice, and stated that:

The principle of State liability must always be interpreted extensively so as to favour the protection of the individual against the actions of the States ... pro-individual interpretation of State liability can be inferred from the fact that the conditions set out by the Court of Justice do not prejudice the application of any less restrictive national rules. This is particularly important in the sphere of our national legal system, in which the institute of State liability is of an objective nature. It is therefore sufficient to be in the presence of an individualized, illegal harm and of a causal link between the infringement act and the harm to engender this State liability.[40]

3.3 Direct Link Between Member State Conduct and Damage to Individual

Finally, there must be a direct causal link between the conduct of the Member State and the damage suffered by individuals. The mere lack of implementation of the ET Directive by the Member State, even if it has been previously declared by the Court of Justice, or the fault or negligence on the part of the organ of the Member State, are not definitive criteria for establishing causality and, thus, the Member State's liability.

It is up to the national court to determine whether there is a direct causal connection between the breach of EC law and the damage.[41] National courts have usually denied the causal connection between the damage and the ordinary operation of public authorities insofar as such a connection was not direct, immediate and exclusive, with the result that the concurrence of any alien circumstance usually meant that the State was not held liable. Today, the situation is slightly different, since most national courts accept the causal connection even if other circumstances intervene or if the damage arose as an indirect or mediate consequence.[42] Similarly, and according to the Court of Justice's *Re-*

chberger (1999) ruling, once the link has been established, the Member State's liability for a breach of EC law 'cannot be precluded by imprudent conduct ... or by the occurrence of exceptional and unforeseeable events'. Such circumstances, 'in as much as they would not have presented an obstacle' to the guarantee system, 'are not such as to preclude the existence of a direct causal link'.[43]

4.   STATE LIABILITY BEFORE THE KYOTO PROTOCOL'S PARTIES

At the international level, the participation of the EU and its Member States in the Kyoto Protocol's Compliance Procedure (art. 18 and Decision 24/CP.7 adopted by the Conference of the Parties in Marrakech in 2001), and with respect to international rules governing State responsibility, poses fairly relevant legal challenges that have to do with the nature of the powers shared between the EC and its Member States regarding the climate change regime. Not only will Member States have difficulties in implementing the ET Directive before the EU institutions and European citizens, but the EU itself also needs to tackle the issue of meeting its international target of an 8% reduction in EU emissions before other Kyoto Protocol Parties. The questions relating to the rules of standing applying to the EC and its Member States appear to be of particular interest.[44]

As for the legal standing to initiate international procedures, from the Community's perspective there are some elements that seem to require joint action by the EC and its Member States. Therefore, unilateral actions by Member States against other non-EC Parties would not fit well with the 'joint fulfilment' agreement subscribed under art. 4 of the Kyoto Protocol. Likewise, these actions could undermine the effect of art. 174.4 of the European Community Treaty (EC Treaty), which establishes that 'within their respective spheres of competence' in the framework of environment, 'the Community and the Member States shall cooperate with third countries and with the competent international organisations'. Furthermore, controversies within the EU should be resolved using the remedies provided by EC law: monitoring and control mechanisms, the infringement action against a Member State, or the action for failure to act against any eventual lack of action of the Community's institutions. In other words, a Member State cannot submit a question of compliance or a claim of responsibility against another Member State before the bodies of the Kyoto Protocol.

In terms of the legal standing for being sued, and in the event that a Member State does not comply with its obligations, it would be feasible to consider that a non-EC Party to the Kyoto Protocol cannot bring any actions against this

Member State individually. Therefore, the 'joint fulfilment' agreement implies a joint responsibility of the EC and its Member States, and this feature cannot be disregarded by other non-EC Parties to the Kyoto Protocol.

However, if the obligations can be separated from the 'European bubble' context, non-EC Parties to the Kyoto Protocol may not be jeopardized by the uncertainties concerning the specific distribution of competences between the EC and its Member States, with the result that they can either bring the case against the EC and its Member States jointly, or against any one of them. Thus, in implementing a mixed agreement, priority should be given to the interest in certainty and predictability[45] of the non-EC Parties.

5. FINAL CONSIDERATIONS

As regards the implementation of the ET Directive and the achievement of the Kyoto Protocol targets, the EU and its Member States are faced with both internal and external challenges that have to do with supervision and enforcement of EC law.

On the internal front, the basis for State liability lies, ultimately, in the special nature of art. 10 EC Treaty, according to which Member States not only have the duty to take all the necessary implementing measures, but 'shall abstain from any measure which could jeopardize the attainment of the objectives of this Treaty'.

The EC Commission ensures the correctness of measures taken by Member States. Should the latter fail to fulfil their obligations the EC Commission may bring the matter before the Court of Justice. Notwithstanding, a number of obstacles have been mentioned concerning the feasibility and effectiveness of the EC infringement procedure. In addition, Member States must nullify the unlawful consequences of a breach of EC law. Despite, however, the abundant case-law of the Court of Justice and the practice of national courts, there is no single, general and sufficiently consolidated path that can be applied to the ET Directive pertaining to the seriousness of the breach or the direct causal link between the breach and the damage caused to individuals involved in the ET system. The assessment of these conditions will, therefore, depend on a case-by-case analysis, and the difficulty of characterizing the breach and of establishing the direct causal link may require the Court of Justice to issue a preliminary ruling. Furthermore, the Court of Justice's case-law does not resolve the practical issues that the actors involved in the ET system may be confronted with, should the Member State fail to implement the ET Directive correctly.

On the external front, the shared nature of the powers conferred on the EU and Member States determines the issues related to the rules of standing within the framework of the climate change regime. These rules must be considered in relation to the 'duty of cooperation' between the Community's institutions and the Member States announced by the Court of Justice in its Opinion 1/94,[46] and without disregarding the preservation of the interests of third parties.

NOTES

1. Professor of Public International Law (EU Law) at the Faculty of Law of the Universitat de Barcelona, E-mail: mcampins@ub.edu. This paper has been conceived as part of the research project 'Nuevos enfoques de la responsabilidad ambiental: Más allá del principio quien contamina, paga' (BJU2003-09190).
2. OJ L 275, of 25 October 2003.
3. No flexibility was accorded to the implementation of the ET Directive by the acceding Member States, as it was not in place at the time that the Accession Treaties were negotiated. The Visegrad countries have all been included as Annex I Parties of the FCCC and as Annex B Parties of the Kyoto Protocol, but they were not considered for inclusion in the 'European bubble'. However, the use of the Kyoto Protocol project-based mechanisms, and particularly Joint Implementation could be a powerful tool for their integration in the EU climate policy strategy. See Fernández Armenteros, M., Michaelowa, A. (2002), *Joint Implementation and EU Accession Countries*, Discussion paper 173, Hamburg Institute of International Economics; Van der Gaast, W. (2002), 'The Scope of Joint Implementation in the EU Candidate Countries', *International Environmental Agreements: Politics, Law and Economics*, **2**, pp. 277–92.
4. OJ L 59, of 26 February 2004.
5. *Ley 1/2005, de 9 de marzo, por la que se regula el régimen del comercio de derechos de emisión de gases de efecto invernadero*, *BOE* of 10 March 2005. Depending on the issue, the Autonomous Communities (ACs) in Spain have normative competence for the development of basic State legislation and executive competence. Moreover, the ACs may adopt additional protection measures in accordance with art. 149.1.23 of the Spanish Constitution (CE) and the respective Statutes of Autonomy of the ACs. Overall, the competence in the field of environmental protection may thus be considered as a shared competence between the State and the ACs. In contrast, the ACs have received exclusive competence in attribution in the field of industry. Notwithstanding, according to the ruling of the Spanish Constitutional Court, the State may intervene in this particular field in the event that the State's measures concern the general planning of the economy (art. 149.1.13 of the CE), as would probably be the case for the issue at stake.
6. See the Federal National Allocation Plan of 24 June 2004 and the Allocations Plans for Brussels: Executive Order of 3 June 2004; Walloon region: Decree of 10 November 2004; and Flemish region: Decree of 2 April 2004 and Executive Order of 4 February 2005.
7. See Mullins, F., Karas, K. (2003), *EU Emissions Trading: Challenges and Implications of National Implementation*, The Royal Institute for International Affairs, at http://www.riia.org/briefing-papers/EUETSSworkshopreport.pdf, at pp. 23–27.
8. According to reports commissioned by governments, Member States have published NAPs allowing companies to increase their emissions. See *Ecofys Interim Report on National Allocation Plans*, September 2004, at http://www.ecofys.co.uk/uk/publications/reports_books.htm.
9. See COM(2004) 500.
10. See C(2004) 2515/6; C(2004) 2515/5; C(2004) 2515/1; C(2004) 2515/8; C(2004) 2515/7; C(2004) 2515/3; C(2004) 2515/2; C(2004) 2515/4.
11. COM(2004) 681.
12. See C(2004) 3982; C(2004) 3982/8; C(2004) 3982/5; C(2004) 3982/3; C(2004) 3982/4; C(2004) 3982/6.
13. See C(2004) 3982/2; C(2004) 3982/7.
14. See C(2004) 5287; C(2004) 5295; C(2004) 5298; C(2004) 5292.

15. See C(2004) 5285.
16. See C(2005) 1083; C(2005) 549.
17. See C(2005) 1527.
18. See C(2005) 1788.
19. See Grant, W., Matthews, D., Newell, P. (2000), *The Effectiveness of European Union Environmental Policy*, London (UK) and New York (USA): and MacMillan Press Ltd., St Martin's Press, at pp. 74–79; and Kramer, L. (2003) *EC Environmental Law*, London: Sweet & Maxwell, at pp. 380–84.
20. ECJ, Judgment of 4 July 2000, Case C-387/97, *Commission* v. *Greece*, ECR (2000), I-5047.
21. ECJ, Judgment of 25 November 2003, Case C-278/01, *Commission* v. *Spain*, ECR (2003) I-14141; and ECJ Judgment of 12 July 2005, Case C-304/02, *Commission* v. *France*, ECR (2005).
22. ECJ Judgment of 12 July 2005, Case C-304/02, *Commission* v. *France*, ECR (2005).
23. *Ibid*, paras 91 and 93.
24. ECJ Judgment of 19 November 1991, Joined Cases C-6/90 and C-9/90, *Francovich*, ECR (1991) I-5357, paras 31, 35 and 37.
25. ECJ Judgment of 5 March 1996, Joined Cases C-46/93 and C-48/93, *Brasserie du Pêcheur*, ECR (1996) I-1029, paras 32 and 33.
26. ECJ Judgment of 30 September 2003, Case C-224/01, *Klober*, ECR (2003) I-10239, paras 50 and 42.
27. ECJ Judgment of 4 July 2000, Case C-424/97, *Haim*, ECR (2000) I-5123, paras 28 and 30.
28. ECJ Judgment of 5 March 1996, Joined Cases C-46/93 and C-48/93, *Brasserie du Pêcheur*, ECR (1996) I-1029, paras 50 and 51; ECJ Judgment of 26 March 1996, Case C-392/93, *British Telecommunications*, ECR (1996) I-1631, paras 39 and 40; ECJ Judgment of 23 May 1996, Case C-5/94, *Hedley Tomas*, ECR (1996) I-2553, paras 25 and 26; and ECJ Judgment of 8 October 1996, Joined Cases C-178/94, C-179/94, C-188/94, C-189/94, *Dillenkofer*, ECR (1996) I-4845, para. 21.
29. See Somsen, H. (1996), 'Francovich and its application to environmental law', in Somsen, H. (ed.), *Protecting the European Environment: Enforcing EC Environmental Law*, London Blockstone Press Limited, pp. 146–7.
30. ECJ Judgment of 7 July 2004, Case C-201/02, *Wells*, ECR (2004) I-723, para. 57.
31. In a different context, related to protection against the acts of institutions and the action for annulment, those are the allegations sustained by Arcelor SA, one of the largest steel groups in the world, to challenge the ET Directive before the Court of First Instance of the European Communities, arguing that the Directive distorts competition and penalizes the European steel industry. Essentially the company alleges the violation of its fundamental rights to property and the pursuit of an economic activity, the infringement of the principles of proportionality by subjecting steel installations to the Directive, and of the principle of equality by not subjecting to the Directive other sectors in direct competition and with comparable or even higher emissions of GHG. The company also contests the Directive because it infringes the freedom of establishment by affecting the company's right to transfer production freely from a less efficient plant in one Member State to a more efficient plant in another Member State. See the Action brought on 15 January 2004 by Arcelor SA against the European Parliament and the Council of the European Union, Case T-16/04, OJ C71, of 20 March 2004, p. 36. Up to now, other companies (Heidelberg Cement and EnBW) have also contested the ET Directive's legality before the national courts, those pleas having been finally rejected by German courts. See ENDS Environment Daily, issues 1678 of 2 June 2004, 1755 of 26 October 2004 1912 of 1 July 2005.
32. ECJ Judgment of 5 March 1996, Joined Cases C-46/93 and C-48/93, *Brasserie du Pêcheur*, ECR (1996) I-1029, para. 55.
33. ECJ Judgment of 8 October 1996, Joined Cases C-178/94, C-179/94, C-188/94, C-189/94, *Dillenkofer*, ECR (1996) I-4845, para. 25.
34. ECJ Judgment of 5 March 1996, Joined Cases C-46/93 and C-48/93, *Brasserie du Pêcheur*, ECR (1996) I-1029, para. 57.

35. ECJ Judgment of 8 October 1996, Joined Cases C-178/94, C-179/94, C-188/94, C-189/94, *Dillenkofer*, ECR (1996) I-4845, para. 25.
36. ECJ Judgment of 30 September 2003, Case C-224/01, *Klober*, ECR (2003) I-10239, para. 56.
37. *Ibid*, paras. 95 and 96.
38. ECJ Judgment of 5 March 1996, Joined Cases C-46/93 and C-48/93, *Brasserie du Pêcheur*, ECR (1996) I-1029, para. 56.
39. As for Spain, *Tribunal Supremo (Sala de lo contencioso-administrativo)*, judgment of 29 May 1991, *RJ Aranzadi* 1991/3901.
40. *Tribunal Supremo (Sala de lo contencioso – administrativo)*, judgment of 12 June 2003, *RJ Aranzadi* 2003/8844. For a comment in English (and partial translation of the judgment) see *Common Market Law Review* (2004), **41**, n. 6, pp. 1717–34 (paragraph cited at p. 1724).
41. ECJ Judgment of 5 March 1996, Joined Cases C-46/93 and C-48/93, *Brasserie du Pêcheur*, ECR (1996) I-1029, para. 65; and ECJ Judgment of 15 June 1999, Case C-140/97, *Rechberger*, ECR (1999) I-3499, para. 72.
42. As for Spain, judgments of the Spanish *Tribunal Supremo (Sala de lo Contencioso-Administrativo)* of 25 January 1997 (*RJ Aranzadi* 1997/226), 26 April 1997 (*RJ Aranzadi* 1997/4307), 5 May 1998 (*RJ Aranzadi* 1998/4625) and 12 May 1998 (*RJ Aranzadi* 1998/4640).
43. ECJ Judgment of 15 June 1999, Case C-140/97, *Rechberger*, ECR (1999) I-3499, paras 75 and 76.
44. For a further discussion, see Campins Eritja, M., Fernández Pons, X., Huici Sancho, L. (2004), 'Compliance Mechanisms in the Framework Convention on Climate Change and the Kyoto Protocol', *Revue Générale de Droit*, **34**, n. 1, pp. 51–105, at 92–7.
45. Heliskoski, J. (2001), *Mixed agreement as a technique for organizing the international relations of the European Community and its Member States*, The Hague Kluwer Law International, para. 5.2.2.
46. ECJ Opinion 1/94, of 15 November 1994, ECR (1994), I-5267, paras 106–10.

5. A level playing field? Initial allocation of allowances in Member States

Bettina Schmitt-Rady[1]

1. INTRODUCTION

The launch in Europe of the largest domestic cap-and-trade carbon market on 1 January 2005 requires a level playing field between competing companies that needs to be supportive of the long-term environmental and economic goals of the European Union (EU). The European Commission (the Commission) emphasized this need from the very beginning of the development of the European-wide emissions trading scheme (EU ETS). The allocation of CO_2 allowances to the almost 12,000 installations concerned and the corresponding mechanisms are key factors in implementing the EU ETS. The economic value of the allowances allocated in the first trading period amounts to well over €100 billion at current market prices.[2] Those installations affected which feel that they have been placed at a disadvantage by the initial allocation in their home countries have already initiated legal actions before the European Court of First Instance (CFI) and before their respective national courts. Hence, Commissioner Stavros Dimas is aware that 'we need to have a serious look at the way we allocate'.[3]

The purpose of this chapter is to have an initial look at the chosen method of allocation as set forth in Directive 2003/87/EC of 13 October 2003, which establishes a scheme for greenhouse gas (GHG) emission allowance trading within the Community[4] (the ET Directive), in particular in Annex III to the ET Directive, and the national transposition of the EU ETS with a focus, from the practitioner's perspective, on Germany.

2. METHOD OF ALLOCATION

Common sense tells us that the method of allocation needs to be fair, consistent, clear and feasible.[5] The ET Directive stipulates accordingly that the National Allocation Plans (NAPs) have to be based on objective and transparent criteria, including those listed in Annex III to the ET Directive. It can already be stated that these requirements have not been met as far as allocation in the first trading period is concerned. Definitions and criteria in the ET Directive and in Annex III are too vague to ensure a level playing field, as mentioned above, between those companies affected that are competing in the internal market.

2.1   Need for Harmonization Among Member States

The Commission has given Member States considerable leeway in choosing the allocation measure.[6] In fact, Member States have generally exercised flexibility in the setting of targets and interpretation of definitions, which causes different market conditions for the affected installations.

In Germany, lobbying has achieved no less than 58 different options for allocation. Rüdiger Schweer, the Head of the Climate Protection Department at the Environmental Ministry of the German State of Hesse, states that:

> The multitude of special rules, particularly the option rule,[7] has made the allocation procedure lacking in transparency, bureaucratic and cost-intensive. The attempt to achieve a high degree of equality in individual cases, with regard to advance performance such as early action, the differing utilization of various plants and differing reduction potential, has placed a particular burden on small plants with low emissions and little possibility of avoiding emissions.[8]

The differences in transposition of the ET Directive are confirmed in an analysis conducted by Ecofys for the United Kingdom (UK) Government.[9] It comes to the conclusion that not all the criteria of Annex III have been adhered to strictly in the NAPs of the Member States. The three following examples illustrate some considerable variations.

2.1.1   Variations in the definition of installation
Three definitions have, for example, been used to determine which installations should be included in the NAP. These are commonly referred to as the narrow, medium and broad definition, respectively. The majority of countries have used the broad definition, but three countries have used the medium definition, while three more countries have used the narrow definition (namely, France, Spain and Italy).[10] The definition of 'combustion installations' with a rated thermal input exceeding 20 MW in particular affects the chemical sector, of which a considerable number of such installations are currently not covered in some Member States. However, with regard to France and Spain the Commission objected and the countries expanded the scheme to such combustion installations.[11] This expansion led to 900 plants being included, or the equivalent of 30 million tons of CO_2 per year, in France alone. In Germany, a court ruling was necessary to clarify whether combustion plants within the chemicals sector (such

as crackers,[12] if permitted as part of the installation delivering heat or electricity) fell under the EU ETS if they exceeded 20 MW.

The different interpretation by Member States of the definition of 'combustion installations' causes distortions of competition if some installations are burdened with new requirements and others are not. The principle of legal certainty also requires that potential investors and operators of installations should know in advance whether or not their installations fall under the EU ETS. It needs to be ensured that the NAPs for the next trading period will be based on a harmonized definition to avoid unequal treatment and unwanted distortions of competition.

2.1.2 Variation in treatment of new entrants

Treatment of new entrants is one of the areas where policies among the Member States differ the most.[13] Most of the NAPs assessed under the Ecofys study provide new entrants (criterion 6 of Annex III) with allowances from a reserve. Allowances from these reserves are usually provided on a first-come, first-served basis. If this reserve is not large enough, then the outcome differs from Member State to Member State. Poland and Italy, for example, intend to purchase allowances from the market for new entrants if their reserves are oversubscribed. Germany even guarantees free allowances to a new entrant for up to 14 years if it is a completely new installation;[14] thereby exceeding the original term of the first two trading periods of the EU ETS.

Such different treatments again potentially give rise to unequal treatment and, again, distortions of competition can arise.[15] However, the Commission has accepted the respective national rules and focused on compliance with the national overall cap during the previous notification procedure without setting up rules on whether or not new entrants should be allocated allowances free of charge. In the second trading period Member States should be required to apply a more harmonized approach.

2.1.3 Protectionist approach in the event of cessation of activities

Another notable variation refers to the international transfer of allowances after an installation ceases to carry out an activity listed under the ET Directive and Annex I.

The vast majority of Member States decided not to issue any further allowances to the operator of that installation during the remainder of the period in respect of that installation. However, in a number of Member States, the owner of a closed installation may transfer the allowances to a new installation instead of losing them altogether. Germany, for example, provides for the cancellation of allowances in the event of a plant closure unless the operator commissions a new installation in *Germany*, but not in another Member State.[16] Similarly, according to French law,[17] an operator may keep the allowances for a plant that is shut down only if the activity is transferred to another installation in *France*. An installation that operates cross-border in the Community would lose its emission allowances for a plant in one Member State in the event of closure and would not be in a position to transfer these allowances to its plants in another Member State – in the event that it plans a corresponding increase in capacity, and thereby CO_2 emissions, over there.

However, the ET Directive states that allowances are transferable between persons within the Community and that issued allowances are to be recognized in each Member State.[18] Consequently, the European Parliament had suggested inserting a clause that states that the operator may keep the allowances if it can demonstrate that the closure is related to a corresponding new investment made within the *Community*.[19] This amendment was not inserted into the final version of the ET Directive. The ET Directive is silent as to the closure of installations and does not provide for any corrective measure for cross-border situations such as the transfer of production between Member States. By omitting any regulation, it fosters the adoption of such measures by Member States, which is not surprising given that the allocation of allowances is carried out on a purely national level in accordance with national reduction targets under the Burden Sharing Agreement on the basis of grandfathering. Of course, Member States have no interest in the transfer of production to another Member State and in the corresponding loss of job opportunities and tax base. Moreover, the cancellation of allowances in the event of a cross-border transfer of production enables the Member State to create spare allowances that can then be reallocated to operators who stay with their installations in the respective Member State. In this way, installations that are international players in the internal market are forced to continue to operate less efficient plants by way of maintaining capacity for the sole purpose of not losing the allowances.

The withdrawal of allowances after closure endangers the integrity of the market and does not take into account the fact that many entities operate cross-border within the Community market. Such an impediment to the free transfer of production within the EU is incompatible with the freedom of establishment and makes no sense by the very fact that it does not matter where the CO_2 emissions get reduced. Consequently, the ET Directive's very approach is to provide for free transfer of allowances between Member States.

However, the question is whether emitters should receive all of their allowances beyond their operational life 'for ever', which would preserve the allowances as property rights of the respective (former) operator in perpetuity. This concept is politically undesirable. Ahman, Burtraw, Kruger, Zetterberg (2005) propose a '10-Year Rule', whereby existing installations that shut down would continue to receive allowances for 10 years. The perverse incentive to continue operation would be diminished. Eventually, the group of installations receiving allowances would slowly shift, as emitters that cease to operate would eventually stop receiving allowances. After that, a transition period with a steadily increasing auction of allowances could begin.[20]

Alternatively, continuing allocation could be limited to installations which can provide evidence that they transfer the corresponding capacity and corresponding CO_2 emissions to another substitute installation, either in the same Member State or in another. Annual allocation would then continue for the operational lifetime of such substitute installations as it would have been carried out for the replaced installation. Since it is hard for national governments to introduce this proposal as a policy on the national level, which would be more favourable to the closure of installation's in their own country and supports investments in neighbouring countries, it should be introduced by way of an amendment of the EU Directive on the international level.

2.2 Grandfathering

For existing installations the ET Directive chose the 'grandfathering' approach, which means that 95% of the allowances in the first trading period and 90% in the second period must be allocated free of charge to them.[21] This approach was, politically-speaking, the easiest and takes into account the existing installations' vested rights to operate their installations, including the right to emit CO_2. Auctioning of allowances is only envisaged as being implemented after 2012.[22]

However, grandfathering led to a fierce battle concerning the amounts allocated between governments and affected sectors, between industries of different sectors and between industries within single sectors. The special rules created for certain industries could not always comply with the principle of equality; challenges by national industries, which claim distortions of competition and even the granting of prohibited state aid, have arisen. The intensive lobbying that took place influenced the political decision-making, proved to be an impediment to a sensible approach for implementing the EU ETS on a national basis and finally led to over-allocation. Similar problems could be observed as regards the implementation of the United States (US) ETS for SO_2 in the US Clean Air Act, which also chose to grandfather SO_2 emission allowances to existing power plants.[23]

The Ecofys analysis revealed that most caps set by Member States are exceeding their Kyoto commitment. Criterion 1 of Annex III was not completely obeyed by the Member States.[24] Criterion 1 states that the total quantity of allowances to be allocated must be consistent with the respective Member State's obligation to limit its emissions, pursuant to the Burden Sharing Agreement[25] and the Kyoto Protocol. While the caps are mostly below the expected 'business as usual' operation, ET is generally not being used to bring industry emissions into line with the reductions needed to meet the Kyoto targets.

Interestingly enough, in order to deliver the shortfall towards the Kyoto target, the focus of Member States is on other sectors, which are not covered by the ET Directive, and on the acquisition of Joint Implementation (JI) and Clean Development Mechanism (CDM) credits. This illustrates, *inter alia*, the short-term lobbying success of affected industries.

2.2.1   German allocation based on grandfathering

The German allocation is a good example of some of the problems that grandfathering creates. As already mentioned, there exist, in all, 58 different allocation options. The German transposition law is extraordinarily complicated. Major polluters were able to achieve a better position for them than were small ones. The German national cap was almost exceeded by all the special requests for special treatment.

In principle, German allocation is based on historic emissions during certain baseline periods. The general rule for existing installations without special conditions is that the number of emission allowances to be allocated corresponds to the result of the calculation of average annual CO_2 emissions during the baseline period, the applicable emission factor and the number of years in the initial allocation period. Compared to the period 2000–02, an existing plant thus either had to reduce its CO_2 emissions by at least 2.91% or it had to purchase emission allowances. There are various other allocation rules whereby no emission factor or – on request – an emission factor of 1 was applied, for example for process emissions in the steel and cement industry.

However, installations could also opt for an allocation on the basis of estimates for the annually expected capacity and incurring emissions instead of the basis of historical data. Of course, many operators used this option. The total number of allowances that all applicants had applied for would have exceeded the German cap of 495 million tons of CO_2 per year for the first trading period. The authorities were forced to reduce the number of allowances for all installations after the original NAP, with the original reduction factor, had been notified to and accepted by the Commission, thereby tightening the emission reduction factor from 0.9709 (as set out in the NAP) to 0.9538, a so-called 'second reduction factor'.

The tightening and use of the second reduction factor was already provided for in the German NAP, which allows for the number of emission allowances allocated to be subject to further proportionate capping, if necessary, in order to ensure that the entire budget of CO_2 emissions to be allocated to existing plants does not exceed 495 million tons per year. Where this cap is exceeded, the allocations subject to an emission factor should also be proportionately capped, thereby adhering to the limit of 495 million tons per year. The proportionate cap did not cover allocations where the law provides for an emission reduction factor of 1 or does not expressly provide for the application of an emission factor at all, such as process emissions or replacement installations. However, all plants, which, instead of allocation on the basis of historical data, had requested allocation on the basis of expected average annual emissions, had proportionate capping applied to them in order to ensure that these plants did not derive any disproportionate benefit compared with other existing plants.

It is questionable whether this step complied with the principle of equal treatment, the principle of legal certainty and the principle of proportionality because it leads to the result that smaller installations are ultimately subject to a reduction of a much higher percentage than the bigger emitters, indeed 7.4%, as compared with the average historic emissions. Thus, smaller installations bear the main burden despite the fact that they are the smallest emitters, usually with the least possibility and financial means for further reduction of their CO_2 emissions.

Thus far, the dispute concerning the second reduction factor has led to 585 legal actions being filed against the allocation decision. Overall, only 1,119 allocations (60%) have become final and unappealable three-quarters of a year after the start of the German ETS.[26]

The risk inherent in such challenges is that trading by the affected companies is halted until the challenges are resolved. A less complicated approach should be chosen for the next trading period, as it is not possible to live up to every expectation of affected operators and sectors. The interaction between different rules for special treatment of particular sectors or companies could partly not be anticipated and lead to inconsistencies that interfere with the ETS's credibility and efficiency.

2.2.2 *Ex-post* adjustments

Since grandfathering entails the risk of over-allocation of allowances, the national authorities are trying to intervene by administrative means in order to modify the number of allowances *after* they were issued to the entitled parties (known as '*ex-post* adjustments'), such as German allocation authorities.

Germany intends to make *ex-post* adjustments to the allocation of allowances to new entrants and to listed installations in certain events. Provisions have even been made for the withdrawal of allowances if the installation experiences lower capacity utilization than foreseen or the installation's annual emissions are less than 40% of its base period emissions.[27] The same applies if the installation is benefiting from an additional allocation for combined heat and power (CHP) and generates a lower amount of power production from CHP than in the base period.[28]

The modification of the number of allowances *after* the allocation decision was issued is a much worse impediment for business planning than the short trading periods of the EU ETS and the uncertainties about the future availability of allowances and their price, which, particularly in the capital-intensive large industry complaints, interfere with their long-term business planning. However, risks about availability and price of allowances can be covered by use of the market instruments like forwards and futures. Nevertheless, Commissioner Stavros Dimas acknowledged that the 5-year periods create an uncertainty for companies investing in long-lasting capital equipment. He has announced that providing certainty about allocations for a longer period of time is an idea which should be explored.[29]

In any event, the principle of legal certainty requires that citizens should be able to know in advance what the legal consequences are that will result from any course of action they are required to commit. *Ex-post* adjustments are therefore unacceptable. Such subsequent adjustments lead to uncertainty among enterprises, a low degree of market liquidity and high indicative prices for small amounts traded.[30]

Consequently, the Commission rejected the German provisions, stating that they contravene criteria 5 and 10 of Annex III.[31] The NAP must state the quantities of allowances intended to be allocated to each installation *ex-ante*. The Commission emphasizes that any later adjustments create uncertainty for business, since the total number of allowances might be modified by intervention from public authorities and hamper the market exchange of allowances.

Nevertheless, the German Government brought the matter before the CFI in September 2004, claiming that the Commission's objection would not be appropriate or covered by the ET Directive. In order to avoid over-allocation on the basis of wrong estimates, *ex-post* adjustments would be an absolutely crucial means of observing the German overall cap. Consequently, the contested provisions were transposed into national law after the Commission had rejected them and are still currently in force and applicable, despite the Commission's decision. The CFI has not yet delivered its ruling on the matter.

2.2.3   Grandfathering and electricity prices

It is also presumed that grandfathering of allowances free of charge has an impact on the German energy market. German electricity prices have recently risen by €10/MWh since the EU ETS started in January 2005.[32] Major energy consumers assume that the steep rise is partly due to the fact that energy producers price the cost of allowances in as opportunity costs, despite the fact that they have received the initial allowances free of charge. They assume that 'fictional' acquisition costs of allowances are factored into energy prices and must be borne by consumers. Four aluminium companies have therefore decided to challenge the German Act on the Allocation of Allowances 2007 (*ZuG 2007*), which transposes the German NAP, before the German Constitutional Court (*Bundesverfassungsgericht*). Although aluminium is not covered by the current version of the EU ETS, it is one of the most energy-intensive industries and it therefore considers the rise of energy prices as posing a danger to its competitiveness and, ultimately, its very survival. This would violate its constitutional property rights.[33]

2.2.4 State aid

Grandfathering also requires tough state aid control.[34] Consequently, the Directorate General Environment (DG) of the Commission forwards all complaints against the EU ETS to DG Competition.[35]

Although the mandatory criterion 5 of Annex III expressly states that the NAPs must not discriminate between companies or sectors in such a way as to unduly favour certain undertakings or activities, the mere reference to the general state aid rules in arts 87 and 88 of the EC Treaty seems insufficient for an effective control in the first trading period. Member States are left considerable freedom in deciding the quantity of allocation of allowances to specific industries, as mentioned above. This can grant the perverse incentive of pursuing other policy goals through their allocation decisions.[36]

It would be helpful if the Commission could draw up more detailed guidelines on which types of allocation would be covered by state aid rules and which allocations could be exempted to reduce the level of uncertainty and to strengthen reliance.

Moreover, an element of state aid may also be seen, not only in the generous initial allocation of allowances, but also if industries have the advantage of being located in a country that either does not conduct the ETS as ambitiously as other countries do from the beginning or makes heavy use of JI and CDM. For instance, in the Netherlands the government intends to acquire 20 million tons of CO_2 per year in the period 2008–12 through CDM and JI projects. Germany, in contrast, will not fund any acquisition of reduction credits with public money. The industry and other sectors in the Netherlands seem to profit from the fact that its government follows such a far-reaching strategy in using the flexible mechanisms of the Kyoto Protocol.

2.3 Benchmarks

Referring to benchmarks instead of historic emissions data for the initial allocation to existing installations could solve some of the problems that are connected with the current implementation of the grandfathering approach. In fact, the Commission and Member States are currently considering their practicality. However, it is difficult to define Community-wide benchmarks for all affected sectors across Europe.[37] Benchmarks are already in place for the allocation to new entrants in combination with forecasted activity in most Member States. However, it is already observed that they differ significantly across Member States, even for identical products such as heat or energy.[38]

Mr Schweer, who is also working on the development of benchmarks, outlines the following problems:

> In the view of various trade associations, benchmarks for existing installations are not an appropriate device for carrying out a proper allocation process, due to the numerous differing products with different emissions, individual plant configurations and different production methods. Furthermore, it is necessary to make an *ex-post* adjustment for existing installations, depending on the amounts actually produced. Although this would solve the problem of differing degrees of utilisation, the fixed sector 'caps' would make a second fulfilment factor necessary to the detriment of the other installations. Furthermore, the *ex-post* correction would have the effect of inhibiting liquidity on the ET market. Industrial companies in particular are waiting until the end of the first period, even if it is clear that they have excess allocation, so as not to fall short.

However, it is to be believed that benchmarks, once set up, will pave the way to a fairer way of allocation that is less prone to influence from the political and lobbying side. Legal certainty and equal treatment should be fostered.

Their correct initial definition and their regular update in coming trading periods must be taken into account, though. The risk arises that it is public authorities again which have to define the best available technology for the respective benchmark without having enough information on the business involved, thereby falling back to the old problems under the current command-and-control scheme which should be solved by the market under the ETS. One of the goals of the ETS is that the definition of the relevant 'benchmark' for their business is left up to the operators themselves who know best about the applicable technology in their installations and market.

In fact, it has already been observed that Member States chose administrative rules instead of market forces where the ET directive allowed a choice. Over-regulation was the answer to the new market-based instrument emissions trading. It can also be stated that implementation was more complex than probably expected, in particular with a view on the existing command-and-control scheme.

3.   GERMAN JUDGMENT ON THE GERMAN ETS'S COMPLIANCE WITH FUNDAMENTAL CIVIL RIGHTS

In relation to the German ETS's compliance with fundamental civil rights, the highest German administrative court (*Bundesverwaltungsgericht*, in Leipzig) dismissed, on 30 June 2005, a cement company's action against the ETS.[39]

The cement company had claimed, *inter alia*, that the German ETS would infringe its constitutional civil rights due to the unbearable burden it would put on the cement industry. Its vested rights to emit CO_2 under an operating permit already granted, pursuant to the German Federal Emissions Control Act

(*Bundes-Immissionsschutzgesetz*), would protect it against the new obligation to hold and surrender allowances for CO_2 emissions. The court found that the new obligations under the German Emissions Trading Act would neither violate European civil rights nor German constitutional civil rights. The court, in particular, referred to the claimant's property right and its freedom to choose an occupation (*Berufsfreiheit*). Both rights are protected under the German constitution and also under the EC Treaty. However, the court found the scheme to be in compliance with these provisions and therefore it did not violate the rights. Neither European nor German law would prevent the new regulating of an area of law for the future. The new ETS would, therefore, not expropriate vested rights to emit.

Insofar as the claimant stressed that the initial allocation of allowances puts the cement sector at an unacceptable disadvantage and that it should have received more allowances, the court did not take into account these pleas in law when giving its ruling, as they would not be connected with the obligation to participate in the ETS as such. Consequently, the court denied to rule on the issue of allocation.

4. FURTHER PRACTICAL PROBLEMS

Some further problems that need a solution for the upcoming trading periods will now be examined.

4.1 Integration with Existing Policies

The ETS needs to integrate with a number of other policies and measures which have been developed over time. For example, the ETS and energy taxation must be seen as complementary instruments. It is, therefore, appropriate to take into account the level of taxation that pursues the same objectives. Neither in the ET Directive nor in the national transposition has this been attained to a satisfactory extent. The same applies to numerous other existing rules and regulations under the command-and-control-schemes.

4.2 Cost-efficient Regulation

The question of how transposition of the ET Directive can be carried out in a cost-efficient manner also needs to be considered carefully.

One way that has been suggested of reducing the administrative burden on both authorities and installations is the idea of implementing a '*de minimis*' rule and reducing the scope of the ET Directive as concerns small installations. The ratio in number of installations versus the number of CO2 allowances that have been allocated is skewed.[40]

In Germany, small installations with annual emissions below 25 kilo tons cover less than 2% of total emissions, but account for 53% of all the installations covered by the ET Directive, while 31% of the installations emit even less than 10 kilo tons and account for only 0.5% of the total emissions. All concerned installations need to comply with legal obligations, such as the potential need to purchase allowances, obtaining a GHG permit, and monitoring, reporting and verification obligations. Commentators have come to the conclusion that their low emissions volumes are not worth the effort, as the administrative costs outweigh the potential economic and environmental benefits of their participation.[41] Rüdiger Schweer shares this view, and states that small installations, especially those from the brick, ceramic and paper industries would be at a disadvantage. He stresses that, conversely, some large installations, with emissions exceeding one million tons per year, are still not included in the EU ETS. He calls for a practical solution to be found, such as that found in the Netherlands. A threshold value of 25 kilo tons would mean that only the approximately 600 relevant installations would participate in the EU ETS.

However, each exemption needs to be measured against the principle of equal treatment and the polluter-pays principle. Larger emitters will most likely refer to these principles if they would be the only ones who fall under the scheme. They will probably argue that a comparable situation exists as all installations emit CO_2, be it in larger volumes or in less, and that the burden should not only be placed on the big emitters. From the economic market point of view, it is also desirable to have as many *participants* as possible.

However, the principles of proportionality and the need for an efficient administration could be relied on in arguing that such exemption is objectively justified. One of the ETS's main targets is to reduce CO_2 emissions in an efficient manner, with less administrative work and expenses than under the existing command-and-control schemes.

4.3 Monitoring and Reporting

Practitioners such as Rüdiger Schweer note that the monitoring and reporting guidelines fail to deal with significant issues concerning the precision requirements imposed on emissions reporting and the criteria for a 'cost-efficient' supervision. This relates in particular to the use of standard factors, the admissible uncertainty margins, the examination of individual batches and the use of accredited laboratories. Numerous enterprises criticize the guidelines for not being 'user-friendly'. The 'Tier approach' is not regarded as viable in practice, especially as other requirements – for example from the fiscal sector – are easier to deal with. Therefore, an adjustment and simplification of the monitoring and reporting guidelines is urgently required.

5.   CONCLUSION

The current regulations in the ET Directive and in Annexes I and III are too vague to ensure a level playing field between affected companies competing in the internal market. The Commission has given Member States considerable leeway in choosing the allocation measure. The ET Directive is strict in relation to many obligations but leaves other important questions unanswered. The Annex I definitions, the Annex III criteria and the Commission's guidance on the implementation of the criteria are insufficient. Member States have used their flexibility in the setting of targets and interpretation of definitions, which causes different market conditions for the affected installations. Administrative over-regulation at national level was the answer to the new market-based instrument emissions trading. The coverage and allocation method also needs serious review. The ET Directive's mere reference to the general state aid rules in arts 87 and 88 of the EC Treaty seems insufficient for effective control during the first trading period, something which is all the more important against the background of the grandfathering approach chosen. Competition distortions and issues of equal treatment need to be addressed by the Commission and Member States in time before the next trading period. An adjustment and simplification of the accompanying monitoring and reporting guidelines is urgently required.

The Phase II NAPs must be notified to the Commission by 30 June 2006. Again, this is a very challenging timetable and it is to be feared that the 'race against the clock', as Commissioner Stavros Dimas put it for the first trading period,[42] might start again and interfere with a sensible harmonization and improvement of the scheme that is so necessary to make the new instrument successful and acceptable to national governments, affected installations and the public. As one of the 'fathers' of emissions trading put it:

> any anti-pollution policy is therefore bound to be in the nature of a social experiment that is neither right nor wrong, but only more or less successful in leading to wise and socially agreed-upon patterns of use of our air and water resources.

(JH Dales, *Pollution, Property and Prices*, Toronto 1968.)

NOTES

1.   The author is an attorney at law with the German law firm Heiermann, Franke, Knipp in Frankfurt am Main, email: rady@kanzlei-hfk.de.
2.   Source: Speech of Commissioner Stavros Dimas at the Green Week, Brussels, 2 June 2005, Public Information Europe, 3 June 2005.
3.   *Ibid.*
4.   OJ L 275, 25.10.2003, pp. 0032–46.

5.   For example, Peeters, Marjan, 'Emissions Trading as a New Dimension to European Environmental Law', *European Environmental Law Review*, March 2003, p. 86.
6.   Also de Cendra de Larragán, Javier, (2005), *EU Greenhouse Gas Emissions Trading and Legal Principles*, University Maastricht METRO Institute, 5 April 2005, p. 32.
7.   Regarding the option rule, see below.
8.   In an interview with the author on 15 September 2005.
9.   Ecofys UK (2004), *Analysis of the National Allocation Plans for the EU Emissions Trading Scheme for 2005–2007*, interim report on Austria, Belgium, Denmark, Estonia, Finland, France, Germany, Ireland, Italy, Latvia, Lithuania, Luxembourg, Portugal, The Netherlands, Slovenia, Spain, Sweden and United Kingdom, August 2004.
10. *Ibid.*
11. See IP/04/1250 of 20 October 2004 and IP/05/9 of 6 January 2004.
12. Chemical plant to break down crude oil into light oil by way of thermal or catalytic cracking of hydrocarbons.
13. Details Ahman, S., Markus/Burtraw, Dallas/Kruger, Joseph/Zetterberg, Lars, (2005), *A Ten-Year Rule to Guide the Allocation of EU Emission Allowances*, IVL Swedish Environmental Research Institute/Resources For the Future Discussion Paper of 23 August 2005, pp. 12–14.
14. Section 11, para. 1 of the German Allocation Act 2007 (*ZuG 2007*).
15. See also de Cendra de Larragán, op cit, n. 6, p. 25; Ahman/Burtraw et al recommend not allocating free allowances to new entrants as this is not an efficient solution for the trading programme as a whole, op cit, n. 12, p. 14.
16. Section C. 3.3 of the German NAP; Sec. 9 para. 4 and Sec. 10 para. 1 of the German Allocation Act 2007.
17. Implementing decree 2004-832 (art. 5 para. 3).
18. Article 12 paras. 1 and 2 of the ET Directive.
19. The EU Parliament legislative resolution on the proposal for an EU Parliament and Council ET Directive (*Moreira da Silva Report*) at the sitting of 10 October 2002, P5_TA-PROV(2002)10-10, amendment 38 to Article 12a.
20. Op cit, n. 12, pp. 16–18.
21. Article 10 of the ET Directive.
22. Article 30 (c) of the ET Directive.
23. ELR Stat. Sec. 401–416 Clean Air Act 1990, Title IV (a), Acid Deposition Control, see the detailed analysis by Schmitt-Rady, Bettina (1999), *Die Umsetzung des Zertifikatsmodells im Luftreinhalterecht der USA*, Baden-Baden: Nomos, pp. 75, 114 and 150.
24. Op cit., n. 9.
25. Decision 2002/358 EC, OJ L 130, 15 May 2002, p. 1.
26. As per 14 September 2005 according to Rüdiger Schweer, op cit, n. 8.
27. Section 7 para. 9 of the German Allocation Act 2007.
28. Section 14 para. 5 of the German Allocation Act 2007.
29. Speech at Green Week, 2 June 2005 op cit, n. 2.
30. Also Rüdiger Schweer, op cit, n. 8.
·31. Commission Decision of 7 July 2004, C(2004) 2515/3 final.
32. On 13 October 2005 the price for one MWh reached EUR 61.40 on the spot market of the German European Energy Exchange in Leipzig (source: http://www.eex.de/spot_market/market_data).
33. Claimants are: Hydro Aluminium, Hamburger Aluminium Werke GmbH, Corus Aluminium Voerde GmbH, Trimet Aluminium AG. The writ was only lodged on 7 September 2005, see http://www.umweltdialog.de/umweltdialog/emissionen/2005-09-07_Verfassungsgericht_prueft_Emissionshandel.php.
34. See also de Cendra de Larragán, op cit, n. 6, p. 36.
35. Confirmed by Peter Zapfel, DG Environment, in July 2005 in an email-correspondence with the author.
36. For example, German energy producer Energie Baden-Württemberg AG (EnBW) challenged, under state aid and competition rules, the German option to transfer allowances of an old existing plant to a new substitute plant with lower emissions. RWE, one of Germany's four major

energy suppliers, operates outdated coal-fired power plants that are in need of replacement. EnBW, in turn, operates many nuclear power plants with no CO_2 emissions and is now obliged to shut them down and replace their capacity with conventional power plants based on a political compromise with the German Government for the phasing out of nuclear energy. The compensation allowances for the phasing out that EnBW receives from a special reserve, with a maximum of 1.5 million allowances per year, are deemed by EnBW to be insufficient in comparison with the number of allowances RWE will receive. EnBW is of the opinion that Sec. 10 of the German Allocation Act ('transfer rule') directly favours its competitor RWE by way of excess allowances, which need to be considered as illegal state aid. EnBW commissioned a study by Bode, Sven/Hübl, Lothar/Schaffner, Joey/Twelemann, Sven (2005), *Discrimination against Newcomers: Impacts on the German Emissions Trading Regime on the Electricity Sector*, HWWA/Universität Hannover/Eduard Pestel Institut für Systemforschung e.V., Universität Hannover, Institut für VWL Discussion Paper No. 316, June 2005, which confirms this assumption.

37. Also see de Cendra de Larragán, Javier, op cit, n. 6, p. 36.
38. See Ahman/Burtraw et al, op cit, n. 12, pp. 12–14.
39. BVerwG 7 C 26.04 of 30 June 2005.
40. van Slobbe, P. and Brinkoff, J. (Dutch Ministry of Economic Affairs), 'Guest Commentary', in *Carbon Market Europe*, 20 May 2005, p. 3.
41. See for example van Slobbe and Brinkoff at Green Week in Brussels.
42. Op cit, n. 2.

6. Linking the project based mechanisms with the EU ETS; the present state of affairs and challenges ahead

Javier de Cendra de Larragán[1]

1. INTRODUCTION

The linking directive,[2] an amendment to the ET Directive,[3] creates the possibility for companies covered under the European Union Emissions Trading Scheme (EU ETS) to use the credits obtained through the project-based mechanisms of the Kyoto Protocol for compliance along with EU allowances.[4] This chapter gives an analysis of this linking directive. First, a survey of the evolution of the EU position on the acceptability of the so-called flexible mechanisms within the Kyoto Protocol is given, to highlight the challenges present in the negotiations leading to the linking directive, and to understand the challenges lying ahead. Then, a brief review of the European Climate Change Programme will follow. The third part focuses on the analysis of the main provisions of the Directive. To conclude, the main challenges are presented in summarized form.

2. THE NEGOTIATIONS OF THE FLEXIBLE MECHANISMS WITHIN THE UNFCCC FROM THE PERSPECTIVE OF THE EU

Upon ratification of the Kyoto Protocol, the 15 Member States of the EU adopted a commitment to reduce its (GHG) emissions by 8% in the period 2008–12 as compared with its emissions in 1990. The Kyoto Protocol (the Protocol) to the United Nations Framework Convention on Climate Change (UNFCCC) lists a series of policies and measures (PAMs), that parties can implement in order to achieve their reduction and limitation commitments listed in Annex B to the Protocol. In addition, the Protocol created three flexible mechanisms, called the Clean Development Mechanism (CDM), Joint Implementation (JI) and International Emissions Trading (IET). They allow parties to use reductions made abroad for compliance purposes. Flexibility reflects the fact that emissions have the same impact on climate change regardless of the location of the source, and reduces compliance costs by achieving reductions where they are cheaper. It turned out to be essential to ensure enough ratifications to make the Kyoto Protocol enter into force.

It is well known that the EU has been historically sceptical of the notion of flexibility mechanisms, favouring instead a climate change strategy on co-ordinated PAMs and pushing for limitations on the use of flexibility mechanisms.[5] However, pressure from the United States (US) and other key players forced the EU to accept the introduction of flexible mechanisms in the text of the Protocol. Still, the EU tried to reduce the extent of their use, to avoid creating a loophole in the regime,[6] by stating that 50% of the reductions at least should be achieved domestically.[7] This fight was, however lost, as is clear from the definition of supplementarity given in the Marrakech Accords.[8]

It has been argued that the EC Commission (the Commission) pushed for EU-wide co-ordinated PAMs in order to increase its competences vis-à-vis Member States.[9] As we will see, this argument is also valid in the negotiations of the linking directive. Supplementarity, environmental integrity and Commission versus Member States competences were recurring topics in the negotiations of the linking directive, are still hot in the implementation process of the Directive into domestic legislation and will keep featuring prominently in the revision of the Directive that will start in June 2006.[10]

3. THE PROJECT BASED MECHANISMS (CDM AND JI)

3.1 The Nature and the Objectives of the CDM and the JI

Both CDM and JI mechanisms were further defined and refined in Marrakech, in COP7.[11] Still, there are large differences as regards both, because the CDM was given a prompt start,[12] whereas the JI must wait until the JI Supervisory Committee is set up, which will happen in COP/MOP1 at the earliest.[13]

3.2 Joint Implementation

Joint implementation allows Annex I Parties to implement projects that reduce emissions, or remove carbon from the atmosphere, in other Annex I Parties, in return for emission reduction units (ERUs). Emission reduction units can then be used towards meeting their emissions targets under the Protocol. Joint implementation projects must have the approval of all Parties involved, and must lead to emission reductions or removals that are additional to any that would have occurred without the project. Projects starting from the year 2000 which meet JI requirements may be listed as JI projects. However, ERUs may only be

issued in relation to periods from 2008 onwards. There are two possible procedures (or 'tracks') for carrying out a JI project.

The 'track one' procedure may be applied when the Annex I Party hosting
the project fully meets all the eligibility requirements to participate in the
mechanisms.[14] In this situation, the host Party may apply its own national rules
and procedures to the selection of JI projects and the estimation of emission
reductions from them. The host Party may also issue ERUs (through converting
existing Assigned Amount Units (AAUs) or Removal Units (RMUs)) and transfer them to project participants.

The 'track two' procedure must be applied if the host Party does not meet all
eligibility requirements. In such cases, the project and the quantity of ERUs it
generates must be verified under rules and procedures supervised by the art. 6
Supervisory Committee, to be elected by COP/MOP1 – but for transfers to take
place, the host Party must meet at least those eligibility requirements relating
to the calculation of its assigned amount and the establishment of its national
registry.[15] A host Party which is eligible to carry out JI projects under track one
may, nevertheless, choose to carry them out under track two.

3.3   The CDM

The CDM defined in art. 12, provides for Annex I Parties to implement project
activities that reduce emissions in non-Annex I Parties, in return for certified
emission reductions (CERs). The CERs generated by such project activities can
be used by Annex I Parties to help meet their emissions targets under the Kyoto
Protocol.

CDM project activities must have the approval of all Parties involved and this
may be gained from designated national authorities (to be set up by each Annex
I and non-Annex I Party). The CDM project activities must reduce emissions
below those emissions that would have occurred in the absence of the CDM
project activity (the so-called 'additionality' requirement). The CDM is supervised by the CDM Executive Board (CDM EB), which itself operates under the
authority of the COP/MOP (a role being performed by the COP until the COP/
MOP meets for the first time).

3.4   Key Challenges Affecting both JI and CDM

Since the inception of the project-based mechanisms in climate change negotiation processes, several issues have featured prominently in the discussions on
their architecture and use. At international level, discussions have revolved
around the CDM, because the JI will not start until 2008. It is expected that the
JI Supervisory Committee will integrate much of the work already done by the
CDM EB.[16]

As the focus of the mechanisms are specific projects, the first challenge is to design a credible baseline from which reductions can be measured. This is a difficult task and it has taken many to defend the need to take a sectoral approach both to JI and CDM projects.[17] Furthermore, the next hurdle is to establish the 'additionality' of the project. The concept raises unending discussions going even to the mere feasibility of using a counterfactual as a realistic benchmark. That almost impossibility is attenuated to some extent in the Marrakech Accords, where it is considered sufficient to make a '*reasonable estimation*[18] of what would have occurred in the absence of a project activity wherever possible and appropriate'.[19] The CDM EB has also addressed the 'perverse incentives' that the concept of additionality was said to give to developing countries in order to avoid adopting environmentally progressive legislation, as its implementation would lower the baseline and reduce the generation of credits. The CDM EB decided that policies or regulations implemented since the adoption of CDM modalities and procedures in the Marrakech Accords may not be taken into account when developing the baseline.[20] Additionality is, of course, very close to the problem of ensuring environmental integrity of the project-based mechanisms, but this is a broader one, for it also includes the decision of the project-types that are allowed. The choice has been to exclude for the moment nuclear energy projects, and has led to set limits on the use of sinks in the CDM.[21]

Related to this issue is the even broader concept of ensuring sustainable development which, according to art. 12 of the Kyoto Protocol, CDM projects must achieve. It is a widely held concern that many CDM projects do not contribute to sustainable development[22] or even go against it,[23] and that those projects that would do most about it are prevented from taking place because: (a) they do not take place in the most deprived countries and areas;[24] (b) the associated transaction costs are so high that they render them unattractive for private investors. It seems that the decisions on small-scale CDM projects are not enough to solve this problem. However, there is a more fundamental critique of the CDM as a tool to promote sustainable development: though the economic and technical data of CDM projects are thoroughly assessed by the CDM EB, the assessment of a project's impact on environmental and social conditions is entirely at the discretion of the host country government.[25]

The CDM regulations do not include any formal way to support host countries in the formulation, monitoring and enforcement of sustainable development criteria. For these reasons, the price of CERs do not reflect the contribution to sustainable development, only the integrity and additionality of the reductions. Together with the fact that the CDM only focuses on specific projects so far has led to some commentators to argue that the CDM is structurally not capable of fostering sustainable development.[26] Last but not least, the determination of 'supplementarity' and the level at which it has to be applied are not easy tasks.

When one looks at all these unresolved challenges, one is left with the impression that the EU has, perhaps, run too fast at the time of linking them with the EU ETS, which is itself filled enough with various challenges. However, as it has been said in the discussions that preceded the adoption of the linking directive, the reduced volumes of credits that will be available in the period 2005–07 leave the linking as a truly 'learning by doing' exercise.[27]

4.   THE EU POLICY AS REGARDS JI AND CDM

4.1   The European Climate Change Programme

In June 2000, the EC Commission, responding to a request from the EU Council of Environment Ministers launched the European Climate Change Programme (ECCP).[28] The goal of the ECCP was to identify and develop all the necessary elements of an EU strategy to implement the Kyoto Protocol in several economic sectors,[29] by establishing working groups. One of them, Working Group 1 (WG1), dealt specifically with the flexible mechanisms of the Kyoto Protocol. Working Group 1 got its mandate on 5 June 2000.[30] WG1 had one sub-group on emissions trading and another on JI and CDM.[31] The latter's overall purpose was to identify Community policy initiatives and actions to facilitate the implementation of the project based Kyoto Mechanisms.

Both sub-groups found it desirable to allow entities participating in the EU ETS to use credits from the JI and the CDM, as this would lower compliance costs and also promote the transfer of clean technologies. Upon presentation of the main findings of the working groups, the EC Commission released a Communication on the implementation of the first phase of the European Climate Change Programme.[32] Among the measures proposed, there was included a proposal for a Directive on linking project based mechanisms, linking JI and CDM to the EU ETS.[33] The aim of the proposal was to specify under which conditions ERUs and CERs can be added to the allowances for compliance purposes. The EC Commission wanted to ensure that the proposal was fully compatible with the provisions of the relevant UNFCCC decisions at a time where the specific rules for JI and CDM were not yet in place[34] – and consistent with the European Community's development policy 'in order that the overall objective of sustainable development in developing countries and countries with economies in transition is maintained'.[35] Last, but not least, the proposal intended to act strategically by alleviating the transaction costs linked to the start-up of the mechanisms, which could lead to a less than optimal demand for credits.

In 2001, the Commission proposed a Directive on GHG emission trading, which gave way to Directive 2003/87/CE,[36] and to the beginning of the EU ETS on 1 January 2005. The proposal did not include a link with the Kyoto Protocol flexible mechanisms. A second phase of the ECCP started in 2002 and covered the period 2002–03. A working group on JI and CDM was again established to look into the manners for linking the EU ETS with the Kyoto Protocol flexible mechanisms. The WG convened four times, and the discussions revolved around three topics: linking, domestic offset projects and the desirability of establishing an EU Fund to support JI and CDM projects.[37]

4.2   The Preparatory Works of the Linking Directive

The Commission issued the first proposal in July 2003,[38] and then transmitted it to the European Council and to the Parliament, following the co-decision procedure. The first reading at the European Parliament took place on 20 April 2004. The Council approved the proposal on 13 September, and the Directive was formally adopted on 27 October 2004. In this remarkably short period of time, a number of important amendments to the original proposal took place. However, a detailed account of the negotiation process has already been completed elsewhere and space does not allow for an expansion of it here.[39]

5.   THE LINKING DIRECTIVE

The linking directive aims to engage private investment in the development of the CDM and JI, because it is estimated that linking will lower the annual compliance costs for companies covered by the scheme by about a quarter and will improve the liquidity of the EU ETS.[40] At the same time, the linking directive aims to promote technology transfer, and thereby contribute to the sustainable development of host countries.[41] Of relevance is the stated political objective to engage non-EU countries in the fight against climate change.

5.1   The Regulatory Scope of the Linking Directive

As just mentioned, the Directive set up the legal regime to authorize companies to surrender credits for compliance with their obligations. This includes establishing the acceptable project types, the mandate to Member States to limit the use of credits by companies, and the conditions for the surrender of credits. In addition, the Directive lays the basis for the regulation of the approval and accounting of JI projects undertaken in the EU Member States and in the candidate countries (Bulgaria and Romania).

The Directive does not regulate Member State policies on the purchase of credits, beyond the requirement that Member States have to comply with the supplementarity requirement. However, these policies will clearly have an effect on the availability and prices of credits that companies can purchase.

The linking directive has four articles. Article 1 introduces several amendments to the ET Directive, adding and modifying some provisions. I will refer always to the provisions of the ET Directive as amended by the linking directive.

5.1.1   The use of credits for compliance

Five issues need to be analyzed: (1) the choice given to Member States as to whether or not to allow the use of credits for compliance; (2) from which projects; (3) up to which limit; (4) the surrendering procedure of the credits; (5) aspects related to the trading of credits within the EU ETS.

5.1.1.1   To allow or not to allow the use of credits?   Article 11(a)1 of the ET Directive regulates two different issues. Paragraph 1 states that Member States *may* allow operators to use CERs and ERUs from project activities in the Community scheme – up to a percentage of the allocation of allowances to each installation. That percentage has to be specified by each Member State in its national allocation plan (NAP) for each period.

Accordingly, Member States have to decide first whether to allow their industry, under the EU ETS, to use credits for compliance.[42] In the period 2005–07 only the use of CERs can be authorized, because ERUs can only be generated from 2008 onwards. A Member State may have several reasons for not allowing the use of CERs in the first trading period; for example it may feel that it will only complicate the domestic legal framework and/or bring about limited benefits to its industry.[43] It may also feel that there would be the risk of an excessive use of credits that could jeopardize the achievement of domestic targets. In addition, a Member State may choose *not* to allow the use of CERs in the first trading period, but to allow the use of both CERs and ERUs in the second trading period. If a Member State forbids use for compliance, it may put its companies at a disadvantage vis-à-vis competing companies from other Member States where use is allowed.[44]

5.1.1.2   The choice of projects   If a Member State does authorize the use of CERs in the first trading period, then it will have to decide from which types of projects it will accept credits. A Member State may take a stricter environmental stance than that adopted in the linking directive as regards admissible projects, given the wording of art. 11(a)3, which simply says that all CERs and ERUs *may be used* in the Community scheme, with some exceptions listed below.[45] Thus, a Member State could decide to introduce additional qualitative limitations to the types of acceptable projects. For example, the United Kingdom (UK) has issued a consultation paper which, in principle, would only allow use for compliance of those CERs coming from projects that reduce CO_2 emissions.[46] The rationale would be that without that limitation companies could rely excessively on the use of CERs for compliance, avoiding undertaking reductions which are considered necessary to achieve the national objective of reducing emissions by 20% in 2010 and by 60% in 2050.[47] Other Member States could take the same choice. Several comments on this seem necessary: (1) the limitation imposed would only be on the use for compliance, not for trade – thus a British company can still trade credits for example, for speculative purposes; (2) what would happen if a Member State imposes a limitation like the one seen above, whereas another Member State accepts credits from all the projects allowed by the linking directive without restrictions? A practical example may cast light on this event; the UK bans non-CO_2 projects, whereas Spain does not impose any such restriction. A British company purchases 10 000 CERs from a methane project in China. The British company subsequently swaps those CERs for EU allowances with a sister company located in Spain. The outcome is the same than if no qualitative limitation had been introduced. This shows that the most effective manner to impose qualitative limitations is by imposing a uniform approach across the EU.

Another issue would be to ask whether there is any obligation for a Member State that decides not to allow the use of credits for compliance to implement the Directive anyway. Since the Directive regulates other aspects than the surrendering of credits for compliance, for example the regulation of several aspects of JI projects when they take place in an installation within the EU ETS, they do have to implement the Directive.

5.1.1.3 The choice of the percentage Once a Member State has decided to authorize the use of credits for compliance, it then has to set a limit to the volume that each installation can use. That volume is calculated as a percentage of the total volume of allowances allocated to each installation.

The Directive gives, in principle, freedom to Member States to choose this percentage, with the limit that has to comply with art. 30.3 and the criteria of Annex III ET Directive, which will be discussed below. The set of a percentage is compulsory only for the second trading period, thus a Member State can decide that companies can use CERs for compliance in the period 2005–07 without any quantitative limit.

When setting the percentage, a Member State also needs to take into consideration the choices made by other Member States, in order to avoid inflicting a damage to the competitiveness of its firms. For instance, the Netherlands has recently issued a draft law implementing the linking directive, which sets a limit of 8%. However, the draft underlines that the Netherlands will only apply such a cap in the second phase NAP if it becomes accepted EU-wide, which does not seem probable, because Sweden seems to land on a cap somewhere between 10% and 20%.[48] Of course this strategy is not without risks, because there will most likely be a number of Member States which will not comply with the deadline for transposition, which was 13 November 2005. One solution could be that the Commission would issue guidance clarifying this aspect, in time to assist Member States in the elaboration of the NAPs for the second trading period. The Commission is most likely to issue guidance by the end of 2005, leaving Member States with 6 months to take it into account.[49]

According to art. 30.3, Member States are required to publish in their NAPs their intended use of CERs and ERUs. The total use of ERUs and CERs shall be consistent with the relevant supplementarity obligations under the Kyoto Protocol. It is interesting to note that supplementarity in the context of the linking directive is not defined, nor appears in the Kyoto Protocol. In any case supplementarity applies at EU and Member State level, not at a company level. Thus, it would not go against supplementarity to set a percentage of, i.e. 70% of the allocated allowances to each installation, as long as the sum of credits used by companies under the scheme and those purchased by the Member State does not go beyond supplementarity.[50] In that case, a company could then decide to purchase credits massively and to sell its allowances, so as to make a profit – provided that the price of credits is lower than the price of allowances, which is presently the case.[51] In the period 2008–12, those credits would be used by each Member State for compliance with its Kyoto targets, and this would balance the reduction efforts between covered and not covered sectors. If the Member State finds that it has too many credits, it could decide to bank them – so as to respect the percentages set in the Marrakech Accords.[52]

Once Member States have included a percentage for the use of credits by companies in the NAPs, it is then for the Commission to check them against the criteria of Annex III, including new criteria 12 on supplementarity, and the guidance on the implementation of those criteria, including the new guidance it may issue.[53] As regards supplementarity, the Commission must ensure that each Member State complies with the supplementarity requirement and that the total use of credits by the EU-15 is in compliance with the supplementarity requirement. This is not a moot point, because the fact that Member States comply with the requirement does not automatically imply that the EU as a whole is in compliance.[54]

Among the criteria of Annex III, number 5 is specially relevant. It states that the NAPs shall not discriminate between companies or sectors in such a way as to unduly favour certain undertakings or activities in accordance with the requirements of the EC Treaty, in particular arts 87 and 88 on state aid. If a Member State would set up a limit on the use of credits for companies within the EU ETS of i.e. 70%, those companies could massively buy credits and sell EU allowances at market price. This would, most definitely, raise an issue of state aid. Although there is no public money involved in the operation, the current price differential between EU allowances and Kyoto credits is such that companies would effectively could swap 70% of the EU allowances for credits.

The Member State would have given, for free, EU allowances worth maybe €20 each, which they can then sell and make a profit, of maybe €10, if the price of credits would be €10.

In addition, such a move would probably go against criteria number 1, because the installations covered by the EU ETS could manage a budget of EU allowances and credits combined that would be far larger than its needs and make a profit with it. The companies would be able to comply with their legal obligation without reducing emissions, and all the reduction effort would fall on sectors outside of the EU ETS. This solution would be inefficient from an economic perspective, and, if in order to avoid competitiveness distortions a large number of countries would also set up a high percentage, then the EU ETS would be environmentally ineffective.

5.1.1.4 The surrendering procedure   The second paragraph of art. 11(a) regulates the surrendering procedure of the credits. It reads: '[the surrendering] shall take place through the issue and immediate surrender of one allowance by the Member State in exchange for one CER or ERU held by the operator in the national registry of its Member State'.

The provision is complex and can lead to misunderstandings. The reason for that wording lies in the original proposal made by the Commission, where it proposed to 'convert' credits into allowances. A Member State would receive a credit from a company and would then decide whether or not to issue an allowance in exchange. The Commission had three reasons to impose conversion: (1) it gave freedom to the Member State to impose additional criteria for the conversion; (2), the Commission wanted that, within the EU ETS, trade would only be done with allowances – to increase the certainty of use and thus the value of credits; (3) to give an incentive to Member States to avoid large swaps of allowances for credits. However, under pressure from the majority of Member States, conversion was deleted. The Directive now allows for the direct use of CERs and ERUs into the Community scheme. The rationale of the 'issuance and immediate surrender', provision is mostly technical in order to ensure simplicity, as only allowances are actually counted as compliance, and to simplify the management of Kyoto units by Member States for compliance with their Kyoto commitments.

5.1.1.5 Trading CERs within the EU ETS   For an operator to use CERs for compliance, several steps need to take place. First, the CER needs to be formally issued. This is to be done by the CDM registry administrator upon instruction of the CDM EB, in accordance with provision 6 of Appendix D of Draft Decision-/CMP.1 (art. 12). Provision 6(c) indicates that the CERs will be issued into a pending account of the CDM EB, and from there forwarded to the registry accounts of the project participants and parties involved.[55] All of this operation need to be completed in accordance with the procedures set out in the modalities for the accounting of assigned amounts. The realization of this sequence requires that the CDM registry, the independent transaction log (ITL) and the national registry are operational. In addition, for a party to transfer or acquire Kyoto units, it has to comply with all the eligibility requirements.[56]

One practical problem is that the UNFCCC Secretariat does not have enough funding for the time being to programme and maintain the ITL software. This may mean that the ITL will not be operative until 2007. Once the CER has been issued, it will be transferred to an account in the national registry. There are no legal obstacles at this point, only some technical difficulties which may delay the process some months after the ITL is operative. Once the CER has been transferred into the national registry, then the ITL and the Community Transaction Log (CTL) will allow it to be freely transferred within the registry. Surrendering for compliance is a transfer within the national registry. However, a Member State has to allow operators to surrender CERs for compliance during the first trading period, and may introduce a maximum percentage, therefore the linking directive has to be implemented.[57] Upon implementation of the Directive, the operator can simply surrender a CER in the same way that it would surrender an allowance. This process means that, in the worst possible scenario, operators would not be able to surrender CERs for compliance until the end of the first trading period. Uncertainty can be reduced for the time being because operators can borrow allowances from the next year.[58] Problems may arise if operators miscalculate the volume of CERs, that they will be able to obtain at the end of the period to compensate for borrowing. From a financial perspective, the risk that CERs cannot be used for compliance by the end of 2007 could make the price of allowances rise without limit.

There is yet another issue regarding tradability. For transfers between national registries to take place, both parties need to be eligible which includes having calculated their assigned amounts in accordance with Draft Decision -/CMP.1 2(B) and the requirements established in the Annex.[59] The EC intends to submit the report to establish its assigned amount by 31 December 2006. At that point, CERs will become fully transferable across national registries. However, not all Member States have operational registries in place at this time, and this undermines liquidity.[60]

5.1.2 Project scope

Article 11(a)3 ET Directive establishes a general authorization of CDM and JI projects with two exclusions and additional conditions for hydro projects. All CERs and ERUs are allowed except those coming from nuclear facilities and from Land Use, Land Use Change and Forestry (LULUCF). All CERs from nuclear power plants are not allowed for compliance during the first and second trading period, but their use after 2012 will depend on the negotiations about nuclear energy at international level.[61]

The linking directive excludes, for the time being, the use of CERs and ERUs from LULUCF. However, art. 30.2(o) leaves – for the review of June 2006 – the possibility of allowing operators to use credits from LULUCF. Arguably, this deferral will not be critical in practice because, for the first trading period only CERs could have been used for compliance, and volumes are still very small. The Directive mentions that the Commission will have to consider technical provisions relating to the temporary nature of credits and the limit of 1% for eligibility for LULUCF as established in Decision 17/CP.7, in accordance with decisions taken pursuant to the UNFCCC and the Kyoto Protocol. In COP9 those decisions were taken, namely Decision 19/CP.9 on afforestation and re-forestation modalities and procedures under the CDM.[62] This decision dealt with the problem of the 'non-permanence'[63] of carbon stored in afforestation and reforestation projects.

Basically, the solution consisted in creating two new types of CERs, namely 'temporary CERs' (tCERs) and 'long-term CERs' (lCERs). The basic difference with 'normal' CERs is that tCERs and lCERs have a shorter life and have to be fully compensated upon the date of expiration. Decision 19 CP.9 was reached before the linking directive was adopted, and there was pressure to introduce the possibility of using credits from sinks in the text of the Directive. However, one of the main reasons to avoid doing this was the perceived incompatibility of expiring CERs with other CERs and EU allowances. The problem is how to achieve fungibility between expiring and non-expiring units. For this, insurance, credit replacement after expiration, and investor state acceptance are needed. Insurance to account for the risk of non-permanence and credit replacement, because otherwise the risk of non-compliance would be too high.[64] Finally, investor state acceptance because the government is accepting a credit that will have to back with AAUs at the end of the commitment period. Thus, it is accepting a liability in international law, namely the risk of non-permanence and the cost of replacement.[65]

In the case of tCERs, it is unlikely that an Annex I government will accept a chain of successive tCERs in exchange for a permanent emission allowance.[66] Moreover, as tCERs expire after one commitment period, a EU Member State may accept them from a company for compliance, and in the next commitment period, upon expiration, the allocation budget of that company will be reduced. The tCERs merely buy time for companies, as they can represent a cheaper way of compliance in the short-term.

As regards lCERs, the Member State may, as Dutschke et al. explain,[67] consider convenient to pass part or all of that liability to the company using the lCER. A government has basically four options: (1) to assume all the liability itself. In this case, it will treat lCERs identically than normal CERs; this option would fail to provide incentives for good project design and implementation of the projects. The government carries the risks of non-permanence and assumes the cost of replacement upon expiry; (2) lCERs can be converted into allowances provided that the company insures against non-permanence. In this case, the government still assumes the cost of replacement; (3) lCERs can be converted into EU allowances if the company purchases insurance against non-permanence and purchases new credits to replace the lCER upon expiry; (4) the government exchanges each lCER against a discounted amount of EU allowances. Then it will bank the remaining portion of each allowance to replace the lCER upon expiry. In this case, the risk of non-permanence is internalized into the price of EU allowances by means of discounting. The cost of replacement is undertaken by the government, but only up to the non-covered share of every lCER. Clearly, it would be up to each Member State to decide the risk management strategy to use. Member State A could decide to assume all the risk, whereas Member State B could decide to transfer all the risk to companies. Furthermore, the manner in which expiring CERs are treated under fiscal legislation in different Member States will influence their attractiveness for companies, if they would be considered liabilities.

Member States have the possibility to use CERs from sinks up to 1% of their assigned amount per year (5% during the 2008–12 period), and it would be strange that a Member State would use this possibility while at the same time forbidding it within the EU ETS. The EU should harmonize this issue for the second trading period.

In addition, the linking directive mentions the need for the Commission to adopt provisions relating to the risk that genetically modified organisms (GMOs) and invasive alien species used in afforestation and reforestation project activities may create. Decision 19/CP.9 leaves it to host parties to evaluate, in accordance with their national laws, potential risks associated with the use of GMOs by afforestation and reforestation project activities and to Annex I parties to evaluate, in accordance with their national laws, the use of tCERS and/or lCERs from afforestation and reforestation project activities that make use of GMOs.[68]

The linking directive does not forbid the use of CERs and ERUs from large hydro plants. However, it requires Member States to ensure that, when approving hydroelectric power production projects with a generating capacity exceeding 20 MW, relevant international criteria and guidelines – including those contained in the World Commission on Dams November 2000 Report *Dams and Development – A New Framework for Decision-Making* – are respected during the development of such project activities.

Some comments can be made: (1) Member States can indeed go individually beyond this requirement when implementing the Directive and state that all projects should be submitted to the criteria and guidelines of the World Commission on Dams (WCD criteria and guidelines) in order to use them for compliance; (2) the Directive establishes that the Commission will only review those projects over 500 MW. This leaves projects between 20 MW and 500 MW out of the scope of the Commission's review. The risk would be that projects not meeting the WCD guidelines could still be approved by non-EU governments and generate ERUs and CERs that would be used for compliance because the Commission will not track them; (3) currently, the largest part of all the CDM projects registered and in the process of validation are large hydro projects (28% in May 2005). Had the proposal from the European Parliament been incorporated into the Directive, the entrance of CERs into the EU ETS could have been largely reduced.[69] In any case, the fact that the Commission will review *ex-post*[70] shows that there seems to be a lack of confidence regarding whether the host countries will apply tough sustainable development standards.

In the original proposal the Commission inserted an article after art.11 (called art.11 (ter)) stating that Member States had to ensure that all the projects in which they participate or authorize private or public entities to participate outside the territory of the Community would take into account the environmental and social impacts of those projects and would ensure their contribution to sustainable development.[71] An article (called art. 17 bis), was inserted after art. 17, which obliged Member States to assess the environmental impacts that may result from their national strategies or programmes for the implementation of projects and consult to the public in accordance with Directive 2001/42/EC on the assessment of the effects of certain plans and programmes on the environment (SEA Directive).[72] Both requirements have been deleted from the final text of the linking directive, to bring the text in line with recital 4 and para. 33 of Decision 17/CP.7, but Member States can decide to put in place extra checks to ensure the environmental quality and sustainability of the project.

Furthermore, the Parliament proposed to exclude the use of CERs and ERUs from power sector installations that emit in excess of 400kg CO_2/MWh electricity or the equivalent for heat output.[73] The Parliament defended that it would be against the objectives of the CDM to allow the generation of CERs from projects that displace a very inefficient coal power plant for other a little bit more efficient. This is not a moot point to the extent that the EU and the world at large will keep relying on fossil fuels, representing an 85% of the increase in world energy demand by 2030 and this will represent an increase of 50 billion tones of CO_2, but was not included in the final version of the Directive.[74]

5.1.3 Specific aspects regarding JI projects

5.1.3.1 Participation in JI projects Although the JI does not have a 'prompt start' as the CDM, the fact that JI projects can exist (since the year 2000) has created a market for ERUs in which both governments and companies have been participating for years now.[75] So far, that participation has been based on the rules established by the Kyoto Protocol and the Marrakech Accords and in memoranda of understanding between countries, without a specific legal base in national law.[76] The linking directive will allow companies within the EU ETS to use ERUs for compliance from the second trading period onwards, provided Member States authorize it, and subject to the percentages laid down in the NAPs. The scope of the Directive is, however, broader in relation to JI, as it establishes the same project exclusions from that of the CDM and regulates the approval and accounting of JI projects that take place within the EU. Having now analyzed the surrender for compliance and project types I will now focus on the approval and accounting of JI projects.

As participation in the project-based mechanisms is not compulsory, Member States first need to decide whether they want to participate in them or not for compliance with their own targets. If they fulfil all the eligibility criteria, they can then participate under JI track one, otherwise they can still participate under JI track two. All EU Member States, including those which acceded on 1 May 2004, are expected to fulfil all their monitoring and reporting obligations before 2006, when the pre-commitment period review starts. Accordingly, they will be able to participate under JI track one. Member States should design their own project cycle for JI and decide upon which methods to use for baseline setting, (in particular when it comes to benchmarking), monitoring, verification and compliance. The following analysis assumes compliance with JI track two.

Once a Member State decides to participate, it has to choose whether to participate only as an investor country or whether to become a host country for JI projects. There are several reasons why a country would not want to become a host country.[77] A Member State that becomes only an investor country would probably decide not to duplicate any functions properly undertaken by project participants, designated or independent operational entities, or the relevant international bodies (the supervisory board for JI or the Executive Board for CDM). However, it retains its right to request a review of the project registration or of the issuance of credits in case it should emerge that false information has been provided. In addition, when the Member State may fear that a project may have negative environmental impacts which will not be reflected in an environmental impact assessment (EIA) either because it is not required, either because it is of mediocre quality, it could decide to require an EIA in accordance with international law or even with its domestic law before giving its approval to the project. The application of this check needs to be carefully assessed and applied only in very specific cases, to avoid the possibility of associated transaction costs rendering sound projects unviable.

5.1.3.2 Harmonizing JI regulations?   If and when a Member State decides to become a host country for projects, then it needs to design the project cycle, methods for the calculation of baselines, monitoring, verification and compliance. Rules and methodologies for setting baselines and developing monitoring plans are only mentioned, in a embryonic way, in the Marrakech Accords. Thus, when setting national guidelines for JI projects, Member States have a large degree of freedom, although they can seek guidance from the work of the CDM EB. For instance, new conditions can be included and the concept of additionality can be further developed. A Member State could produce a JI Project Design Document (JI PDD) to be done in accordance with the CDM-PDD, could set identical requirements for approval of projects and participants that those required by the CDM EB and could make approval of projects valid only until 31 December 2012, thus rejecting late crediting through allocation of AAUs to the project.[78]

Furthermore, a Member State has to decide whether to allow private entities to participate in JI. In case it does, it assumes a liability for the credits purchased, as ultimately it has to make sure that it complies with its Kyoto target. For this reason, the Member State will want to establish a legal procedure to ensure that liability is minimized. Neither the Kyoto Protocol nor the JI Guidelines or CDM Modalities and Procedures set out any procedural requirements for obtaining approval of and authorization to participate in project activities, and the linking directive is equally silent on this respect.[79] Each Member State has, therefore, the freedom to establish the criteria it considers convenient.[80]

If a number of Member States would authorize their private entities to participate in projects, and if a number of them would decide to become host countries, it is possible that the regulations of those Member States would vary widely. The question that can be raised is what would happen if some parties prescribe lax requirements on baseline, monitoring and verification while others adopt very stringent rules, taking into account that Annex I parties compete with each other in order to benefit from JI as much as possible. A risk of a race to the bottom may arise. On the contrary, it could well be that some Member States set unilaterally very strict requirements for JI projects that could hinder the development of the mechanism. For this reason, it would seem beneficial to agree on simplified procedures for JI EU-wide in order to facilitate the approval of projects and to ensure their environmental integrity.

Possibly this does not seem to be a serious issue in the EU-15, since Member States do not have much room to implement JI projects within their territories, and since the policy objective is to increase the size of the EU ETS as soon as possible to cover more sources and gases. Here the role of JI could rather be to seize opportunities to reduce emissions of GHG other than CO_2, when they cannot be included in the EU ETS due to difficulties and inaccuracies in the monitoring of emissions.[81]

However, in the case of the countries that recently acceded to the EU there are, in general, large volumes of 'hot air' that could be sold through JI projects. As it is probably not enough to have the *acquis communataire* implemented on paper, the harmonization of JI modalities could be desirable.

Most JI projects are already taking place in central and eastern European countries. This pattern will more than likely to intensify in the mid-term. Taking into account that many of those countries will join the EU sooner or later, it would be worthwhile to establish standardized guidelines for the implementation of JI projects in those countries. By so doing, it would be possible to ensure their environmental integrity and to help those countries to come closer to the environmental *acquis*.

In general, those issues that could be harmonized are participation requirements for legal entities, eligibility criteria, baseline requirements including sectoral benchmarks, monitoring and verification requirements and fast-track procedure for small scale projects. The advantages would be guaranteed environmental integrity, equal access of EU companies to JI, prevention of incompatible state aid, ensuring consistency and synergies with EC legislation on environment and energy, in particular with respect to baseline setting – including Best Available Technologies (BAT) required by the IPPC Directive – and accreditation and verification requirements.

5.1.3.3   Compliance with the acquis communataire   Article 11(b)1 ET Directive requires Member States to ensure that the baselines for project activities undertaken in countries that have signed a Treaty of Accession with the Union fully comply with the *acquis communataire*, including the temporary derogations set out in that Treaty. The EU pre-accession policy requires that all new investments in candidate countries comply with the *acquis*. The rationale is that a baseline lower than the *acquis* might: (1) serve as an incentive for candidate countries to delay implementation of investment-heavy parts of the *acquis* till after 2007; serve as a disincentive for investors to engage in necessary high-tech projects, if greater credits are available from low-tech investments (the 'low-hanging fruit').

The *acquis communataire* relevant to climate change is very extensive.[82] The *acquis* includes minimum environmental standards (such as emission limit values for large combustion plants), and more flexible legislation like market based instruments (i.e. the EU ETS). There are pieces that affect the baseline directly, such as the IPPC Directive[83] or the Landfill Directive,[84] and others that affect it more indirectly, such as the directives on fuel quality and vehicle emissions and the Large Combustion Plant Directive.[85] The *acquis* is implemented over time and can follow a different pace in different Member States for two reasons: first, transition measures can be negotiated with the Commission for some reasons, e.g. when a substantial adaptation of infrastructure is needed; secondly, EU environmental legislation is often defined in terms of BAT. Best Available Technologies can differ between countries and even within countries and can be subject to negotiations. The scope for negotiation is however limited, because the IPPC develops constantly Best Available Technique Reference

Documents (BREFs) that, although not legally binding, narrow considerable the room for manoeuvre when setting the baselines. However, even BREFs differ widely between Member States. In general it is safe to say that the implementation of the *acquis* has minimized the scope for JI projects in all new Member States, not least because it has already covered the 'low-hanging fruit', leaving the expensive reductions available. In the case of Romania and Bulgaria the scope is higher. However, there is the other side of the coin. The *acquis* lowers the level of emissions of the Member State, which for new Member States means enlarging the gap between the AAU and the actual emissions, thereby making IET more attractive. Further, when these countries join the EU bubble they will bring that 'hot air' into it, making compliance with Kyoto commitments easier.

5.1.3.4   Double counting   Double counting in the context of linking JI with the EU ETS may arise when emission reductions generated by a JI project are counted twice and give rise to the issuance of an ERU and an EU allowance at the same time. There are two cases where double counting may happen: (1) when the JI project activity takes place in an installation included in the coverage of the EU ETS; (2) when a JI project takes place out of the EU ETS but has an indirect effect on the emissions of installations under the EU ETS, like projects dealing with renewable energy or energy efficiency. The linking directive has sought to give a solution to each of these two cases. It does so by establishing a general prohibition for Member States to issue credits (ERUs and CERs, the latter when the project would be a CDM project taking place in Cyprus or Malta), from reductions or limitations of GHGs emissions from installations within the scope of the EU ETS.[86] However, the Directive provides for two exceptions, which can apply until 2012. For the case where the JI project activity reduces or limits directly the emissions of an installation within the EU ETS, the Member State can issue credits, provided that an equal number of allowances is cancelled from the account of the operator of that installation. For the case where the project activity reduces or limits indirectly the emissions of an installation within the EU ETS, the Member State can issue credits as long as it cancels an equal number of allowances from the national registry of the Member State where the credits are originated.

This provision has several consequences. In case an installation under the scope of the EU ETS seeks to host a JI project, it has to make sure that it receives a number of allowances in line with the baseline of that project. In other words, the allocation of allowances has to be done according to the baseline of the project, otherwise the installation will not have enough allowances to cover the ERUs generated.

A Member State that considers to host JI projects, may need to set up a reserve pool in its NAP from which to cancel allowances as ERUs are generated.

The Member State needs to calculate in advance the volume of that reserve pool, taking into account the JI projects estimated to take place. As it is difficult that the estimations are precise, the pool may be too short or too large. The situation can be looked at from the perspective of the project developers, from the perspective of installations covered by the EU ETS and from the perspective of the Member State. For the former, it is important that they present early in advance to their government their project development document, with accurate analysis of the baselines and the indirect impact on emissions within the EU ETS, otherwise they may find that there are not enough allowances available to cover their project. From the perspective of the companies covered by the EU ETS the risk is that the reserve pool is too large, because that means that their allocation is necessarily shorter. Finally, from the perspective of the Member States, a short reserve pool may create legal challenges from project developers that cannot meet their contractual obligations, or from companies within the EU ETS which feel that their allocated volumes are too short.[87] The Member State should put in place clear application procedures for allowances in the reserve pool. The solution of establishing reserve pools is the one recommended by the Commission.

A Member State may decide to avoid this issue altogether by simply forbidding JI projects which directly or indirectly reduce or limit emissions within the EU ETS, although in that case existing projects need to be taken into account.[88] The Directive gives freedom to Member States to do so.[89]

There is yet another case of JI projects which fall out of the scope of the linking directive: those which generate emissions reductions and do not directly or indirectly limit or reduce emissions within the EU ETS. These projects are not regulated under the Directive and therefore can be developed as usual provided they comply with the *acquis*. This type of project includes capturing methane from landfills, energy efficiency, nitrous oxide reduction projects and non-grid connected renewable generation. The implementation of the Landfill Directive, however, will exclude the potential of the first ones to a great extent and the others will, in general, fall under the scope of the EU ETS.

It is hoped that JI will help renewable energy to increase its share in the energy mix. However, there is a financial problem associated; the Renewable Energy Directive[90] is part of the baseline, as well as all the incentives given by Member States to developers; if it is cheaper to purchase CERs and ERUs from outside the EU, then JI will not help to increase the percentage of green electricity within the EU. The promotion of green electricity depends more on regulatory certainty for new investments in power generation – which is not likely to arise within the EU ETS until the international climate change regime is decided – in the good functioning of promoting mechanisms such as markets for tradable certificates or feed-in tariffs, and in the development of a market for biomass.

5.1.4 Domestic offset projects

Article 30.2(n) leaves for the review of 2006 the consideration of the modalities and procedures for Member State's approval of domestic project activities and for the issuing of allowances in respect of emission reductions or limitations resulting from such activities from 2008.

These projects are also called 'domestic offset projects' (DOPs). They are project-based activities. However, they are not part of the flexible mechanisms of the Kyoto Protocol. Thus the 'domestic' label attached to them. Accordingly, they can generate neither CERs nor ERUs, but would generate EU allowances if and when introduced. The main difference with JI projects is that the latter is bilateral, as it requires participation by two countries, whereas the DOPs would be unilateral. However, in practice that distinction is of little relevance, because unilateral JI projects could be possible and because the bilateral character of a JI project can be easily circumvented in terms of entity involvement. They were proposed by some Member States as a way to provide an economic incentive for sectors not covered by the EU ETS to engage in abatement efforts at an early stage. On the other hand, some considered that the first best option for EU climate change policy would be to expand – as fast as possible – the EU ETS. Allowing DOPs could made this more difficult. In addition, setting baselines for these projects could shield the installations where they take place from further reductions in the future, whereas the EU ETS would see the volume of allocated allowances further reduced. Moreover, because baselines could be shielded from newer and tougher environmental policies, it was seen as unattractive from an environmental perspective. This concern went to the point of fearing that DOPs could create perverse incentives for governments to avoid adopting new policies and measures, in order to attract more DOPs.[91] The alleged benefits of DOPs are that they can be implemented quickly, have a low risk profile, focus on domestic action and increase the availability of options for compliance within the EU ETS. The Commission and the European Parliament opposed them on the ground that they offered a piecemeal approach to some sectors instead of the comprehensive approach that the EU ETS seeks to maximize efficiency. However the Parliament offered as a compromise in the first reading of the proposal that domestic projects could start on an experimental basis in the rail sector.[92] This raises the question of the definition of 'project'. Neither the Kyoto Protocol nor the Marrakech Accords give a definition of project and, as I have shown above in relation with the CDM, there is some margin for interpretation. In any case, being a domestic measure, DOPs could define project in a way that could include the rail sector. Clearly macro-economic or sector-wide definitions may raise serious methodological and practical problems, particularly when it comes to measuring their impact, determining environmental additionality and establishing ownership of reductions. On the other hand, it can be argued that those difficulties may be overcome[93] and that it is worth considering the great potential for emission reductions in some sectors, specifically the transport sector that should receive incentives. Be that as it may, the solution adopted by the Directive clearly excludes, for the moment, the possibility that Member States implement them unilaterally.

5.1.5 Support of capacity building activities

Article 21(a) of the linking directive requires Member States and the Commission to 'endeavour to support capacity-building activities in developing countries and countries with economies in transition' so that they may profit from CDM and JI projects and proceed with their sustainable development strategies. The Community, through its annual budget, and the European Development Programme is already supporting programmes that can be utilized in the context of JI and the CDM.[94] Those programmes cover all geographical areas and are accessible to all stakeholders. The different programmes cover specific themes and activities. Most of the funds are allocated through competitive bidding processes, such as calls for proposals or public tenders. The themes include capacity building for JI/CDM in host countries, research and technology development for JI/CDM and JI/CDM investments including, for the preparatory phases, sectoral/regional policy formulation, regulatory frameworks, exchange of experiences, including private sector meetings, feasibility and investment studies and joint ventures. Nevertheless, this is a topic which will also be subjected to the revision of June 2006.[95]

6. CONCLUSIONS AND CHALLENGES AHEAD

Several conclusions can be ascertained. First, the CDM is plagued with challenges, but is already up and running – and there are several projects already registered and delivering reductions.[96] Joint implementation will start in 2008, and is expected to cover a niche market for credits, but the size of it is unknown. Despite the existing challenges, the EU has decided to link them to the EU ETS already in the first trading period of the EU ETS. This is part of its 'learning by doing' strategy.

Secondly, Member States are now busy preparing the implementation in national law of the linking directive. However, many are still lagging behind in the full implementation of the ET Directive. In addition, they are starting to prepare the NAPs for the second trading period. Further, consultations to make amendments to the ET Directive are already in place. All these factors combined may be too much too grasp for many Member States, already overwhelmed with the speed of the process.

Thirdly, the implementation of the linking directive can differ widely in different Member States, ranging from those that want to implement it 'as it is' to those which want to further restrict the market for credits. This could fragment the nascent emissions market even further. In the case of JI track two, it would be interesting to agree on EU-wide harmonized rules and guidelines for implementation, reaching even Central and Eastern European Countries (CEEC).

As there are increasing calls to further harmonize the EU ETS, there will probably be calls – there are already from industry – to further harmonize the implementation of the linking directive. It seems important to let the market work as freely as possible. However, the radical innovation that the CDM and the JI create in international law, and the uncertainties that surround them, advises against letting the market work without restraints. Learning by doing and the full application of the precautionary principle (specially in relation to GMOs and alien and invasive species) seems indeed a reasonable way to proceed.

Finally, in the future the EU should remain a key actor in the international negotiations, sending input to the UNFCCC as regards the project based mechanisms, and ready to shift to a sectoral or even a policy approach if it transpires that that is the best way to proceed. The role of the mechanisms in promoting sustainable development probably depends on such an active role from the EU.

NOTES

1. Junior Researcher, Institute for Transnational Legal Research (METRO), and Ius Commune Research School; email Javier.decendra@facburfdr.unimaas.nl.
2. Directive 2004/101/EC of the European Parliament and of the Council of 27 October 2004 amending Directive 2003/87/EC establishing a scheme for GHG emission allowance trading within the Community, in respect of the Kyoto Protocol's project mechanisms, OJ L 338, 13.11.2004, pp. 0018–0023. *I will refer to it either as the Directive or the linking directive.*
3. Directive 2003/87/EC of the European Parliament and of the Council of 13 October 2003 establishing a scheme for greenhouse gas emission allowance trading within the Community and amending Council Directive 96/61/EC, OJ L 275, 25.10.2003, pp. 0032–0046.
4. Article 12.3 ET Directive reads in part: 'Member States shall ensure that, by 30 April each year at the latest, the operator of each installation surrenders a number of allowances equal to the total emissions from that installation during the preceding calendar year'.
5. For a good account of this evolution, see Christiansen, A.C. 'The Role of Flexibility Mechanisms in EU Climate Strategy: Lessons Learned and Future Challenges', *International Environmental Agreements: Politics, Law and Economics*, **4**, pp. 27–46. The author suggests that the EU favoured this approach for three reasons: (1) because earlier calculations of costs within the EU to achieve its commitment were rather low (lower than those for the US, Canada and Japan); (2) the EU relied on the fact that some Member States had enjoyed strong emission reductions for reasons totally separated from climate change considerations, namely the UK with the 'dash for gas' and Germany with the re-unification; (3) the EU had achieved flexibility through the inclusion of art. 4 Kyoto Protocol, a provision created for the specific case of the EU and which enabled the creation of the EU 'bubble'. Thus, the EU position could be considered as enlightened self-interest and strategic manoeuvring.
6. Christiansen, ibid., notes that the loopholes in this context refer to opportunities for windfall gains due to the existence of hot air from Annex I countries, the use of sinks and the 'additionality test' problem.

7. More precisely, the EU formulation stated that each party should acquire and surrender no more emission certificates than the equivalent of 50% of the difference between five times the emission in one of the years between 1994 and 2002, on the one hand, and its number of AAUs on the other. See Draft Set of Guidelines for Joint Implementation, EU Submission, para. 8, and Draft Modalities and Procedures for Clean Development Mechanism, EU Submission, para. 9. Both can be found in UNFCCC (1999): Principles, Modalities, Rules and Guidelines for the Mechanisms under Articles 6, 12 and 17 of the Kyoto Protocol: Submissions from the Parties: Note by the Secretariat: Addendum: Bonn: UNFCCC Secretariat (Official Documents; FCCC/SB/1999/MISC.3/Add.3).

8. Adopted in COP-7, held in Marrakech in November 2001. Supplementarity is defined as 'the use of the mechanisms shall be supplemental to domestic action and [...] domestic action shall thus constitute a significant element of the effort made by each Party included in Annex I to meet its quantified emission limitation and reduction commitments under Article 3, paragraph 1'.

9. Christiansen, A.C., 'The Role of Flexibility Mechanisms in EU Climate Strategy: Lessons Learned and Future Challenges', p. 30, referring to Edward, J.H. (2002), 'Has the European Union Exercised Leadership in the International Climate Change Regime Since The Hague Conference?', Brussels: College of Europe, Department of European Political and Administrative Studies.

10. Article 30.2 ET Directive.

11. Decision FCCC/CP/2001/13/Add.2, including Decision 16/CP.7 on JI, Decision 17/CP.7 on CDM and Decision 18/CP.7 on EIT.

12. Decision 17/CP.7, para. 1.

13. COP11/MOP1 will take place in November–December 2005.

14. Those laid down in para. 21 of the Annex on Guidelines for the implementation of Article 6 of the Kyoto Protocol, Draft Decision -/CMP.1.

15. Ibid, para. 24.

16. The Guidelines for the Implementation of Article 6 of the Kyoto Protocol, state in part C, para. 3(c) that the JI Supervisory Committee will have given consideration in its work to the relevant decisions of the CDM EB, although clearly is not obliged to follow them.

17. See for instance a recent paper from Sterk, W. and Wittneben, B. (2005), 'Addressing Opportunities and Challenges of a Sectoral Approach to the Clean Development Mechanism', *JIKO Policy Paper*, 1. See also Cosbey, A., et al. (2005), 'Realizing the Development Dividend: Making the CDM Work for Developing Countries', *Report of the International Institute for Sustainable Development*. For a specific analysis on baselines for JI projects, see Illum, K., Meyer, N.I. (2004), 'Joint Implementation: Methodology and Policy Considerations', *Energy Policy*, **32**, 1013–23.

18. Emphasis added.

19. Appendix C on the terms of reference for establishing guidelines on baselines and monitoring methodologies, 17/CP.7, FCCC/CP/2001/13/Add.2, 46.

20. Executive Board of the Clean Development Mechanism, 16th Meeting, Report, Annex 3, Clarifications on the treatment of national and/or sectoral policies and regulations in determining a baseline scenario.

21. Decision 17/CP.7, Preamble and para. 7, respectively.

22. See Cosbey, A., et al. (2005), op. cit., n. 17.

23. One of the reasons that environmental groups used to oppose the inclusion of CDM projects on sinks was that they would be used by large-scale industries to establish vast mono-species plantations, possibly displacing and further marginalizing local and indigenous populations. This would go against any notion of sustainable development.

24. For instance in the case of sink projects, the populations living in the area of the project, in order to profit from it would need to have tenure rights on the land, which they often lack. In addition, the definitions of afforestation and deforestation impede that populations living in areas deforested after 1990 profit from them.

25. According to para. 40 of Draft decision -/CMP.1 (art. 12), in order to submit the validation report of the CDM project to the CDM EB, the project participants need to obtain from the host country confirmation that the project assists it in achieving sustainable development. The

host state approvals have, since the beginning, invariably included a statement very similar to the following: 'the project contributes to sustainable development in country X'.

26. Pearson, B. (2004), 'Market Failure: Why the Clean Development Mechanism Won't Promote Clean Development', *CDM Watch*, available at http://cdmwatch.org/files/market%20failure. pdf or www.cdm.watch.

27. See the Summary Record of the Second Meeting of the ECCP Working Group on JI and CDM, Brussels, 27th March 2002, on the chapter regarding discussion on issues raised in Background Document N.1.

28. Communication from the Commission on EU policies and measures to reduce GHG emissions: Towards a European Climate Change Programme, COM (2000) 88.

29. Ibid., pp. 12–13.

30. Mandate of the WG on flexible mechanisms, available at http://www.eu.int/comm/environment/climat/pdf/wg1_mandate.pdf.

31. Mandate approved by the ECCP Steering Committee on 18th February 2001. All the material from the meetings is available at http://www.eu.int/comm/environment/climat/jicdm/jicdm. htm.

32. EC Commission, Communication from the Commission on the implementation of the first phase of the European Climate Change Programme, COM (2001) 580 Final.

33. Ibid, 8.

34. Those rules were adopted in the Marrakech Accords, November 2001.

35. Op. cit., n. 32.

36. Proposal for a Directive of the European Parliament and of the Council establishing a scheme for greenhouse gas emission allowance trading within the Community and amending Council Directive 96/61/EC, COM (2001) 581 Final.

37. Meeting agendas, meeting minutes and background documents are available at http://www. eu.int/comm/environment/climat/ji_cdm.htm.

38. EC Commission, Proposal for a Directive of The European Parliament and of the Council amending the Directive establishing a scheme for greenhouse gas emission allowance trading within the Community, in respect of the Kyoto Protocol's project mechanisms, COM (2003) 403, adopted 27 July 2003.

39. See Langrock, T., Sterk W. (2004), 'Linking CDM & JI with EU Emission Allowance Trading', *Policy Brief for the EP Environment Committee*, EP/IV/A/2003/09/01, available at http://www.europarl.eu.int/comparl/envi/pdf/externalexpertise/ieep/flexible_mechanisms_ brief.pdf, 12 January. For an analysis of the full evolution of the negotiations, see Lefevere, J. (2005), 'Linking Emissions Trading Schemes: the EU ETS and the 'Linking Directive', in Freestone, D. and Streck, C. (eds), *Legal Aspects of Implementing the Kyoto Protocol Mechanisms – Making Kyoto Work*, Oxford: OUP, pp. 511–33.

40. MEMO/03/154, 23 July 2003.

41. Recital 3 of the Preamble, linking directive.

42. For instance, the Ministry of Environment of Slovakia has passed a draft law which forbids Slovakian companies covered by the EU ETS to use credits for compliance. See Carbon Market Europe of 23 September 2005, 6.

43. One of the reasons that the Slovakian Government gives to forbid the use of credits for compliance is that a survey among companies has shown that only one company would be interested in doing CDM projects.

44. Industry consistently puts pressure to allow the use of credits without limitations, to increase the cost-effectiveness of the EU ETS.

45. Article 11(a)3 of the ET Directive.

46. Consultation paper on the linking directive, p. 18, available at http://www.defra.gov.uk/corporate/consult/euets-linkingdir/consultation.pdf. Last accessed 30 September 2005.

47. As stated in the White Paper 'Our Energy Future – Creating a Low Carbon Economy', available at http://www.dti.gov.uk/energy/whitepaper/ourenergyfuture.pdf, last accessed 30 September 2005.

48. See Carbon Market Europe, 23 September 2005, 6.

49. National Allocation Plans have to be submitted to the Commission by 30 June 2006. Article 11.1 ET Directive.

50.   A Member State could decide not to buy credits itself, and increase within the NAP the volume of credits that installations within the EU ETS can purchase. A possible problem that may arise is that at the end of the compliance period there are not enough credits at its disposal to comply with its commitments. In practice, most Member States have set up funds to purchase themselves credits in the international markets.

51.   The current price of the EU allowance is of around €21, whereas CERs do not trade at prices higher than €10.

52.   Decision 19/CP.7 regulates the carry-over of Kyoto units. It sets a limit of 2.5% of the Assigned Amount for ERUs and 2.5% of the Assigned Amount for CERs. See Decision 19/CP.7, 61.

53.   Communication from the Commission on guidance to assist Member States in the implementation of the criteria listed in Annex III to Directive 2003/87/EC establishing a scheme for GHG emission allowance trading within the Community and amending Council Directive 96/61/EC, and on the circumstances under which force majeure is demonstrated, COM (2003) 830 Final.

54.   See the Policy Brief for the European Parliament Environment Committee EP/IV/A/2003/09/01, 'Linking CDM & JI with EU emissions allowance trading', 15 and Annex II. The linking directive as such cannot ensure that supplementarity is complied with, because the linking proposal does not regulate all channels through which the EU Member States can acquire credits. In particular, it does not regulate the CERs and ERUs that Member States can acquire in the international markets, nor does it regulate the AAUs that Member States can acquire through international emissions trading. Due to this limitation, the EU has approached the supplementarity issue by a combination of two measures: (1) the establishment of a mechanism for monitoring EU GHG emissions and evaluating progress towards meeting international commitments; (2) the inclusion of the supplementarity question in the elaboration of the NAPs.

55.   After forwarding the quantity of CERs corresponding to the share of proceeds to cover administrative expenses and to assist in meeting the costs of adaptation.

56.   As laid down in provision 2 of the Modalities, rules and guidelines for emissions trading under Article 17 of the Kyoto Protocol, FCCCP/CP/2001/13/Add.2, 52.

57.   The deadline is 13 November 2005.

58.   Articles 11.4 and 12.5 of the ET Directive.

59.   Draft decision -/CMP.1, Chapter II Registry Requirements, points 17–22.

60.   In July 2005, there were only nine national registries in place: Austria, Denmark, Finland, France, Germany, Netherlands, Spain, Sweden and the United Kingdom. Those of Belgium and Ireland were close to becoming operational. Transfer between registries is essential to allow the development of CER exchanges in exchange platforms. There is already one player offering forwards whose buyers can elect to take CERs or cash at delivery and which takes the risk that CERs cannot be transferred should the ITL not be linked to the CER registry before delivery. This risk largely explains that guaranteed CERs do not trade at or above EUA price levels.

61.   According to the Marrakech Accords, nuclear CDM and JI projects are not prohibited, only to use CERs and ERUs from those projects before the year 2012. See the Preamble of Decision 16/CP.7 on Guidelines for the implementation of Article 6 of the Kyoto Protocol.

62.   Decision 19/CP.9 on 'Modalities and procedures for afforestation and reforestation project activities under the clean development mechanism in the first commitment period of The Kyoto Protocol, Document FCCC/CP/2003/6/Add.2. the Kyoto Protocol recognizes that parties can rely on afforestation, reforestation and deforestation to meet their targets. In addition, the Marrakech Accords introduced four additional activities: forest management, cropland management, grazing land management, and re-vegetation. In relation to their validity for CDM and JI projects, the Marrakech Accords regulate the issue, establishing some exceptions and a cap. As regards the CDM, the Marrakech Accords limit CERs to CDM projects in afforestation and reforestation.

63.   Non-permanence refers to the risk that, for some reason, the carbon contained in the ground or in biomass, is released back into the atmosphere.

64.   Decision 19/CP.9 requires that both tCERs and lCERs are replaced before their expiry date. A tCER expires at the end of the commitment period following the one during which it was

issued. A lCER expires at the end of the crediting period of the afforestation or reforestation project activity under the CDM for which it was issued. The crediting period for a proposed afforestation or reforestation project activity under the CDM shall be either: (a) a maximum of 20 years which may be renewed at most two times, provided that, for each renewal, a Designated Operational Entity (DOE) determines and informs the Executive Board that the original project baseline is still valid or has been updated taking account of new data where applicable; or (b) a maximum of 30 years. See Draft decision -/CMP.1, para. 23.

65. These ideas and the possible solutions mentioned are taken from Dutschke, M. et al. (2005), 'Value and risks of expiring carbon credits from afforestation and reforestation projects under the CDM', *Climate Policy*, **5**, pp. 109–25.

66. A chain of successive tCERs is created when a project investor replaces tCERs that expire with tCERs from the same sink project.

67. Ibid., pp. 120–22.

68. Preamble to Decision 19/CP.9.

69. See the Report prepared by Mr Alexander de Roo for the European Parliament, A5-0154/2004, Final. In amendment 8, Article 1, point 2, Article 11(a), para. 3, point (b a) (new) (Directive 2003/87/EC), it was proposed that hydro projects larger than 10 MW would be excluded from the scope of the Directive, and those of 10 MW or smaller should be in full compliance with the criteria laid down by the World Commission on Dams.

70. Article 30.2(l) ET Directive.

71. Article 11 (ter).4, Commission, Proposal for a Directive of the European Parliament and of the Council amending the Directive establishing a scheme for greenhouse gas emissions allowance trading within the Community, in respect of the Kyoto Protocol project mechanisms, COM (2003) 403.

72. Directive 2001/42/EC of the European Parliament and of the Council of 27 June 2001 on the assessment of the effects of certain plans and programmes on the environment, OJ L 197, 21/07/2001 pp. 30–37.

73. See the Report prepared by Mr. Alexander de Roo for the European Parliament, A5-0154/2004, Final. In amendment 9, Article 1, Point 2, Article 11(a), para. 3, point (b) (new) (Directive 2003/87/EC), it is proposed to exclude power sector installations that emit in excess of 400kg CO_2/ MWh electricity or the equivalent for heat output.

74. An added fear is that these types of investments could create a lock-in effect in developing countries and economies in transition for the whole lifetime of the new power plants. This can be serious because investment in the global energy sector will amount to $16 trillion under a Business as Usual scenario (BAU scenario). These observations become even more relevant at a point in time where it is acknowledged that the CDM and the JI are not leading to significant technology transfer, and there is an ongoing debate on this issue. Member States can decide to impose more stringent conditions as regards the type of energy efficient projects they approve and can put more emphasis in technology transfer.

75. See for instance Streck, C. (2005), 'Joint Implementation: History, Requirements and Challenges', in Freestone, D., Streck, C. 'Legal Aspects of Implementing the Kyoto Protocol Mechanisms—Making Kyoto Work', Oxford: OUP, pp. 107–26.

76. Ibid., p. 118.

77. The UK has recently adopted the Greenhouse Gas Emissions Trading Scheme (Linking Directive) Regulations 2005. Article 5 para. 6 states that '[t]he Secretary of State may not approve a proposed project activity to be carried out in the United Kingdom.

78. As an early proposal of the German regulation implementing project based mechanisms suggested.

79. Article 11(b) para. 5 merely restates the provisions of the Marrakech Accords by stating that Member States that authorizes private or public entities to participate in project activities shall remain responsible for the fulfilment of its obligations under the UNFCCC and the Kyoto Protocol.

80. See for instance Part 3 of the Greenhouse Gas Emissions Trading Scheme (Linking Directive) Regulations 2005, on project approval and authorization to participate in project activities. The application has to be made to the Secretary of State, which can require any information to be independently verified, and may require further information.

81. Some utilities have proposed this solution in the context of the Climate Change Policy Review undertaken by DEFRA. See the Final Report on the Analysis of Responses to the Analysis of Responses to the Consultation for the Review of the UK Climate Change Programme to Defra, available at http://www.defra.gov.uk/corporate/consult/ukccp-review/responses.pdf.
82. Directorate General of the Environment publishes information sheets with the climate relevant pieces of the *acquis communataire*.
83. Council Directive 96/61/EC of 24 September 1996 concerning integrated pollution prevention and control, OJ L 257, 10.10.1996, pp. 0026–0040.
84. Council Directive 1999/31/EC of 26 April 1999 on the landfill of waste, OJ L 182, 16.07.1999, pp. 0001–0019.
85. Directive 2001/80/EC of the European Parliament and of the Council of 23 October 2001 on the limitation of emissions of certain pollutants into the air from large combustion plants, OJ L 309, 27.11.2001, pp. 0001–0021.
86. Article 11(b)2 ET Directive.
87. As allocation *ex-post* is not allowed except for cases of force majeure, legal challenges may arise if the size of the pool leads to distortions of competition or to default of projects.
88. Which was the reason to introduce the exceptions in the linking directive in first place.
89. Slovakia has issued a proposal for a draft regulation implementing the linking directive which prohibits JI projects that reduce or limit, directly or indirectly, emissions of installations within the EU ETS. Other countries could decide to do the same.
90. Directive 2001/77/EC of the European Parliament and of the Council of 27 September 2001 on the promotion of electricity produced from renewable energy sources in the internal electricity market, OJ L 283, 27.10.2001, pp. 0033–0040.
91. This problem arose in the context of the CDM, and the CDM EB tackled it as explained in Chapter 3.3 above. See also n. 29 above. Another approach to solving these problems can be to establish dynamic baselines that take into account trends and objectives of environmental policies, as opposed to static baselines.
92. Amendment 16 Article 1, Point 9, point A (new), A5-0154/2004, 15. Article 30, para. 2, point (d a) (new) (Directive 2003/87/EC).
93. The methodological problems could be solved by setting a baseline which refers to the national energy system from which the project is part, so as to account for the whole reductions generated. See Illum, K., Meyer, N.I. (2004), 'Joint implementation: methodology and policy considerations', *Energy Policy*, **32**, pp. 1013–23.
94. The PHARE programme provides funds to CEEC countries to meet the requirements of accession and to prepare the countries for the Structural Funds. The Pre-Accession Structural Instrument (ISPA) has been funding transport and environmental schemes in all the CEECs since early 2000, along the same lines as the Cohesion Fund model designed for the least prosperous EU members. The Special Accession Programme for Agriculture and Rural Development (SAPARD) has also been in operation since 2000, helping the applicants prepare for the common agricultural policy, in particular for its standards of food quality and consumer and environmental protection.
95. Article 30.2(m) ET Directive.
96. There are currently 25 CDM projects registered; see http://cdm.unfccc.int/Projects, last accessed 30 October 2005.

7. Emissions trading and the Aarhus Convention: A proportionate symbiosis?

Karen MacDonald and Zen Makuch[1]

1. INTRODUCTION

1.1 Aim of this Chapter

The aim of this chapter is to examine the extent to which some central elements of the European Union Emissions Trading Scheme (EU ETS) comply with the objectives of the Aarhus Convention. The Aarhus Convention represents a regional revolution in environmental democracy and this chapter determines whether or not the EU Emissions Trading Directive (hereinafter, the ET Directive), is equal to the legal demands of this groundbreaking Convention.

The EU ETS is arguably an innovative attempt to balance the goals of the sustainable development nexus – frequently understood to comprise environmental, economic and social concerns. However, the fact that the ETS concerns the emissions of pollutants and provides an incentive for the relocation of industries or changes in industrial practice which may positively or adversely affect the environment of a particular locale raises issues of accountability. These issues of accountability should be dealt with under the ETS itself, but if they are not, the provisions of the Aarhus Convention may provide a 'fallback' position.

Calls for accountability in climate change-related matters emanate from art. 4(1)(i) of the 1992 United Nations Framework Convention on Climate Change (UNFCCC) 'Commitments' section. This provides that all Parties shall: 'Promote and cooperate in education, training and public awareness related to climate change and encourage the widest participation in this process, including that of non-governmental organizations'. This language is reinforced in key provisions of art. 6 on 'Education, Training and Public Awareness' which requires (in accordance with national laws and regulations, and within their respective capacities), public awareness programmes on climate change and its effects; Public access to information on climate change and its effects; and public participation in addressing climate change and its effects and developing adequate responses.

Furthermore, the 1997 Kyoto Protocol to the UNFCCC, which provides a basis for the ET Directive, reminds States of their UNFCCC art. 6 obligations, though in even less obligatory terms and language. Article 10(e) requires States to 'facilitate at the national level public awareness of, and public access to information on, climate change'.

1.2   The Aarhus Convention

The Aarhus Convention on Access to Information, Public Participation in Decision-Making and Access to Justice in Environmental Matters (the Convention) was signed by the European Community and its Member States on 25 June 1998, with entry into force on 30 October 2001. It was ratified in the European Community by Council Decision 2005/370/EC on 17 February 2005. If they did not already fully ratify the Convention earlier, Member States are now required at national level to meet or exceed the implementing standards set by EU law.

The Convention constitutes a groundbreaking environmental law agreement[2] that aims to ensure that members of the public benefit from procedural rights advanced under its three core 'pillars': access to environmental information; public participation; and access to justice. It places obligations on key actors and provides members of the public with certain rights that have not readily been available in the past. Of significance, Annex I of the Convention lists sector-specific activities for which public participation procedures should be in place, as per art. 6(1)(a) (eg, for permitting of an activity). The sectors, including those which could impact on emissions production and trading, are: energy; metals; minerals; chemicals; waste; waste-water; infrastructure; and, extraction. Their relevance is discussed further below.

1.3   ET Directive 2003/87/EC

The Emissions Trading Scheme (ETS), created under Directive 2003/87/EC, commenced operation on 1 January 2005. Though the ETS and the Directive are discussed elsewhere in this book, some key provisions are examined herein in view of their relevance to this chapter.

Pursuant to art. 4 (requirement to obtain a greenhouse gas (GHG) emission permit) of the ET Directive, Member States are required to ensure that all installations listed under Annex I are permitted to emit GHG. The installation/activity operator[3] is required to obtain a GHG emissions permit issued by a competent authority in accordance with arts 5 (application to obtain a GHG emission permit) and 6 (conditions and content of permit) of the Directive.

Activity/installation operators that are eligible to participate in the ETS – in connection with the Directive Annex I categories of activities – can apply for GHG emissions permits as per the procedure identified in art. 4, art. 5, art. 6 and art. 7 (concerning changes in operation of installation). With respect to art. 27, an installation may be temporarily excluded from the Community ETS until 31 December 2007.

Under art. 30, the European Commission (the Commission) is required to undertake a review of the functioning of the ET Directive and to submit a report and any proposals on this to the European Parliament and Council by 30 June 2006. This review process (currently underway at the time of writing) comprises, in part, a stakeholder consultation in the form of a survey[4] to enable key stakeholders and organizations to contribute to the potential improvement of the EU ETS. The survey identifies stakeholders within its scope, as: Companies within or affected by the ETS and associations representing these companies; Governments and government bodies of all the 'old' and 'new' Members of the EU; Market intermediaries such as banks, law firms and independent trading organizations; and, NGOs and 'others' such as environmental organizations, academic organizations, and so on. An extensive definition of 'others' is not provided by the Commission but if 'other' persons can justify why the survey is relevant to them they may also participate.

1.4 An Ethical Dimension

It is important to note that Aarhus-type rights and obligations are conventionally sourced in traditional command and control legislation. Hence, regulators should pay special attention to ensuring that these such rights and obligations are included and strengthened in economic instruments such as emissions trading schemes. After all, such instruments entail some transfer of regulatory powers and actions to self-regulating firms. In this sense, the ET Directive is being implemented by businesses but will have no less an impact upon civil society and the ecosystems upon which it depends.

The concept of emissions trading 'relies on the privatization of formerly publicly held rights'.[5] This has attracted an element of criticism of an ethical nature. Arguably, 'privatization of the atmosphere' results in 'an abuse in the global commons that [it] is the responsibility of the State to protect, not to give away'.[6] Such an approach of regarding the 'sky as though it were a common resource free of any substantial public interest' is what has, arguably, led governments as guardians and trustees to neglect their broader environmental and other obligations to citizens, of what are in essence, public resources and public interests.

A decision has been taken to continue to pollute: to emit GHG, rather than striving for what is, at this moment in time, largely an aspirational zero-tolerance approach to harmful emissions. Whether such a decision has been taken on the basis of a sufficient consensus (and thus meets governments' 'legitimation function') requires debate that is beyond the scope of this chapter. Still, the observation that the global atmosphere is being 'used' in a manner that could conflict with other uses requires further policy option-based scrutiny; for example, one arguable 'use' of the global atmosphere is as a cooling system for the earth. Thus, conflicts over 'use' and 'ownership' are evident.

As a result, it is apparent that there has to be some level of governmental accountability to society even though it may favour polluters to some extent. It is clear that a policy consensus on emissions trading does not exist because there will be societal groups that oppose the permissible pollution policy inherent in an emissions trading regime. Perhaps there is a sense that emissions trading represents something of a compromise as an unavoidable consequence of sustainable *development*, though it is hoped that the said compromise will result in the 'ratcheting down' of GHG emissions over time.

It is clear that there are many willing participants in the ETS and that this fact adds legitimacy to this policy instrument. Still, if 'polluters'/activity/installation operators apply for a GHG emissions permit and benefit from emitting or sending their pollution elsewhere through trading their emissions allowances, then the public should have a right to the relevant information, to be able to participate in decision-making procedures regarding the issuing of a permit and the allocation of emissions allowances. Equally, stakeholders should be given access to data, reports and other information related to such a regime. The public should be informed, involved and included in the process, on the basis that a stable 'atmosphere' is a common concern, is the common heritage of humankind and is necessary to maintain life on this planet.

Evidence[7] from experience with emissions trading in the USA over the past 20 years is positive and indicates both environmental improvement and an overall cost reduction for industry in reducing pollution through the use in particular of cap-and-trade policies on emissions. The three Aarhus Convention pillars and subsequent EU implementation of these offer a way for governments and polluters to be accountable in relation to such policy matters as they apply to the emissions trading instrument.

2.   IMPLEMENTATION OF THE AARHUS CONVENTION IN THE EU

The first pillar of the Convention on public access to information has been implemented at Community level through Directive 2003/04/EC on public access to environmental information (Access to Information Directive). The second pillar on public participation in environmental procedures has been implemented through Directive 2003/35/EC (Public Participation Directive). Also, at the time of writing, the implementation of a Directive to transpose the third pillar which guarantees public access to justice in environmental matters (Access to Justice Directive) was proposed in 2003 and is now in the EU codecision procedure.

2.1 Directive 2003/04/EC on Public Access to Environmental Information[8]

Since 1990, there has been public access to environmental information requirements in the EU, though these have been subsequently repealed in favour of the current Directive 2003/04/EC (as of 14 February 2005).

The Access to Information Directive grants a general right of access to information to *any* person, including non-EU citizens, to request environmental information held by or for public authorities and from any public authority within the EU. It sets out the basic terms and conditions and practical arrangements for access to environmental information[9] and implements the first 'pillar' of the Aarhus Convention on public access to environmental information.

A summary of the Directive art. 2 definition of 'environmental information' indicates that the public shall have access to any information in written, visual, aural or electronic or other material form on the state of elements of the environment such as air, atmosphere, water, land, wetlands, genetically modified organisms etc.;[10] factors such as noise, emissions and discharges etc.;[11] measures such as policies, legislation, environmental agreements and others;[12] reports on the implementation of environmental legislation;[13] cost-benefit and other economic analyses;[14] and, the state of human health and safety, including any effects or consequences of the aforementioned.[15] A 'public authority' is defined as government or other public administration including national, regional and local level bodies;[16] any natural or legal person performing public administrative functions in relation to the environment under national law[17] or having public responsibilities or functions in relation to the environment.[18]

'Information held by a public authority' refers to environmental information in its possession, which has been produced or received by that authority,[19] and 'information held for a public authority' means that environmental information physically held by a natural or legal person on its behalf.[20] An 'applicant' is the natural or legal person that requests environmental information,[21] while the 'public' means one or more legal persons, and in accordance with national legislation or practice, their associations, organizations or groups.[22]

In line with the Aarhus Convention, the Directive governs 'passive' access to environmental information upon request (art. 3) and 'active' dissemination of environmental information (art. 7).

'Active' dissemination of environmental information requires, *inter alia*, that Member States ensure that public authorities organize the environmental information that they hold (which must be kept up to date, accurate and comparable[23]), which is relevant to their functions, in electronic/computer-based form where possible, and that such environmental information 'progressively' be made available in easily accessible, electronic, public databases (para. 1). Examples of information that has to be made available includes: data or summaries of data derived from monitoring activities[24] and progress reports in the implementation of international treaties and Community environmental law.[25] Thus, the UNFCCC, the Kyoto Protocol and the ET Directive, as amended, come under its scope.

A 'passive' information request may be refused, in line with the Aarhus Convention provisions, if any of the art. 4 'exceptions' apply. Some examples of art. 4 'exceptions' to supply environmental information could arise say, because the public authority in question does not hold the information requested;[26] the request is too general;[27] or, the request would 'adversely affect' public security[28] or intellectual property rights.[29] However, the grounds for refusal have to be interpreted in a 'restrictive way', requiring a narrow view to be taken of the scope of each exception. Efforts, as far as possible, should be made to supply the information requested. Thus, on a practical legal point, it is acceptable to 'extract' the relevant piece of requested information from any sources that come under the scope of the exceptions rule[30] as the emphasis is on making as much information available as possible.

Though all exceptions are subject to the 'public interest test', of particular relevance for our purposes is the penultimate closing paragraph of art. 4(2) which provides that, in cases where the information request relates to emissions into the environment, Member States may not apply the following exceptions and are thus required to disclose the relevant information: the confidentiality of the proceedings of public authorities;[31] the confidentiality of commercial or industrial information;[32] the confidentiality of personal data;[33] the interests or the protection of any person who has supplied the requested information on a voluntary basis;[34] or, the protection of the environment to which such information relates.[35] This public interest test in relation to emissions is not as unusually transparent as it would appear to be at first blush. It only applies to the specific emissions data itself (understood purely in numerical and unit terms).

Article 6 concerns access to justice in cases where a request for environmental information may have been wrongfully refused, wrongfully ignored, inadequately answered or otherwise not dealt with in accordance with arts 3, 4, and 5 of the Directive. This right of access to justice will be supported by the Directive on Public Access to Justice in Environmental Matters (see below) once it enters into force.

2.2 Directive 2003/35/EC on Public Participation in Environmental Procedures[36]

This EU Directive was supposed to have been implemented by Member States by 25 June 2005.[37] It implements the second 'pillar' of the Aarhus Convention on public participation in environmental decision-making and amends, *inter alia*, the IPPC Directive[38] (more on which is below in section 3.3) and the EIA (Environmental Impact Assessment) Directive.[39]

The key purpose of the Public Participation Directive is to provide for public participation in respect of the drawing up of certain plans and programmes relating to the environment[40] and to improve public participation and access to justice in relation to EIA approvals and the issuance of permits under the IPPC Directive. The latter Directive is of particular relevance in the context of this chapter as art. 8 of the ET Directive requires the co-ordination of industrial permitting procedures.

In relation to IPPC permit issuance procedures, art. 4 of the Public Participation Directive contains provisions for the amendment of the IPPC Directive in relation to its public participation requirements. In addition, there are amendments to the art. 2 definition of 'substantial change'[41] and the definitions of 'the public' and 'the public concerned';[42] as well as requirements for the consideration of the main alternatives in the permit application.[43] The public concerned are to be given early and effective opportunities to participate in the permit issue procedures for new installations, for substantial changes in installation operation and for the updating of an installation permit or permit conditions.[44] There is also a requirement to have access to justice in cases where the public participation procedures are inadequately followed where there is sufficient interest or impairment of a right.[45]

2.3 Proposed Directive 2003/624 Final on Public Access to Justice in Environmental Matters[46]

Notwithstanding that the Access to Information Directive and the Public Participation Directive contain implicit access to justice dimensions, the Commission adopted a proposed Directive on Public Access to Justice in Environmental Matters (hereinafter Access to Justice Directive) so as to ensure that the third pillar of the Aarhus Convention is fully implemented.

Article 1 of the proposed Access to Justice Directive seeks 'to ensure access to justice in environmental proceedings for members of the public and for qualified entities'. It covers access to administrative or judicial proceedings against acts or omissions by private persons as well as public authorities.

Under the proposal, 'members of the public', defined in art. 2(1)(b) as 'one or more natural or legal persons and in accordance with national law, associations, organisations or groups made up by these persons', are given the legal right to challenge acts and omissions by private persons which are (potentially) in breach of environmental law (art. 3) if they demonstrate legal standing in accordance with art. 4 requirements. These requirements include passing the 'sufficient interest test'[47] or 'maintaining the impairment of a right, where the administrative procedural law requires this as a precondition'.[48] It is the responsibility of Member States to determine what would constitute the appropriate procedural rights for members of the public on these grounds[49] though, by analogy, arts 6 and 13 of the European Convention on Human Rights may provide guidance to its observers.

The art. 2(1)(g) definition of 'environmental law' is wide in scope and is indicative only. It includes 'emissions, discharges and releases in the environment'[50] and 'access to environmental information and public participation in decision-making'.[51] As the ET Directive is a law covering emissions it likely falls within the definition of 'environmental law'.

Legal persons that are 'qualified entities' are defined in art. 2(1)(c) as 'any association, organisation or group, which has the objective to protect the environment and is recognised according to the procedure laid down in Article 9' and which meets the art. 8 criteria.[52] (For illustrative purposes, in the UK such entities as Friends of the Earth and Greenpeace will likely qualify under this Article.)

Under art. 5(1) legal entities do not have to demonstrate a sufficient interest or maintain the impairment of a right, 'if the matter of review in respect of which an action is brought is covered specifically by the statutory activities of the qualified entity and the review falls within the specific geographical area of activities of that entity'.

Members of the public and qualified legal entities are given the right under art. 6 to request an internal review of a public authority that might have undertaken an administrative act or administrative omission in breach of environmental law. Where this is inadequate, environmental proceedings may be instituted in accordance with arts 7[53] and 10.[54]

As the Aarhus Convention is the legal basis for the implementation of the three EU Directives, we shall now examine the relationship between the ET Directive, Aarhus Convention provisions and the key EU Directives, including the proposed Access to Justice Directive.

3. THE EU EMISSIONS TRADING DIRECTIVE AND AARHUS INTERFACES

The ET Directive, as amended,[55] requires mandatory access to environmental information and public participation via several articles. Recalling in para. 13

of the Recital provisions (those provisions that signal the intent and purpose of the Directive): '(i)n order to ensure transparency, the public should have access to information relating to the allocation of allowances and to the results of monitoring of emissions, subject only to restrictions provided for in the Access to Information Directive 2003/4/EC on public access to environmental information'.

First, it is important to determine exactly who 'the public' are if they are to have access to environmental information and to participate in environmental decision-making in relation to emissions trading. Here, art. 3(i) of the ET Directive defines 'the public' as: 'one or more persons and, in accordance with national legislation or practice, associations, organisations or groups of persons'.

When contrasted with the definition of 'public' under the Aarhus Convention, this is essentially the same: 'the public' means one or more natural or legal persons, and, in accordance with national legislation or practice, their associations, organizations or groups (art. 2(4)). This definition is also used in the EU Directive on public access to environmental information (2003/4/EC).

However, caution should be directed to the scope for discretion here, particularly for 'new' Member States that have not traditionally functioned with a culture of NGOs that are powerful in policy and law making processes as well as litigious in nature. In some Member States, unincorporated or unregistered organizations may not have NGO status, and hence the status of legal persons. The result may well then be that, in challenging some aspect of national ET Directive implementation, if not the Directive itself, an 'NGO' can only appear in dispute settlement procedures as 'one person', and not collectively as a representative group or organization.

The Aarhus Convention goes one step further than the ET Directive in defining 'the public concerned' as the public affected or likely to be affected by, or having an interest in, the environmental decision-making. For the purposes of this definition, non-governmental organizations promoting environmental protection and meeting any requirements under national law shall be deemed to have an interest (art. 2(5)). The Aarhus Convention term 'public concerned' is also applied in the Public Participation Directive (2003/35/EC).

It is clear then that the ET Directive does attempt to provide for both access to information and public participation in the ETS. However, this provision is not as broad in scope as that of the Aarhus Convention.

3.1 The ET Directive Public Participation Provisions

We address the public participation provisions of the ET Directive prior to the access to information provisions on the basis that public participation in establishing the EU ETS is a vital first implementation step, after which access to information requests will become regularized. There are two main elements of decision-making in which public participation is anticipated: (1) in that of decisions related to the allocation of emissions allowances in the National Allocation Plan NAP (including subsequent decisions made in accordance with art. 11(3) of the ET Directive) and (2) in decisions related to the issuance of GHG emission permits. Without the creation of the NAP and allocation of allowances, there would arguably be no emissions trading – and thus there would be no environmental information on emissions trading available for public access.

3.1.1 The NAP

The development of an art. 9 NAP for the periods referred to in arts 11(1) and (2) of the ET Directive, stating the quantity of allowances a Member State intends to allocate and how, has to be based on 'objective and transparent criteria' and involve public consultation.[56] The list of installations forms part of the NAP and is thus a public document. If installations are removed from that list, i.e., because they opt out of the ETS via art. 27, this information also has to be made public.

In relation to NAPs, the first explicit reference to public participation is found in art. 9, para. 1. This simply states that comments from the public are to be taken into 'due account' in creating a national emissions allocation plan (NAP), which itself is to be based on the 'objective and transparent criteria' listed in Annex III. Annex III has since been amended by Directive 2004/101/EC[57] to require that the NAP also specify the maximum amount of CERs (Certified Emissions Reductions) and ERUs (Emission Reduction Units) which may be used by operators in the ETS as a percentage of the allocation of allowances to each installation. 'The public' are defined under art. 3(i) of the ET Directive as 'one or more persons, and, in accordance with national legislation or practice, associations, organisations or groups of persons'.

In developing the art. 9 NAP, Member States have to decide – as per the art. 10 allocation method – upon the total quantity of emissions allowances it will allocate for a 3-year period commencing from 1 January 2005 and subsequently for a 5-year period commencing 1 January 2008, as per art. 11. Here, Annex III, para. 9 requires that the NAP itself shall include provisions for comments to be expressed by the public, and contain information on the arrangements by which due account will be taken of these comments before the art. 11 decision on the allocation of allowances is taken. The NAP had to be prepared and published by 31 March 2004 (1 May 2004 for the 10 new Member States). Subsequent art. 11 decisions on the allocation and issue of allowances based on the art. 9 NAP require public consultation.

However, the ET Directive *per se* does not outline the procedure by which the public views are to be obtained and taken into consideration in developing the NAP. Nor is the term 'due account' defined. It is possible that the drafters were relying on Member States to follow the art. 7 requirement of the Aarhus Convention concerning plans, programmes and policies relating to the environment, which refers back to the procedure outlined in art. 6 of the Aarhus Convention on public participation in decisions on specific activities. Submitting a NAP to an environmental impact assessment would also appear to be an ideal use of the (Strategic Environmental Assessment) Directive 2001/42/EC on the assessment of the effects of certain plans and programmes on the environment.

Thus, in the absence of explicit provisions in the ET Directive on the procedure for public consultation over the development of a NAP (ET Directive art. 9) or the allocation of allowances (ET Directive art. 11) the Aarhus Convention requirements can come into play. In particular, key art. 6 Aarhus Convention (Aarhus) public participation issues that would need to be considered when consulting the public include: setting reasonable time frames for the different stages of the public participation process – art. 6(3) (Aarhus); providing early public participation – art. 6(4) (Aarhus); and, ensuring that in making a decision due account is taken of the outcome of the public participation process – art. 6(8) (Aarhus).

These public participation provisions are mandatory under the transposing EU Directive 2003/35/EC on public participation in environmental matters which amends the IPPC Directive and can be found in the respective Articles of this Directive: arts 3, 2 and 2(c).

However, though these Aarhus provisions have to be applied in developing public consultation procedures, art. 9(1) (on the creation of a NAP) of the ET Directive required that the European Commission develop guidance on the implementation of the criteria listed in Annex III of the ET Directive. This includes the development of guidance on public consultation procedures, more on which is below.

3.1.2   NAP Commission Guidance

As noted above, the ET Directive *per se* provides little insight into the scope of its obligations to, and the rights of, the public. Nevertheless, Commission Guidance[58] developed to assist States in implementing the Annex III criteria[59] for NAPs provides some insight into the mandatory obligation to allow public participation in implementing the NAP and appears to consider relevant Aarhus Convention requirements. Although the Guidance is not legally binding, it specifies those criteria of the ET Directive that are mandatory or optional.

Annex III(2) indicates that the total quantity of allowances to be allocated shall be consistent with Decision 93/389/EEC (as amended) on monitoring Community CO_2 and other GHGs. This Decision has since been repealed and replaced by Decision No. 280/2004/EC on monitoring Community GHG emissions and for implementing the Kyoto Protocol, though one of the requirements of the original monitoring mechanisms was, and still is, to achieve consistency with the assessments and the publicly available and objective assessment of actual and projected emissions.[60] As per art. 6(1) of the ET Directive, an operator has to be capable of monitoring and reporting emissions as a condition of GHG permit issuance. Operators are then required to monitor and report on emissions as per art. 14(3) of the ET Directive and Commission Decision of 29 January 2004. Emission reports held by the competent authority are thus to be made available to the public by that authority subject to the rules laid down in the Access to Information Directive. With regard to the application of the 'confidentiality' exception laid down in art. 4(2)(d) of that Directive, operators may indicate which information they consider to be commercially sensitive in their report. Of course, a Member State may well determine that such information on emissions data, or extracts thereof, should be made public. Data that are not part of the annual emissions report shall not be required to be reported or to be made public.[61]

Data on GHG emissions and removals are sent by Member States to the UNFCCC, to the EU Greenhouse Gas Monitoring Mechanism/Inventory and to Eurostat, with copies also to be sent to the European Environment Agency (to EIONET – European Environment Information and Observation Network – and the European Topic Centre on Air and Climate Change (ETC/ACC)). The EIONET assists the Commission in data compilation and monitoring and makes data publicly available on its website.

Criterion 9 of Annex III of the ET Directive explicitly concerns involvement of the public and requires that 'the [national allocation] plan shall include provisions for comments to be expressed by the public, and contain information on the arrangements by which due account will be taken of these comments before a decision on the allocation of allowances is taken'. This is a mandatory requirement, which requires the Member State to describe in the NAP how it will make the plan available to the public for comments. Criteria in providing for public involvement that are expressed in the NAP Guidance[62] thus, summarily, include: (1) making the plan (including the draft plan) available in a timely manner so that the public can comment at an effective and early stage. This requires that the public be informed by public notice, the Internet or other appropriate means, and that contact and other relevant information is included; (2) allocating a reasonable time-frame for receipt of responses so that 'due account' can be taken of the comments; (3) providing feedback to the public on the decision taken and the main considerations upon which it is based; and holding a second consultation round if necessary. The NAP shall thus contain a list of the installations covered by the ETS Directive with the quantities of allowances intended to be allocated to each.[63]

The NAP, once determined, has to be notified to other Member States, which may well add another public participation requirement in the notified State (art. 9, ET Directive). Such public participation procedures do not only advance democracy and serve as an environmental protection mechanism but they also offer a means of ensuring fairness among installations and installation operators in terms of allocated allowances. Information on the Member States' proposed NAPs and those approved by Commission Decision is available on the European Commission Environment Directorate-Generale webpage of 'Europa'.[64]

Article 11 of the ET Directive, on allocation and issuance of allowances, builds upon art. 9 by requiring that comments from the public are taken into consideration when allocating and issuing emissions allowances. However, the Directive itself goes no further in terms of outlining the relevant procedural features or steps. It is a logical assumption that such Directive provisions should be interpreted in accordance with Aarhus Convention art. 6 (public participation in decisions on specific activities – which outlines the public participation procedure requirements) and art. 7 (public participation concerning plans, programmes and policies relating to the environment – which cross-references to the art. 6 procedure). As referred to above, the NAP Guidance is indicative in this regard.

3.1.3   Temporary exclusion from the ETS

Finally, as noted above, art. 27 requires that comments by the public should be taken into consideration when deciding to 'temporarily exclude' certain installations from the Community ETS Phase I. Article 27(1) requires that any Member State application submitted to the Commission for such exclusion has to list each installation and has to be published. The public are to be invited to comment on this application for 'opt-out' from the ETS, though the Directive does not outline any procedure for such consultation.[65] Again, the relevant Aarhus Convention recommendations for such consultation are to be followed as indicated in the NAP Guidance, and their comments are to be taken into account by the Commission before a Decision is made, as per art. 27(2).

The Commission Decision to exclude an installation or company will require that the installation or company, as a result of national policies: 'limit their emissions as much as would be the case if they were subject to the provisions of [the] Directive';[66] 'be subject to monitoring, reporting and verification requirements which are equivalent to those provided for pursuant to Articles 14 and 15 [of the Directive]';[67] and, 'be subject to penalties at least equivalent to those referred to in Article 16(1) and (4) [of the Directive] in the case of non-fulfilment of national requirements'.[68] Thus, the public are to be convinced that a firm case demonstrating environmental, administrative and penalty 'equivalency' has been made to the Commission. If the public have comments as to the equivalence of measures, then the Commission will consider such information in its assessment of approving, or not, the application for exclusion.

Article 27 has been applied in a few cases – including in the UK and the Netherlands. Installations covered by Climate Change Agreements (CCAs) –

which may be met through emissions trading – can also apply for temporary exclusion under art. 27.

A possible reason for 'opt-out' in these cases might be that companies are covered by several policies with the same purpose and wish to avoid 'overlaps'. This situation might well change if and when the national policies expire or are amended. Also of note in the context of this chapter is that any installations that breach their 'opt-out' conditions will be notified to the national public under the national scheme. For example, in the UK, Part G(2) of the Emissions Trading Scheme Rules contains the access to information provisions which participants in the scheme are required to consent to, including, for example, publication of the following:

> Part G(2)(2)
> (a) the name of the participant;
> (b) the account name;
> (c) the account number; and
> (c) the name and contact details of the principal account users; and
> (d) the transaction log described in rule B11.
>
> Part G(2)(3) A direct participant consents to the publication of—
> (a) his overall and annual targets;
> (b) the number of allowances allocated to him in each commitment year under rule C5;
> (c) his verified baseline in aggregate form;
> (d) his aggregate emissions in tCO_2e for each commitment year; and
> (e) whether he has complied with his annual emissions limitation commitment under rule C6(1) or to what extent he has failed to do so.
>
> (3A) A CCA (Climate Change Agreement) participant consents to the publication of—
> (a) the number of allowances allocated to compliance accounts held in his name under rule D4; and
> (b) the number of allowances retired from compliance accounts held in his name under rule D5, [etc.].

3.2   The ET Directive Access to Information Provisions

There are four key Articles that deal with access to information under the ET Directive: 16, 17, 19 and 21 (the latter three Articles as amended by 2004/101/EC).

3.2.1   'Naming and shaming'
Article 16(2) of the ET Directive requires that Member States penalize operators who breach art. 12(3) of the ET Directive, the requirement to surrender a number of emissions allowances. Member States are to 'ensure publication of the names of the operators who are in breach of the requirements' (art. 16(2)). This is, of course, in addition to imposing penalties such as an excess

emissions penalty of €100 for each excess tonne of CO_2 equivalent (CO_2e) (art. 16(3)).

'Naming and shaming' is a useful exercise in that when it comes to permit renewals, applications for additional developmental consents, EIAs and so forth, the public will be aware of the past records of installation operators and will be more able to respond with the appropriate level of scrutiny. While the potential for this public response is there, only time will tell whether it becomes effective in practice. After all, the nature of the information is highly technical and would benefit from more easily digestible additional non-technical summaries.

Furthermore, although dependent on the nature of the installation, consumer influence could play a part in the future prospects of the installation that has been 'named and shamed'. For example, members of the public would be in a position to boycott the goods or services offered by the installation or at least bring greater attention to the need to improve company performance, though the extent to which a boycott would actually occur in practice is possibly rather low.

3.2.2 Access to information and the 'linking' directive
Of more significance for its centrality to the ET Directive, and the discussion at hand, is art. 17 on access to information, as amended by the 'linking' directive (so-called as it 'links' the Kyoto Protocol project mechanisms of joint implementation (JI) and clean development to the Community ETS). This Article requires that 'Decisions relating to the allocation of allowances, information on project activities in which a Member State participates or authorises private or public entities to participate, and the reports of emissions required under the greenhouse gas emissions permit and held by the competent authority, shall be made available to the public in accordance with Directive 2003/4/EC' (the Access to Information Directive) except – as stated above – in cases where there are incomplete requests or confidentiality requirements.

The 'linking' Directive art. 1(8) introduces replacements to sub-art. 30(3) of the ET Directive in a manner that also impacts on access to information for the public. Sub-article 30(3) now requires that, in advance of the 5-year period beginning 1 January 2008 and each subsequent 5-year period, Member States are required to publish their intended use of ERUs and CERs in their NAP as well as the percentage of the allocation issued to each installation (expressed as a percentage of the maximum allowable ERU and CER in that period).

While it is commendable that this provision exists within the Directive, it is arguably of limited utility. As noted, the scope of the information that is to be provided is limited to decisions solely relating to allocations of allowances and reports of the emissions allowed under the GHG permit and not to the actual origin, type or amount of emission itself. It is possible to argue that this reference to 'project activities' covers IPPC-related activities. However, as this latter element of art. 17 does not specify precisely what 'information' on project activities has to be made available, recourse should be had to the Access to Information Directive and the IPPC Directive access to information and public participation requirements.

To this end, it appears that there may have to be considerable reliance on related Directives, such as the Access to Information Directive and the IPPC Directive in order to sufficiently provide the public with adequate information on the particular project/industrial activities *per se* and not just their designated emissions allowance. Emissions reports submitted to the European Pollutant Emission Register (EPER)[69] and Pollutant Release and Transfer Registers (PRTR)[70] databases (more of which follows) might also contain additional relevant information.

Perhaps it is assumed that Member States will implement this provision in synchronization with the requirements under the Aarhus Convention, and thus provide more detailed information on the installation or activity by this means. This is not to mention the potential role that can be played by related emissions registers but, of course, there are no such explicit requirements in the ET Directive art. 17 Access to Information provisions (which covers allocation allowances and reports of emissions under the GHG permit). It appears to be the case that art. 17 allocation decisions are thus available through the NAP and then subsequently through the art. 19 Registry requirements.

3.2.3 Information on the GHG Permit

It is observed that, according to art. 4 of the ET Directive, holding a GHG emissions permit is a 'condition *sine qua non*' for undertaking an activity listed in Annex I to the Directive.[71] This Annex more or less mirrors Annex I of the Aarhus Convention list of activities referred to in art. 6(1)(a) of the Aarhus Convention. Thus art. 6 of the Aarhus Convention as it relates to public participation in decisions on specific activities, is applicable here.

Furthermore, the public participation requirements of the IPPC Directive also offer an alternative route for obtaining information on permitting particularly in light of its amendment by the Public Participation Directive through its implementation of Aarhus Convention art. 6 provisions.

Perhaps evolving Member State 'best practice' will determine the extent to which the ET Directive GHG permitting procedures and IPPC permitting procedures are assimilated and thus clarify the public participation elements in this way.

3.3 Emissions Trading Directive Links with IPPC Directive Public Participation

Article 8 of the ET Directive requires co-ordination with the IPPC Directive 96/61/EC: Member States shall take the necessary measures to ensure that, where installations carry out activities that are included in Annex I to Directive 96/61/EC, the conditions of, and procedure for, the issuance of a GHG emissions permit are co-ordinated with those for the permit provided for in that Directive (see art. 26 of the ET Directive, which amends IPPC art. 9(3)).

While ET Directive arts 5 (GHG emissions permits), 6 (conditions/content of GHG permits) and 7 (changes relating to installations) on GHG permitting are not explicitly concerned with access to information and public participation, it is possible to address the content of these articles under the public participation/access to information elements of the IPPC Directive via art. 8 of the ET Directive.

The ETS will cover a total of more than 12 000 installations in the EU-25 (combustion plants, oil refineries, coke ovens, iron and steel plants, and factories making cement, glass, lime, brick, ceramics, pulp and paper) that will be required to obtain GHG emissions permits, just as IPPC installation operators are required to obtain integrated environmental permits – which may also stipulate GHG emission limits.

To this effect, the requirements of arts 5, 6 and 7 of the ET Directive may well be integrated into the procedures provided for in Directive 96/61/EC and this means that the access to information and public participation provisions of the IPPC Directive (particularly as amended by Directive 2003/35/EC) are applicable to the issuance of GHG emissions permits for those installations that come under the remit of the IPPC Directive.

Again, noting the art. 8 IPPC Directive co-ordination requirements in relation to those of the ET Directive, the IPPC Directive was amended to ensure that it is fully compatible with the provisions of the Aarhus Convention, in particular art. 6 (public participation in decisions on specific activities) and art. 9(2) and (4) thereof on access to justice. Hence, because public hearings are a feature of the Aarhus Convention, notably through art. 6(2)(d)(iii) (public participation in decisions on specific activities), its treatment of installations and activities under the IPPC Directive would cover those installations that produce GHGs.

The public should endeavour to have access to environmental information so that they can obtain data on the activities of the installation or activity before the permit is issued. Such data will relate to the conditions of the permit. The conditions of the permit, one would imagine, should also be made publicly available, though arts 17 and 19 of the ET Directive (the latter Article concerns Registries) are not explicit on this point. Again, art. 17 refers to decisions relating to the allocation of allowances, the reports of emissions and now (as amended by the 'linking' Directive), information on project activities. However, would this latter element include the permit conditions themselves? If this is not the intention, the access to information provisions of the IPPC Directive would cover this issue for IPPC permitted installations.

3.4   The ET Directive and Procedural Links to Environmental Impact Assessment and Strategic Environmental Assessment

Noting that the 'linking' Directive links the Kyoto Protocol's project mechanisms of JI and the clean development mechanism (CDM) to the Community ETS and allows credits earned from JI ERUs and CDM CERs, JI and CDM projects should be subject to environmental impact assessment (and even social impact assessment), with meaningful public participation and consultation procedures, prior to the grant of project approval. It would be unacceptable to benefit from emissions credits generated through unsustainable projects and hopefully Environmental Impact Assessment (EIA) and Strategic Environmental Assessment (SEA) processes would assist in preventing this situation from occurring. On this point, information on project activities in which a Member State participates – or authorizes private or public entities to participate – should be made available to the public in accordance with the Directive on public access to environmental information.[72] Noting that art. 12 of the Kyoto Protocol (on the CDM) does not explicitly mention public (or NGO) participation in the CDM project approval process, EIA approval would likely be mandatory in most jurisdictions, though this cannot be taken for granted at the international level. Another consideration, and one that is not explicitly advocated, is that public consultation, prior to the issuance of credits should also take place.

In addition, Recital para. 25 of the ET Directive stresses 'that policies and measures should be implemented at Member State and Community level across all sectors of the European Union economy, and not only within the industry and energy sectors, in order to generate substantial emissions reductions. The Commission should, in particular, consider policies and measures at Community level in order that the transport sector makes a substantial contribution to the Community and its Member States meeting their climate change obligations under the Kyoto Protocol'. As noted above, amendment of Annex I as per art. 30 requirements will elaborate upon this provision in the context of actual emissions trading. Notwithstanding this observation, the point here is that 'expansion' of emissions reductions and even trading to sectors other than those related to industry and energy would require strategic planning and public consultation. Of particular interest and influence will be public responses to integrated transport planning towards a possible carbon free transport economy for the future.

3.5 Data Registries, Kiev Protocol on Pollutant Release and Transfer Registers and the European Pollutant Emission Register

3.5.1 Introduction

In accordance with art. 15(3) of the IPPC Directive, in 2000, an EPER was created in order to ensure effective public access to information regarding emissions. Furthermore, the Aarhus Convention Kiev Protocol on Pollutant Release and Transfer Registers[73] was adopted in Kiev on 21 May 2003 and simultaneously signed by the EU. The PRTR is the first legally binding international instrument on pollutant release and transfer registers. There are evident links with emissions trading, the PRTR, the IPPC Directive and the EPER that are explained below.

3.5.2 ET Directive art. 19 data registries

Under art. 19(1) of the ET Directive, data registries[74] containing accounting, holding, transfer and cancellation information in relation to allowances are to be maintained in a consolidated system by Member States – based on a procedure determined by the Commission as per para. 3. Paragraph 3 amendments to the ET Directive under Directive 2004/101/EC[75] also require that information concerning the identification and use of CERs and ERUs within the ETS shall also be made available on the registry. Paragraph 2 states that '(a)ny person may hold allowances. The registry shall be accessible to the public and shall contain separate accounts to record the allowances held by each person to whom and from whom allowances are issued or transferred'.

Electronically linked national registries are central to both the EU ETS and wider international emissions trading that will take place under the Kyoto Protocol from 2008.

The scope of the right of access to the registry appears to be quite clear: that the public can obtain information on allowances held by each person and the transfer of those allowances and their cancellation. This should provide for additional accountability in the system and may even lead to an informal monitoring role played by the NGO community. However, there is one questionable element. The registry requirements under the ET Directive do not actually appear to require emissions data *per se* (i.e., type of industrial emission release, source, need for the release, etc.) or analysis of how emissions might be reduced over time, which creates the impression that the registry serves a pure accounting interest rather than an environmental protection interest for the public. Of course, NAPs will contain some of this information and should, thus, have a prominent focus in public awareness raising activities.

Transfers and acquisitions of the allowances are to be tracked and recorded through the registry systems under the Kyoto Protocol. These include a national registry to be established and maintained by each Annex I Party. Each EU

Member State will thus have its own national registry containing accounts which will hold the allowances. These registries interlink with the publicly accessible Community transaction log[76] operated by the Commission, which will record and check every transaction and which will provide information on registered account operators, the types of installations and, access to Member States registries. It is noted that some form of register 'user access' registration is required in order to access certain elements of these databases, so access to information is restricted by the requirement to register and/or to meet the criteria which allows one to be registered. Hopefully, this will not give rise to any unacceptable restrictions on public access. Whether the fact that public access to the ET registry is limited to information on the allocation of emissions rather than their actual transfer is an impediment to full environmental transparency and accountability remains to be seen.

Each of the 12 000 installations that participate in the ETS is required to have an 'operator holding account' in its national registry, into which its own allowances will be issued. This can be contrasted with a 'person holding account' which an individual or organization wishing to participate in the market will be able to open in any of the registries.

Only certain types of information will be available via the Community transaction log, which will show that national registries are 'live', such as contact details for the accounts and information on allowances allocated in accordance with the final NAP. Other types of information on transactions will be made available after a period of 5 years has passed, so access to information is again limited. However, the compliance position of each installation across the EU–25 will be available on 15 May of each year from 2006 onwards.

Article 19(3) calls on Member States 'to provide for ... confidentiality as appropriate' in the context of the registry, though there is no guidance as to how to do this or how to determine what is confidential. It would be preferable if the discretion afforded by this provision were exercised in the interests of transparency in circumstances that are within the scope of the art. 4(2) exceptions to Access to Information Directive as it applies to emissions. In this regard, transparency and environmental protection will be the priorities to be treated proportionally vis-à-vis traditional confidentiality considerations.

Finally, in art. 21(1) there is the requirement for Member States to submit information on the operation of the registries to the Commission. This, it is hoped, will ensure that the Member States comply with this element of the access to information obligations, something that is clearly in the interests of the public, even if it does appear to be somewhat limited in scope. Commission Decision 2005/381/EC of 4 May 2005 establishes a detailed questionnaire to assist Member States in their reporting on the application of an ET Directive.

With regard to these questionnaires, the public can make a 'passive' request for access to environmental information as per Directive 2003/04/EC either to the Member State or the Commission as regards questionnaire content and responses. It is also anticipated that the reports on the Members States' replies will be made public when the Environment Directorate-Generale prepares the compilation of all the replies.[77] As noted above, reports on the implementation of environmental legislation come under the scope of the Access to Information Directive.

3.5.3   The Kiev Protocol PRTR

Noting the limitations of ET Directive registry requirements, other databases offer enhanced access to emissions-related data, one of which is the PRTR. The Kiev Protocol PRTR will replace EPER with a European PRTR (E-PRTR) by 2009 as proposed in the European Commission Proposal concerning the establishment of an E-PRTR, which implements the provisions of the Kiev Protocol.[78]

Article 1 of the Kiev Protocol (the PRTR) outlines the objective of the PRTR as follows: 'to enhance public access to information through the establishment of coherent, nationwide pollutant release and transfer registers' (in accordance with the requirements of art. 11). Thus, enhanced public participation in environmental decision-making and related contribution to reducing and preventing environmental pollution in the EU, the Member States and Accession Countries is a clear objective.

Under art. 4 of the PRTR, Parties are required 'to establish and maintain a publicly accessible national pollutant release and transfer register' and to meet set requirements therein on core elements, design and structure and scope, as identified in arts 4, 5 and 6. The latter is a welcome step towards data harmonization. Parties to the Protocol may also introduce a 'more extensive or more publicly accessible' PRTR than that required by the Protocol (art. 3(2)). Article 3(5) states that the PRTR systems may be integrated with other reporting systems (i.e., those contained in licence or permit conditions) in order to avoid duplicative reporting. Such integration may also encourage more streamlining. Furthermore, Annexes I and II provide a list of Activities and Pollutants (including transfers of pollutants) for which 'best available information' (art. 9(2)) must be reported annually (art. 8) by owners or operators of facilities in accordance with art. 7 reporting requirements. The 'best available information' should be, where appropriate, collected (i.e., calculated, monitored and so forth) in accordance with internationally approved methodologies (art. 9(2)).

Linked to the above is art. 10 PRTR on quality assessment. Paragraph 2 indicates that data contained in each Party's register shall be subject to quality assessment by the competent authority, particularly to ensure 'completeness, consistency and credibility'. This is a positive step forward in order to achieve some quality control and uniformity in data reporting and interpretation so that members of the public and other parties might be accurately informed.

However, by way of a broad comment on the PRTR and the EPER and the ET Directive Registry requirement (and other pollutant release databases in general), there is a common concern among industries that data registers will expose their trade secrets to competition. By listing the releases into air (or water), competitors may be able to identify commercially sensitive 'ingredients', technologies or processes used in the formulation of certain products – the knowledge of which may even extend to revelations about intellectual property.

Hence, art. 12 of the PRTR contains the confidentiality-related provisions within para. 1, providing that 'each Party may authorize the competent authority to keep information held on a register confidential where public disclosure of that information would adversely affect [... (among other interests listed in paragraphs (a)–(e), for example, national defence) ...], the confidentiality of commercial and industrial information, where such confidentiality is protected by law in order to protect a legitimate economic interest' (art. 12(1)(c)).

The PRTR is quite innovative in dealing with the 'confidentiality problem'. In accordance with the above, the register can still provide *some* form of data. For example, the register 'shall indicate what type of information has been withheld, through, for example, providing generic chemical information if possible, and for what reason it has been withheld' (art. 12(3)). This is an advanced approach as more relevant information may be gleaned for PRTR purposes, while providing a comfort level to firms, that can apply greater flexibility and creativity in the provision of information. Hence, reliance on the PRTR and not the ET Directive may have an additional public information benefit.

Note also that para. 2 provides further that within the framework of paragraph 1(c) of the PRTR 'any information on releases which is relevant for the protection of the environment shall be considered for disclosure according to national law'.

Article 12 thus appears to constitute a compromise between respecting true commercial confidentiality while striving to retain the openness and access to information that the Protocol, Aarhus Convention and the Access to Information Directive require. However, the extent to which art. 12 provisions will be open to abuse in the GHG emissions trade remains to be seen.

3.5.4 Proposed European PRTR

As it appears that the ET Directive art. 19 Registry requirement may not be sufficient in terms of publicly available information on GHG emissions, the PRTR offers an alternative source. Noting further that several EU databases and the EPER exist, and that art. 30(2)(f) of the ET Directive refers to 'review and further development provisions' for the ETS, in 2004, the Commission proposed a single Community registry/database, the E-PRTR. Accordingly, the public have the opportunity to be involved in further developing the E-PRTR and preparing the necessary amendments. They will also be able to access the register free of charge on the Internet and find information using various search criteria such as type of pollutant, geographical location, affected environment, source facility, and so on.

The benefits of having one Community registry based on the far-reaching provisions and workability of the Kiev Protocol means that Member States' administrative and monitoring responsibilities may be reduced – as they would not have to duplicate staff time for GHG emissions reporting for several databases.

Another benefit to a single database is that the public would be able to find emissions information all in one place. Once the register public access points have been determined (say, on the Internet, in libraries, in town halls, in shopping malls and even in fast food restaurants, etc.) public access to information will thus be made easier to obtain. It may even become second nature for members of the public to regularly browse the database, depending on how access is managed and encouraged.

Centralizing environmental information in one registry will make the links between relevant Directives and related environmental issues clearer; an additional form of environmental education and awareness-raising. This is an advantage, as most public participation procedures concern inter-related environmental issues or overlapping approval procedures such as those related to EIA and IPPC.

Furthermore, noting the cumulative effects of GHG emissions, they should all be reported on an annual basis subject to a *de minimis* rule. In this way, the State of the environment reports being submitted would be able to draw upon this data in determining total pollution loads on the environment, including their cumulative and synergistic effects. For those that would object to the broader emissions reporting mandate referred to here, it should be pointed out that such emissions data is 'environmental information' within the meaning of the Aarhus Convention and would have to be disclosed upon request in any event (without a duty to give reasons in connection with the disclosure request).

4. CONCLUSION

The ET Directive has an important integrative dimension vis-à-vis the Aarhus Convention. This chapter has demonstrated that on selected issues the ET Directive does not address explicit obligations that comply with the requirements of the Aarhus Convention. However, for the most part, the NAP Guidance (though this is not strictly legally binding) and the enhanced Aarhus-related amendments introduced through the 'linking' Directive resolve these matters and there is a more-or-less proportionate symbiosis.

Where elements of full adherence to the Aarhus Convention are lacking, this chapter also makes it clear that a wealth of other legal instruments exist in relation to public participation and access to environmental information. These instruments will contribute to enhanced access to information and public participation in emissions trading procedures. In particular, by 2009, the E-PRTR will represent greater consistency and uniformity in emissions data. The streamlining of access to information and public participation provisions would clearly advance the objectives sought.

Potential positive impacts of public participation in the emissions trading regime include the following: the public will be more informed as to potential environmental impacts at the local levels; industry might be encouraged to reduce their GHG emissions and adopt more 'environmentally friendly' technologies; increased public scrutiny may result in improved environmental 'housekeeping' by firms if mandatory reporting creates competition for favourable consumer opinions in relation to GHG performance by rival firms (though perhaps this is an optimistic view); if competent authority reviews of reporting activities could be bolstered by other interested stakeholders such as NGOs, then emissions reporting will be more precise in the context of NAPs. The broad availability of said data will be a spur to technical progress through sustainable technological system and process innovation as informed stakeholders contribute to efforts to ratchet down emissions over time. The said research will also yield new analytical insight to improve the emissions trading system itself as professional and academic inquiry on the subject is enhanced by the transparency of information.

However, in order to reap these benefits potential abuses should be discouraged. A public information campaign should be developed in order to stimulate stakeholder involvement in what is otherwise a technical and inaccessible field. Otherwise, the 'real' public may not wish to become involved, relying instead on the 'professional' public.[79] Such a tendency should be curbed because if economic instruments (e.g., emissions trading) are to gain a permanent foothold in the arsenal of environmental policy instruments then they will require public support and active stakeholder participation. This is a vital point if often hard-fought for Aarhus-related 'rights' are to be fully utilized. It is, without doubt, that a feature of emissions trading is that governmental and public interest is concerned with the total amount of pollution emitted in a certain area and not with individual decisions related to the permitting of IPPC installations. However, and the following remarks are suited to further research beyond the scope of this chapter, this does not guarantee increased stakeholder involvement, increased public understanding of the issues surrounding emissions trading, nor an increased desire of the public to participate in the emissions trading process.

Nevertheless, the procedures are in place for the public that wish to use them – and this is a very positive step forward in environmental democracy.

NOTES

1. Karen MacDonald, Environmental Lawyer, Imperial College London, E-mail (k.e.macdonald@ imperial.ac.uk) and Zen Makuch, Reader in Law, Barrister, Imperial College London, E-mail (z.makuch@imperial.ac.uk).

2. Hailed by Kofi A. Annan, Secretary-General of the UN as 'a giant step forward in the develop-ment of international law in this field', in Stec, S., Casey-Lefkowitz, S. and Jendrośka, J. (2000), *The Aarhus Convention: An Implementation Guide*, Regional Environmental Center for Central and Eastern Europe and Danish Environmental Protection Agency, New York and Geneva: United Nations / Economic Commission for Europe, http://www.unece.org/env/pp/ acig.pdf.

3. 'Operator' is defined in art. 3 as 'any person who operates or controls an installation or, where this is provided for in national legislation, to whom decisive economic power over the technical functioning of the installation has been delegated'.

4. The ETS Survey is a tool being applied by DG Environment of the EU Commission to review the EU Emissions Trading Scheme (EU ETS). The Commission is being supported by McK-insey & Company and Ecofys to assist the EU in this review in 2005 and 2006 by providing a fact base for the discussion. See the opening page of the survey at: http://www2.perseus. com/mckinsey/prod/eu/eumain.htm#%20User%20name.

5. Wemaere, M. and Streck, C. (2005), 'Legal Ownership and Nature of Kyoto Units and EU Allowances', in David Freestone and Charlotte Streck (eds), *Legal Aspects of Implementing the Kyoto Protocol Mechanisms*, Oxford, UK and New York, US: OUP, pp. 35–53.

6. Ibid., pp. 37–38.

7. Ellerman, A. Denny, (2005) 'US Experience with Emissions Trading: Lessons for CO_2 Emis-sions Trading', in Bernd Hansjürgens (ed.), *Emissions Trading for Climate Policy: US and European Perspectives*, Cambridge University Press, UK, and Cambridge University Press, New York, pp. 78–95.

8. Directive 2003/4/EC of the European Parliament and of the Council of 28 January 2003 on public access to environmental information and repealing Council Directive 90/313/EEC, OJ L 041,14.02.2003, pp. 26–32.

9. Ibid. art. 1.

10. Op cit, n. 8, art. 2(1)(a).

11. Ibid., art. 2(1)(b).

12. Op cit, n. 8, art. 2(1)(c).

13. Ibid., art. 2(1)(d).

14. Op cit, n. 8, art. 2(1)(e).

15. Ibid., art. 2(1)(f).

16. Op cit, n. 8, art. 2(2)(a).

17. Ibid., art. 2(2)(b).

18. Op cit, n. 8, art. 2(2)(c).

19. Ibid., art. 2(3).

20. Op cit, n. 8, art. 2(4).

21. Ibid., art. 2(5).

22. Op cit, n. 8, art. 2(6).

23. Ibid., art. 8(1).

24. Op cit, n. 8, art. 7(2)(e).

25. Ibid., art. 7(2)(c).

26. Op cit, n. 8, art. 4(1)(a).

27. Ibid., art. 4(1)(c).

28. Op cit, n. 8, art. 4(2)(b).

29. Ibid., art. 4(2)(e).

30. Op cit, n. 8, art. 4(4).

31. Ibid., art. 4(2)(a).

32. Op cit, n. 8, art. 4(2)(d).

33. Ibid., art. 4(2)(f).

34. Op cit, n. 8, art. 4(2)(g).
35. Ibid., art. 4(2)(h).
36. Directive 2003/35/EC of the European Parliament and of the Council of 26 May 2003 providing for public participation in respect of the drawing up of certain plans and programmes relating to the environment and amending with regard to public participation and access to justice Council Directives 85/337/EEC and 96/61/EC, OJ, L 156, 25.06.2003, pp. 17–25.
37. Ibid., art. 6.
38. Council Directive 96/61/EC of 24 September 1996 concerning integrated pollution prevention and control, as amended, PbL 257, 10 October 1996, pp. 26–40.
39. Council Directive 85/337/EEC of 27 June 1985 on the assessment of the effects of certain public and private projects on the environment, as amended, PbL 175, 5 July 1985, pp. 40–48.
40. Ibid., art. 1(a).
41. Op cit, n. 39, art. 4(1)(a).
42. Ibid., art. 4(1)(b).
43. Op cit, n. 39, art. 4(2).
44. Ibid., art. 4(3).
45. Op cit, n. 39, art. 4(4).
46. Proposal for a Directive of the European Parliament and of the Council on access to justice in environmental matters, Brussels, 24 October 2003, COM(2003) 624 final 2003/0246 (COD). This proposed Directive has to be adopted in the co-decision procedure since it is based upon art. 175(1) EC Treaty. The European Parliament completed its first reading of the proposal in March 2004. The proposal is now pending before the Council of Ministers. At the recent Environment Council meeting of 24 June 2005, the outgoing Luxembourg Presidency indicated the state of play of the proposed Directive (see the Council website: http://ue.eu.int/ueDocs/cms_Data/docs/pressData/en/envir/85575.pdf). No progress on the substance of the file has been reported and it is not known when the Council will advance the co-decision procedure. (Personal communication with Directorate-Generale Environment, 13.07.05.)
47. Ibid., art. 4(1)(a).
48. Op cit, n. 46, art. 4(1)(b).
49. Ibid., art. 4(2).
50. Op cit, n. 46, art. 2(1)(g)(x).
51. Ibid., art. 2(1)(g)(xii).
52. Article 8 Criteria for recognition of qualified entities.
53. Commencement of environmental proceedings.
54. Requirements for environmental proceedings.
55. Amended by Directive 2004/101/EC, PbL 338, 13 November 2004, pp. 18–23.
56. Article 9(1) Directive 2003/87/EC of 13 October 2003, PbL 275, 25 October 2003, pp. 32–46.
57. Ibid., art. 1(9).
58. Communication from the Commission on guidance to assist Member States in the implementation of the criteria listed in Annex III to Directive 2003/87/EC establishing a scheme for greenhouse gas emission allowance trading within the Community and amending Council Directive 96/61/EC, and on the circumstances under which *force majeure* is demonstrated, COM(2003) 830 final, 7.1.04, Brussels.
59. Of Directive 2003/87/EC.
60. See s 2.1.2 of the Guidance, COM(2003) 830 final, 7 January 2004, Brussels.
61. Annex I Part 6 – Retention of Information of Commission Decision of 29.01.04.
62. See s 2.1.9 of the Guidance, COM(2003) 830 final, 7 January 2004, Brussels.
63. Directive 2003/87/EC Annex III(10).
64. To consult the initial NAPs notified to the Commission, and the Commission Decision on each plan follow the link (accessed 29 July 2005): http://europa.eu.int/comm/environment/climat/emission_plans.htm.
65. DG Environment of the European Commission provides access on its 'Europa' website to documents concerning Member State art. 27 exclusion requests. See for example Commission Decision of 22/III/2005, COM 2005/866 Final concerning the temporary exclusion of certain

installations by the Netherlands from the Community emissions trading scheme pursuant to art. 27 of Directive 2003/87/EC of the European Parliament and of the Council: http://europa. eu.int/comm/environment/climat/pdf/netherland_dec2.pdf#search='article%2027%20emissi ons%20trading%20directive'. However, it is questionable as to whether a member of the public would be able to find such documents or even know that they are publicly available, without there being an awareness raising programme.
66. Article 27(2)(a) COM 2005/866 Final.
67. Ibid., art 27(2)(b).
68. Op cit, n. 66, art.27(2)(c).
69. European Pollutant Emission Register, created by Commission Decision 2000/479/EC, 17 July 2000 on the implementation of a European Pollutant Emission Register (EPER) (OJ L 192, 28 July 2000, pp. 36–43).
70. Pollutant Release and Transfer Registers, as per the 2003 Kiev Protocol to the Aarhus Convention.
71. Jendrośka, J. (2004), 'Public Participation and Information in the Emission Trading Directive', *Environmental Law Network International Review*, **1**, pp. 7–11.
72. As per the new art. 17 introduced by art. 1(3) of Directive 2004/101/EC, 27 October 2004, PbL 338, 13 November 2004, pp. 18–23.
73. See http://www.unece.org/env/pp/prtr.htm.
74. Commission Regulation 2004/2216/EC of 21 December 2004, PbL 386, 29 December 2004, pp. 1–77, which provides for a standardized and secured system of registries pursuant to the ET Directive.
75. Ibid., art. 1(5).
76. See http://europa.eu.int/comm/environment/ets/, (accessed 27 July 2005).
77. Personal communication with European Commission, DG Environment Unit C2: Climate, Ozone and Energy.
78. Proposal for a Regulation of the European Parliament and of the Council concerning the establishment of a European Pollutant Release and Transfer Register and amending Council Directives 91/689/EEC and 96/61/EC, COM/2004/634 final, COD 2004/231.
79. Lee, M. and Abbot, C. (2003), 'The Usual Suspects? Public Participation Under the Aarhus Convention', *Modern Law Review Limited*, **66**(1).

REFERENCES

Commission Decision of 22/III/2005 concerning the temporary exclusion of certain installations by the Netherlands from the Community emissions trading scheme pursuant to Article 27 of Directive 2003/87/EC of the European Parliament and of the Council: http://europa.eu.int/comm/environment/climat/pdf/netherland_dec2.pdf#search='art icle%2027%20emissions%20trading%20directive', 4 August 2005.
Council Decision 2005/370/EC of 17 February 2005 on the conclusion, on behalf of the European Community, of the Convention on access to information, public participation in decision-making and access to justice in environmental matters.
Commission Decision of 29 October 2004 concerning the temporary exclusion of certain installations by the United Kingdom from the Community emissions trading scheme pursuant to Article 27 of Directive 2003/87/EC of the European Parliament and of the Council.
Commission Decision of 29/01/2004 establishing monitoring and reporting guidelines pursuant to Article 14(1) of the Greenhouse Gas Emission Allowance Trading Directive, 2003/87/EEC.
Commission Decision 2000/479/EC (of 17 July 2000 on the implementation of a European Pollutant Emission Register (EPER OJ L 192, 28.07.2000, pp. 36–43): http://www.eper.cec.eu.int/eper/default.asp, 4 August 2000.

Council Directive 85/337/EEC of 27 June 1985 (as amended by 97/11/EC) on the assessment of the effects of certain public and private projects on the environment.

Decision No. 280/2004/EC of the European Parliament and of the Council of 11 February 2004 concerning a mechanism for monitoring Community greenhouse gas emissions and for implementing the Kyoto Protocol, OJ L 049, 19.02.2004, pp. 1–8.

Directive 2003/35/EC of the European Parliament and of the Council of 26 May 2003 providing for public participation in respect of the drawing up of certain plans and programmes relating to the environment and amending with regard to public participation and access to justice Council Directives 85/337/EEC and 96/61/EC, OJ L 156, 25.06.2003, pp. 17–25.

Directive 2003/4/EC of the European Parliament and of the Council of 28 January 2003 on public access to environmental information and repealing Council Directive 90/313/ EEC, *OJ L 041,14.02.2003, pp. 26–32.*

Directorate-Generale Environment, European Commission, Press Release, 'Industrial pollution: new European register adopted', http://www.eper.cec.eu.int/eper/documents/ press%20release,%20first%20reading%20agreement%20E-PRTR.pdf, 4 August.

Ellerman, A. Denny, (2005), 'US Experience with Emissions Trading: Lessons for CO_2 Emissions Trading', in Bernd Hansjürgens (ed.), *Emissions Trading for Climate Policy: US and European Perspectives,* Cambridge University Press, UK and Cambridge University Press, New York, pp. 78–95.

Hancock, J. (2003), 'Environmental Human Rights: Power, Ethics and Law', Critical Security Series, Aldershot, UK and Burlington, US: Ashgate. See in particular, pp. 115–19.

Jendrośka, J. (2004), 'Public Participation and Information in the Emissions Trading Directive', *Environmental Law Network International Review,* 1/2004, pp. 7–11.

Kiev Protocol on Pollutant Release and Transfer Registries, 21 May 2003.

Lee, M. and Abbot, C. (2003), 'The Usual Suspects? Public Participation Under the Aarhus Convention', *The Modern Law Review Limited,* **66** (1), (January).

Proposal for a Regulation of the European Parliament and of the Council concerning the establishment of a European pollutant release and transfer register and amending Council Directives 91/689/EEC and 96/61/EC, COM (2004) 0634 final 2004/0231 (COD).

Proposal for a Directive of the European Parliament and of the Council on Access to Justice in Environmental Matters (presented by the Commission) Brussels, 24.10.2003 COM(2003) 624 Final 2003/0246 (COD).

Stec, S., Casey-Lefkowitz, S. and Jendrośka, J. (2000), *The Aarhus Convention: An Implementation Guide,* Regional Environmental Center for Central and Eastern Europe and Danish Environmental Protection Agency, New York and Geneva: United Nations/Economic Commission for Europe, http://www.unece.org/env/pp/acig.pdf.

UK Greenhouse Gas Emissions Trading Scheme Regulations 2005, SI No. 2005: 925: http://www.opsi.gov.uk/si/si2005/32, 4 August 2005.

UK Greenhouse Gas Emissions Trading Scheme Rules 2002 Updated and Re-Issued, 14 October 2002 (Ets(01)06.Rev2).

United Nations/Economic Commission for Europe Convention on Access to Information, Public Participation in Decision-Making and Access to Justice in Environmental Matters, done at Aarhus, Denmark, 25 June 1998.

Wemaere, M. and Streck, C. (2005), 'Legal Ownership and Nature of Kyoto Units and EU Allowances', in David Freestone and Charlotte Streck (eds), *Legal Aspects of Implementing the Kyoto Protocol Mechanisms,* Oxford, UK and New York, US: OUP, pp. 35–53.

8. The IPPC permit and the greenhouse gas permit

Birgitte Egelund Olsen[1]

1. INTRODUCTION

The development of the environmental law of the European Community (the Community) is characterized by the use of a range of regulatory instruments. This is in accordance with the Community's Sixth Environment Action Programme, which states that the aims and objectives set out in the Programme need to be pursued through the use of a mix of regulatory instruments, including market-based and economic instruments.[2]

Another characteristic of the more recent development of Community environmental law, which is not mentioned in Community policy programmes and strategies, is a tendency for the Community no longer to restrict itself to setting the regulatory framework. To some extent the Community also fills out the framework and thereby leaves less to the discretion of the Member States in the implementation process. Thus, the Community is perhaps moving away from a highly flexible regulatory approach to environmental law, which puts great emphasis on the principle of subsidiarity, and towards a deeper Community involvement and in setting the framework which narrows the competences of the Member States, and which may reduce the scope for the Member States to ensure a form of implementation that reflects national characteristics and priorities within the overall framework of Community legislation and policy. The Directive establishing a scheme for greenhouse gas (GHG) emission allowance trading (hereinafter the 'Emissions Trading Directive') is an example of this.[3] The Directive establishes a Community GHG emissions trading scheme as from 1 January 2005 and contributes to meeting both international and European Union (EU) obligations in response to the severe implications of climate change.

A major regulatory challenge for the Community and the Member States is to ensure that the emissions trading scheme complements and is compatible with other environmental policies and measures. Within the EU there are many such measures already in place, for example quality norms or technical standards, energy taxes and environmental agreements. Accordingly, it is essential that the adoption of a Community emissions trading system reinforces and does not weaken the objectives of each of these.[4] However, the implementation of the Directive reveals several ambiguities. The interplay between the emissions trading scheme and the Directive on integrated pollution prevention and control (hereinafter the IPPC Directive) is of special interest.[5]

This chapter addresses how and to what extent the new market-based approach of the Emissions Trading Directive interacts with the traditional permit system of the IPPC Directive, and whether it is a substitute for or is complementary to the holistic, cross-cutting and flexible approach of the IPPC Directive. The focus of this chapter is narrow. It is not a study of the economic or legal efficiency of the measures in question. Nor does it expound in detail the permit procedures of these directives. Instead, it explores some of the implications of the adoption of a new instrument which affects the scope of the existing regulatory measures. It addresses the specific problems of co-ordinating the procedures for IPPC permits and GHG permits and investigates how GHG emissions that are not covered by the Emissions Trading Directive are dealt with, and what legal problems or questions may arise in this respect. The legal implications depend to a large extent on differing regulatory perspectives, so there is a brief description of the characteristics of the different approaches of Community environmental law.

2. REGULATORY APPROACHES IN COMMUNITY ENVIRONMENTAL LAW

From an overall perspective, Community environmental law can be divided into broad categories based on those to whom the regulations are addressed or the matters that are regulated. The first category includes legislation using a quality-oriented single medium approach to protect the quality of a given medium. It does not directly involve the concerns of other media. An example of this is the basic EU instrument on air pollution, the Air Framework Directive, which establishes the legal framework for setting quality standards for air.[6] This approach is frequently used, but it is generally difficult to deal with this kind of approach using traditional legal methods because the environmental quality of a medium is often the result of cumulative actions and not the isolated action of one identifiable responsible party.

The second category includes legislation which uses a product-oriented approach. Product-oriented legislation is often highly technical and detailed and it generally reflects a piecemeal approach.[7] Historically, it formed the basis of the first Community measures on air pollution, including maximum values for emissions from cars, bans on the marketing of leaded petrol and obligations to make sulphur-free fuels available within the EU. These directives are based on art. 95 of the EC Treaty, as their purpose is not only to achieve a high level of health and environmental protection, but also to ensure the establishment and functioning of the internal market and the free circulation of products.

The third category includes legislation based on a source or sector-oriented approach. The Large Combustion Plants Directive is a good example of sector-oriented regulation that focuses on limiting emissions of certain pollutants into the air, but it is only applicable to large combustion plants.[8] The Emissions Trading Directive is a key element of the Community climate change policy. It is a pollutant-specific regulatory instrument that controls emissions from large industrial installations. The Directive establishes a scheme for GHG emission allowance trading within the Community, based on a traditional permit system under which the emission of GHG is prohibited unless the operator holds a permit.[9] The objective of the Directive is to promote reductions of GHG emissions in a cost-effective and economically efficient manner.[10] It is a typical example of a source-oriented approach. However, as will be seen below, it differs from the existing approaches in Community environmental law in other respects.

The IPPC Directive introduced the process-oriented approach to Community environmental law. Integrated pollution prevention and control takes place within the context of a permit regime. The Directive covers multiple-media pollution control from major industrial installations. The concept of integration is expressed in two ways, reflecting both a procedural and a substantive approach. Substantive integration concerns the integrated approach to all pollution of various media from a specific installation. The objective is to establish a balance across media and to avoid the problem of introducing media-specific solutions whereby one pollution problem may be solved at the expense of creating another.[11] Procedural integration refers to the integration of different permit schemes into one procedure leading to an integrated permit.[12] The key aim is to assist substantive integration.

The IPPC Directive contains no overall targets for the reduction of pollution. It refers to the use of best available techniques (BAT), taking into account the local environmental conditions and an integrated approach to emissions to air, water and soil as a basis for setting individual permit conditions. The Directive deals with the problem of the single-medium approach, which might encourage shifting pollution between the various environmental media rather than protecting the environment as a whole. Thus, the IPPC Directive addresses impacts on all media on the principle that the efforts to reduce emissions to one media should not lead to increases in emissions to other media. However, it only regulates emissions and it does not prescribe the technology to be used in specific cases.[13]

In recent years, the regulatory technique of using the combined approach has been introduced, though not yet for combating air pollution or climate change.[14]

This approach uses a combination of the process-oriented and media-oriented approaches. It is characterized by the introduction of two independent types of measures: the control of emissions at source, and the setting of quality standards as a means of measuring the success of the control of emissions.

3. EMISSIONS TRADING V. IPPC

The introduction of an emissions trading regime has required clarification of the relationship between emissions trading and the IPPC Directive, and in particular the relationship between the permits and the permit procedures of the directives.[15]

Under the IPPC Directive, installations are required to hold an operating permit giving emission limit values based on BAT, relating to pollutants that are likely to be emitted in significant quantities. However, there is an exception. Under the amendment in art. 26 of the Emissions Trading Directive, a permit issued in accordance with the IPPC Directive must not include emission limit values for direct emissions of GHG if the installation in question has obtained a GHG permit. Also, the IPPC permit must contain conditions concerning the energy efficiency of the specific technology, but this is not a requirement if an installation is required to obtain a GHG permit. This means that installations that are subject to the Emissions Trading Directive are not required to implement BAT for the specific GHG pollutant, and may also be exempted from requirements concerning energy efficiency measures. Instead, they can choose whether to reduce their emissions to the level allocated to them or below that level, or they may choose to keep emissions above that level and buy allowances from other undertakings, only having to consider the economic rationale of their actions.[16]

The logic behind the approach is that, in principle, a cap-and-trade system can offer guaranteed results in terms of total emissions, which is something that a permit system based on BAT cannot achieve.[17] Another compelling argument in favour of emissions trading is that it means that there is always an incentive for operators to reduce emissions – and this allows operators to achieve emission reductions in a more cost-effective way. Those who face high pollution abatement costs can continue to pollute by buying additional allowances. Those facing low costs can take action to reduce emissions and sell their surplus allowances for profit. In this way, each source can trade off the cost of controlling pollution against the cost or revenue of buying or selling allowances.[18] Market-based regulation does not involve the regulators in individual decisions on abatement actions.

This approach is supported by the argument that industry often has more information and know-how than regulators and is therefore in a better position to identify cost-effective measures to reduce emissions, provided that there is an incentive to do so. The system put in place by the IPPC Directive does not offer any direct financial incentive to reduce emissions below the BAT-based limit values set in the permit.[19] Accordingly, regulators and industry play very different roles under each of the directives in assessing and deciding where and when technical measures should be taken to achieve the necessary reductions.

4. SCOPE OF APPLICATION

Both the Emissions Trading Directive and the IPPC Directive cover emissions of carbon dioxide and all the other GHG listed in Annex A to the Kyoto Protocol. However, the emission of GHG, including carbon dioxide, is not mentioned explicitly in the IPPC Directive which has an indicative list of polluting substances to be taken into account when granting an IPPC permit. Carbon dioxide is, however, clearly covered by the IPPC Directive's broad definition of pollution.[20] The Emissions Trading Directive explicitly covers all GHG, but initially it applies only to carbon dioxide resulting from the activities specified in Annex I.[21] As a result, other GHG emissions are still subject to the requirements of the IPPC Directive, and indeed also to the BAT standards.[22]

The directives cover similar installations.[23] They are concerned with the most significant carbon dioxide emitting activities, including for example energy industries, production and processing of metals, mineral industries.[24] In one respect, the Emissions Trading Directive has a wider coverage than the IPPC Directive.[25] It applies to combustion installations with a rated thermal input exceeding 20 MW rather than 50 MW. The reason for this is that smaller combustion installations are significant sources of carbon dioxide emissions and their number is likely to increase in the future.[26]

5. PERMITS AND ALLOWANCES

The Emissions Trading Directive has been drafted so as to ensure that the emissions trading scheme and the IPPC Directive are compatible, that they work well together and that synergies between them can be exploited.[27] The most obvious indication of this is their identical use of language. They share basic definitions like those of 'operator' and 'installation'. Both directives call for co-ordination of the issuing of permits. Article 8 of the Emissions Trading Directive explicitly states that the Member States are obliged to co-ordinate the permit procedure under that Directive with that of the IPPC Directive.

The concept of a 'permit' is well established in environmental policy, particularly for the application of technical standards in the field of waste, water and air pollution.[28] The GHG emission permit has to be distinguished from GHG allowances.[29] The permit is a prerequisite for installations to participate in the trading regime as it sets out the conditions for monitoring and reporting emissions. However, it is not emission permits that are traded; it is allowances that are traded.[30]

According to art. 4 of the Emissions Trading Directive, Member States are obliged to ensure that, from 1 January 2005, no installation undertakes any listed activity resulting in carbon dioxide emissions unless its operator holds a GHG emissions permit, or unless the installation is temporarily excluded from the Community scheme pursuant to art. 27 of the Emissions Trading Directive.

6.   PERMIT SETTING

A GHG permit gives permission to emit certain greenhouse gases, whereas the IPPC permit grants authorization to operate larger industrial installations.[31] Article 5 of the Emissions Trading Directive lays down the requirements for applying for a permit. As with the application for an IPPC permit, the operator has to give a description of the installation and its activities, its sources of emissions and the measures intended to be taken to comply with monitoring and reporting obligations.

Article 6 of the Emissions Trading Directive sets out the permit procedure and the conditions and contents of the permit. The first requirement is that a permit must specify the measures to ensure compliance with the monitoring and reporting requirements of the Directive.[32] The interpretation of this requirement and the legal guidance for its implementation are to be found in the Commission Guidelines for monitoring and reporting, which are based on the principles set out in Annex IV to the Directive.[33] The second requirement is that the permit shall contain an obligation to surrender allowances that are equal to the total emissions of the installation.[34] Operators must inform the competent authority of any planned changes to an installation to ensure that permits are updated.[35]

Both the IPPC permit and the GHG permit are aimed at reducing non-local effects on the environment. The primary aim of the IPPC permit is, however, to take local conditions into account, for example noise, odour, discharges into water and soil, and to deal with the accumulation of these effects locally. In addition, the IPPC permit must take account of the possibilities of promoting sustainable production, including the economic use of raw materials, use of environmentally friendly materials, waste prevention, energy efficiency, and so on. In principle, both permit procedures require an installation-by-installation determination, though one permit can cover more than one installation provided they are located on the same site and operated by the same operator.[36] It is a moot point whether environmental impacts can be addressed more effectively in some cases if, for example, a permit is issued to an industrial estate or maybe to a company with installations at different sites.[37] This is not possible under the existing legislation.

The IPPC Directive precludes sector-wide interpretation, due to the definition of BAT.[38] Best available technologies concerns not merely the level of emissions of one pollutant, but also the design, building and operation of an installation as a whole, covering all its emissions and their impact on the environment.[39]

In contrast, the GHG permit does not, to the same extent, require an assessment of the individual aspects of the operation of an installation and its impact on the local environment. The GHG permit focuses on compliance, including the ability of the operator to conduct the required monitoring and reporting of emissions. The constraints on a sector-wide approach allowing single permits for companies with installations on different sites ('company bubbles') are less prevalent when the focus is on the global environment.[40] The instrument of pooling has already been introduced for GHG allowances, which means that Member States may allow operators to form a pool of installations carrying out the same activity so that they can meet their emission targets jointly, by pooling their individual allowances and emissions in a common 'bubble'.[41]

The wording and content of the provisions for making applications and granting permits laid down in the Emissions Trading Directive are similar to the corresponding provisions of the IPPC Directive, taking into account the differences in the scope and aims of the directives. However, in their general approach, the two permit systems are essentially different from one another. They reflect two very different approaches to regulation and the regulatory role of the Member States. The IPPC permit regime is characterized by a flexible approach which gives the regulators broad discretion in applying the basic requirements of the Directive, taking into account the traditions and priorities of the Member State in question. In contrast, the content of the GHG permit is more strictly defined. It does not leave as much room for flexibility or diversity of national characteristics in the interpretation of the permit conditions. To some extent the granting of permits under the Emissions Trading Directive diminishes the regulatory role of the Member States. This is not unusual in an EU context, but it is unusual in the context of environmental law and may be seen as an indication of what could be a change of approach to environmental regulation.[42]

7.   CO-ORDINATING PERMIT PROCEDURES

The GHG emission permit and the IPPC operating permit are parallel but mutually exclusive schemes. Article 26 of the Emissions Trading Directive provides that, where emissions of GHG from an installation are covered by the emissions trading scheme, the IPPC permit relating to that installation shall not set a parallel limit for direct emissions covered by the scheme. In principle, the existing requirements of the IPPC Directive for the efficient use of energy in the form of electricity, steam, hot water and so on, remain in force, for installations covered by the Emissions Trading Directive. However, the competent authorities under the IPPC Directive may choose not to impose energy efficiency measures, provided the emissions do not cause problems of local pollution.[43]

The integrated approach of the IPPC Directive implies an integrated approach to the issuing of permits, which requires the conditions and procedures for granting permits to be fully co-ordinated where more than one competent authority is involved.[44] Accordingly, art. 8 of the Emissions Trading Directive states that the Member States must ensure that the procedures for issuing permits under the Emissions Trading Directive are co-ordinated with the procedures provided for in the IPPC Directive. This implies that Member States may have a single procedure for issuing permits under both schemes. However, Member States are not required to establish a single integrated permit system. The Member States may have separate systems for granting different permits, as long as they ensure that the procedures are sufficiently co-ordinated. In Member States with different competent authorities for issuing IPPC permits and GHG permits, a GHG permit may be issued at a different time, provided that the competent authority for granting the IPPC permit is consulted. The information required for the grant of an IPPC permit tends to include the same information as is required for a GHG permit, and it would be useful for the relevant authorities to be able to check the consistency of applications.[45]

The Commission expects that Member States will want to take advantage of the possibility of having a single procedure.[46] Nevertheless, there is one obvious obstacle to the establishment of a single permit system. This concerns public participation in the granting of permits. In accordance with art. 15 of the IPPC Directive, the Member States are obliged to enable the public to comment on applications for permits for new installations or for substantial changes to existing installations for an appropriate period before the competent authority reaches its decision. The GHG permit is not subject to the same public consultation procedures, which makes it difficult to co-ordinate with the procedures stipulated in art. 8 of the Emissions Trading Directive.

From an environmental law perspective, it is not clear why the GHG permit procedure in itself is not subject to public consultation, considering that emissions of GHG may cause local problems or local concern if an installation keeps increasing its GHG emissions because it is more cost-effective to buy allowances than to reduce emissions. From a trade law perspective, it is perhaps more understandable why public participation is not included in the procedure for granting permits, as this may lead to trade distortion. However, the opportunity of the public to participate seems to be of particular importance for a regulatory scheme such as the Emissions Trading Directive as in many ways it relies on the actors in the market and not the Member States to regulate emissions.

8.   LEGAL IMPLICATIONS OF INTEGRATING EMISSIONS TRADING AND THE IPPC

The Emissions Trading Directive has been drafted to ensure that the emissions trading scheme and the IPPC Directive are compatible and work together. However, the amendment to the IPPC Directive, under which a permit to operate must not include an emission limit value for direct GHG emissions, does not sufficiently clarify the relationship between the directives, and in practice the co-ordination of the regimes is left to the Member States. The potential for conflict is obvious. Two legal problems in particular leap into view. The first problem concerns the introduction of a market-based incentive into a traditional command and control approach with direct regulation, statutory standards and emission limit values. The installations covered by the Emissions Trading Directive may have existing operating permits under the IPPC Directive, and these may provide a legal right to emit carbon dioxide.[47] Although the GHG permit is just the basic requirement for allowing an installation to emit greenhouse gases under the Directive, it also contains an obligation to surrender allowances equal to the total emissions of the installation, which may restrict the right to emit. Consequently, the legal right to emit may be partly withdrawn by obligations under the Emissions Trading Directive, and the introduction of the new market-based measure may amount to a taking, which leads to the question of compensation.

The directives only partly deal with the problem. On the one hand, the provision of the Emission Trading Directive amending the IPPC Directive states that the competent authorities must amend the IPPC permit if necessary. On the other hand, the IPPC Directive prescribes that the competent authorities shall amend permit conditions if new Community legislation so dictates.[48] However, these provisions do not deal with the consequences of withdrawing existing rights under national law. The emissions trading system is not introduced in a vacuum but has to be integrated into an existing legal system, and the legal requirements for revoking existing permits, whether wholly or in part, will have to be examined by each Member State when issuing the GHG permits.[49]

The second problem is that the introduction of a new measure to regulate emissions which are already covered by an existing measure risks duplicating the burden on the affected operators. This problem is expressed in several ways in practice.

First, it is necessary to make a distinction between direct and indirect emissions, as the GHG permit only applies to direct emissions of carbon dioxide on the site, in combustion or by other processes. Indirect emissions are still subject to the requirements of the IPPC Directive and the IPPC permit obligations based on BAT and the requirements for energy efficiency. It is difficult to draw this distinction in practice. Emissions of carbon dioxide can be monitored either by calculation or on the basis of measurement.[50] In principle, emissions can be measured directly at the chimney, but the calculation will normally be based on the amount of fuel and raw materials used, which may have implications for installations that consume all or part of the electricity, steam, and so on, they produce.[51] The end result may be an increased burden for certain installations.

Secondly, local pollution caused by greenhouse gases is not dealt with in the Emissions Trading Directive. Significant local pollution continues to be dealt with under the IPPC Directive. As a result, emission limit values may still be necessary in order to avoid local environmental pollution or 'hot spots'.[52] Consequently, it may be necessary to lay down BAT-based emission limit values for some greenhouse gases in order to ensure that direct emissions do not give rise to significant local pollution.[53] According to current scientific knowledge, there are no problems related to carbon dioxide emissions as they do not have local effects, but it is possible that other greenhouse gases may have significant local effects. If this is the case, an installation may be able to participate in the emissions trading scheme, but will not be able to increase their emissions above the level set in the IPPC permit, regardless of the amount of allowances the installation may hold.[54]

The indirect influence of the IPPC Directive on the functioning of the emissions trading market may also be expressed in other ways, for example where a national competent authority sets an emission limit value for another pollutant, such as sulphur dioxide, or imposes BAT on indirect emissions of greenhouse gases, or lays down energy efficiency standards which, if met by the operator, will result in the incidental reduction of direct GHG emissions from the installation. In such cases, the operator of the installation will incur costs to meet the emission limit value for sulphur dioxide or the standards for indirect carbon dioxide emissions or energy efficiency, and thus indirectly pay for a reduction of direct GHG emissions. In this situation, where the operator has to take measures to reduce other emissions, as a result of which the direct GHG emissions increase or decrease incidentally, the operator has no choice other than to take these measures.[55]

A third problem, which is not however a legal problem, is the lack of scientific knowledge about 'hotspots', for example. The term 'hot spots', which relates to pooling, refers to the interaction of a given pollutant with the same kind of pollutant, in a pattern of rising marginal damage with increased concentrations.[56] Thus, small amounts of GHG may be harmless, whereas larger amounts may be dangerous. Under these circumstances, trading can be highly problematic, because trading may cause concentrations of pollutants in a specific location.

This illustrates a major problem for environmental law in general and for the regulation of climate change in particular: there is a need for scientific knowledge. This is not just a problem for a market-based approach, such as the emissions trading scheme, but the market-based approach of the Emissions Trading Directive brings the knowledge problem to the fore.

We cannot know what the actual effects of emissions trading are or whether the approach is really working without solving the basic problem of monitoring greenhouse gases, and without understanding the synergies, interactions and trade-offs caused by the emissions trading scheme. Although the local effects of an installation buying extensive extra allowances and increasing its emissions considerably are scientifically uncertain, it is perhaps not as uncertain as the reaction of the public if pooling were to be used extensively. This may not be a legal problem, but the signals which the law sends to the citizens who live near to installations with intensified emissions of greenhouse gases are none the less important.

9. A NEW REGULATORY APPROACH?

The problems of environmental protection are addressed through a complex and wide range of regulations and instruments. So far, relatively few non-traditional tools have been applied in Community environmental policy.[57] The Emissions Trading Directive is a landmark in EU environmental law as it is the first overarching market-based instrument to be used throughout the EU for environmental purposes.[58] However, there are several other defining features of the Directive. Compared to the IPPC Directive and EU environmental law in general, there are two main characteristics of the Emissions Trading Directive.

The first characteristic is its unusual reliance on economic thinking. Indeed, no other field of environmental law relies more on economic instruments than climate change law.[59] The key economic rationale behind emissions trading is to use market mechanisms to ensure that the emissions reductions necessary for achieving a predetermined environmental outcome take place where the cost of those reductions is lowest.[60] Emissions trading allows an individual company to emit more than its allowance, on condition that that it can find another company that emits less than its allowance and is willing to transfer its surplus allowance. The overall environmental outcome is presumably the same as if both companies used their allowances fully, but with the important difference that both buying and selling companies benefit from the flexibility offered by trading.[61]

Obviously, it is important to ensure that the emissions trading scheme has a positive environmental outcome. What is less obvious is whether the functioning of the market and cost effectiveness are more important than environmental improvements The answer to this is not quite straightforward, despite the fact that the legal basis of the Directive is art. 175 of the EC Treaty. Emissions trading presumably does not undermine the environmental objective, since the overall amount of allowances is fixed, whereas regulatory instruments, such as emission limit values based on the best available techniques (BAT), cannot ensure a predetermined environmental outcome (for example, the number of new plants may be greater than foreseen) so the total emissions may be greater than foreseen even if all installations use the best technical standards.[62] By comparison, emissions trading provides certainty about the environmental outcome, as the emission allowances represent the overall limit on emissions allowed by the scheme. However, the benefits of emissions trading can only be realized in practice if it is accompanied by a reliable monitoring and compliance regime at a reasonable cost.[63]

The second characteristic of the Emissions Trading Directive is that, in some respects, it goes further in its efforts to approximate the laws of the Member States than has previously been seen in environmental law. Consequently, this Directive may be seen as reflecting a new regulatory tendency in Community environmental law where the framework is less flexible and the culture, legal traditions and diversity of the Member States are not taken into account to the same degree. The functioning of an emissions trading scheme implies that more uniform criteria will have to be used both for setting the standards for permits and for monitoring emissions and the verification and certification of emission reports.

This is also reflected in art. 16 of the Emissions Trading Directive, which establishes a uniform regime for penalties.[64] The provision stipulates that if any operator of an installation does not comply with the requirement to surrender sufficient allowances, the Member States are obliged to impose both a fixed financial penalty and to publish the name of the operator in question. It is questionable whether such penalties will always be effective and proportionate and whether a uniform regime is compatible with the subsidiarity principle. The fact that environmental damage, such as climate change, has a trans-border effect does not in itself explain why penalties are introduced at a European level.[65] The explanation may be that a Community-wide emissions trading system can only function properly with a strict emissions monitoring regime consisting of a reliable credit registration system and a uniform approach to enforcement and compliance. The legal basis of emissions trading must be fair, transparent and practicable, and leave no room for negotiation or consideration for individual circumstances, as this could distort competition and lead to uncertainty.

One the one hand, the Emissions Trading Directive is characterized by provisions that lay down quite detailed substantive requirements as opposed to merely setting the framework. The role of the Member State is not that of a regulator but more that of an administrator and enforcer. This is emphasized by the fact that there are no provisions in the Directive for public participation with regard to the GHG permit as institutionalized public participation would undoubtedly require that the role of the Member States be stronger and more decisive.[66] On the other hand, the approach of the Emissions Trading Directive gives companies wider scope to decide what technical measures to adopt than is feasible under the IPPC Directive. The method by which pollution is reduced is up to the installations and their trading partners, and the emissions trading regime meets the demand for flexibility and cost-effective pollution control because it allows for any number of pollution-reduction methods to be used.

10. CONCLUDING REMARKS

The permit schemes of the Emissions Trading Directive and the IPPC Directive are parallel and complement each other, but they are also potentially in conflict and there is often ambiguity about which code to apply.

The potential for conflict is mainly due to the fact that the emissions trading scheme only includes direct emissions of greenhouse gases, and only to the extent that they cause no significant local pollution. Otherwise, emissions are subject to an operating permit under the IPPC Directive. However, conflict may also arise from the possibility that initiatives taken under the IPPC Directive may indirectly affect the emissions trading scheme. From an environmental perspective it is also a major problem that BAT is not applicable to the Emissions Trading Directive, which may be seen as the main environmental drawback of the new regulatory approach. This may lead to situations in which the overall standard of an individual installation is lowered, because BAT is not fully applied.

The Emissions Trading Directive reflects a different approach in Community environmental law. The Directive lays down (with respect to permitting and monitoring) clearer and less ambiguous criteria, leaving the Member States with less discretionary power. Presumably, more uniform regulation and enforcement will lead to a higher level of compliance and, ultimately, to higher environmental standards. Perhaps no other field of environmental law relies more on compliance than the emissions trading scheme, as it can only function under a strict emissions monitoring regime with a reliable credit registration system and a uniform approach to enforcement and compliance. A uniform regime may lead to the successful functioning of the system from a market and economic perspective, but such harmonization disregards other operative legal measures and ignores the different legal traditions of the Member States.

It is important to recognize that there are some inevitable trade-offs between the impacts addressed by an emissions trading scheme and a permit scheme prescribing emission limit values for individual installations based on the application of BAT. The inevitable question is whether the emissions trading approach, which gives companies a large degree of freedom to decide for themselves the nature, planning and distribution of technical measures, is more effective, also from a legal perspective, than the more traditional command and control permit scheme. Emission trading does not, of itself, reduce emissions. It simply provides incentives to find the lowest cost for achieving a given level of emissions reduction.

From a legal perspective the answer depends on several factors such as the content of the national allocation plans, the choice of allocation method, the technical accuracy of monitoring, the level of compliance, and so on. It implies that the system is fair, predictable and transparent, and that the public understands and reacts positively to this new regulatory instrument that turns the right to pollute into a commercial product that can be bought and sold. The legal outcome cannot easily be foreseen, and the use of this new regulatory instrument will, no doubt, raise new legal questions.[67]

NOTES

1. Associate Professor in EU and Environmental Law, PhD, LLM, Department of Law, Aarhus School of Business, Denmark: E-mail beo@asb.dk.
2. COM(2001) 31 final, Communication from the Commission to the Council, the European Parliament, the Economic and Social Committee and the Committee of the Regions on the sixth environment action programme of the European Community 'Environment 2010: Our future, Our choice' – The Sixth Community Environment Action Programme, p. 13.
3. Directive 2003/87/EC establishing a scheme for greenhouse gas emission allowance trading within the Community and amending Council Directive 96/61/EC as amended by Directive 2004/101/EC.
4. COM(2000) 87 final, Green Paper on greenhouse gas emissions trading within the EU, p. 6.
5. Directive 96/61/EC concerning integrated pollution prevention and control.
6. Directive 96/62/EC on ambient air quality assessment and management.
7. Louka, Elli (2004), *Conflicting Integration: The Environmental Law of the EU*, Intersentia Publishers, pp. 134–38.
8. Directive 2001/80 of the European Parliament and of the Council of 23 October 2001 on the limitation of emissions of certain pollutants into the air from large combustion plants (LCP). Plants covered under the LCP Directive are also covered under the IPPC Directive.
9. For a discussion of permit systems, see Lefevre, Jürgen (2005), 'The EU Greenhouse Gas Emission Allowance Trading Scheme' in Farhana Yamin (ed.), *Climate Change and Carbon Markets*, Earthscan Publications , pp. 81–86.
10. Article 1 of the Emissions Trading Directive.
11. Lefevre, Jürgen, Andrew Farmer and Patrick ten Brink (2002), 'Assessment of the Relationship between Emissions Trading and EU Legislation, in particular the IPPC Directive', Ministry of Housing, Spatial Planning and the Environment Directorate of Climate Change and Industry, p. 29.
12. Ibid., p. 29.
13. Article 2(5) of the IPPC Directive.
14. A good example of this approach is Directive 2000/60/EC establishing a framework for Community action in the field of water policy (the Water Framework Directive).
15. Smith, Adrian and Sorrell, Steve (1999), 'How to avoid a policy mess? The case of carbon emissions trading and IPPC', *ELNI Newsletter*, **2**, pp. 4–8.

16. COM(2003) 354 final, On the road to Sustainable Production. Progress in implementing Council Directive 96/61/EC concerning integrated pollution prevention and control, p. 20.
17. For more on the cap-and-trade approach, see Lefevre, Jürgen (2005) in Farhana Yamin (ed.), pp. 86–88.
18. Smith, Adrian and Sorrell, Steve (1999), *ELNI Newsletter*, p. 5.
19. COM(2003) 354 final, p. 30.
20. Article 2(2): '"pollution" shall mean the direct or indirect introduction as a result of human activity of substances, vibrations, heat or noise into the air, water or land which may be harmful to human health or the quality of the environment, result in damage to material property, or impair or interfere with amenities and other legitimate uses of the environment'. See also Lefevre, Jürgen (2005) in Farhana Yamin (ed.), p. 123.
21. Annex II of the Emissions Trading Directive.
22. DG Environment (2002), 'Non-paper on synergies between the EC emissions trading proposal and the IPPC Directive', p. 3.
23. The IPPC Directive applies to a wider range of installations.
24. The sources of pollution covered are listed in Annex I of the IPPC Directive and the Emissions Trading Directive.
25. Article 27 of the Emissions Trading Directive contains an opting out possibility, whereby the Member States may apply to the Commission for installations to be temporarily excluded until 31 December 2007 at the latest.
26. DG Environment (2002), 'Non-paper on synergies between the EC emissions trading proposal and the IPPC Directive', p. 2.
27. Ibid., p. 1.
28. COM(2000) 87 final, p. 8.
29. Peeters, Marjan (2003), 'Emissions Trading as a New Dimension to European Environmental Law', *European Environmental Law Review*, p. 84.
30. Lefevre, Jürgen (2005) in Farhana Yamin (ed.), pp. 109–10.
31. Article 9 of the IPPC Directive.
32. Article 6(2) of the Emissions Trading Directive.
33. Article 14 of the Emissions Trading Directive, cf. Commission Decision C (2004) 130 final of 29/01/2004 establishing guidelines for the monitoring and reporting of greenhouse gas emissions pursuant to the Emissions Trading Directive 2003/87/EC of the European Parliament and of the Council.
34. The allowances are tradable rights that can either be held or transferred to anyone, companies or persons, within the Community and in third countries provided that an agreement on mutual recognition of allowances has been concluded, cf. art. 25 of the Emissions Trading Directive.
35. Article 7 of the Emissions Trading Directive.
36. IPPC Directive art. 2(9) and Emissions Trading Directive art. 6(1). On the many different approaches to implementing the IPPC Directive in the Member States, see ENAP Study (2004), 'Analysing View, Policies and Practical Experience in the EU of Permitting Installations under IPPC and Alternative ways of Regulation that go beyond Installation Permitting'.
37. ENAP Study, 2004, pp. 63–5.
38. Lefevre, Jürgen, Andrew Farmer and Patrick ten Brink (2002), p. 16.
39. Ibid., p. 16.
40. ENAP Study (2004), pp. 52–54.
41. Article 28 of the Emissions Trading Directive.
42. The reduced regulatory role of the Member States is also remarkable considering the legal basis of the Directive, cf. art. 175 of the EC Treaty. However, it must be noted that a great deal of discretion has been given with regard to the allocation of allowances.
43. Article 26 of the Emissions Trading Directive amending art. 9(3) of the IPPC Directive.
44. Article 7 of the IPPC Directive.
45. DG Environment (2002), 'Non-paper on synergies between the EC emissions trading proposal and the IPPC Directive', p. 2. Since October 1999, the IPPC Directive has applied to new installations as well as existing installations where the operator intends to carry out changes that may have significant negative effects on human health or the environment. Member States

have a transition period until October 2007 to ensure that other existing installations fully comply with the Directive.

46. Ibid., p. 2.
47. See n. 45 above.
48. Article 26 of the Emissions Trading Directive amending art. 9(3) of the IPPC Directive and art. 13 of the IPPC Directive.
49. German companies have already announced their intention to challenge the new requirement in court on the ground that their operating permits provide legal protection against reduction of emission rights. See also Peeters, Marjan (2003), p. 87.
50. Annex IV of the Emissions Trading Directive.
51. COM(2003) 354 final, pp. 20–21.
52. Lefevre, Jürgen, Andrew Farmer and Patrick ten Brink (2002), p. 19.
53. COM(2003) 354 final, pp. 20–21.
54. DG Environment (2002), 'Non-paper on synergies between the EC emissions trading proposal and the IPPC Directive', p. 3.
55. COM(2003) 354 final, pp. 20–21, and the Non-paper on synergies between the EC emissions trading proposal and the IPPC Directive, 2002, p. 3.
56. Rose, Carol M. (2005), 'Environmental Law Grows Up (More or Less), and What Science Can Do to Help', Draft: Final version to appear in 9 *Lewis & Clark Law Review*, pp. 20–24.
57. Another example is the Eco-Management and Audit Scheme (EMAS).
58. COM(2003) 354 final, p. 28.
59. Richardson, Benjamin J. (2004), 'Climate law and economic policy instruments: a new field of environmental law', *Environmental Liability*, p. 19.
60. COM(2000) 87 final, p. 8.
61. Ibid., p. 8.
62. Ibid., p. 8.
63. Ibid., p. 9.
64. Lefevre, Jürgen (2005) in Farhana Yamin (ed.), pp. 118–212.
65. See also the doubts of Peeters, Marjan (2003), p. 85.
66. On 'institutionalized public participation', see Louka, Elli (2004), pp. 78–80.
67. Peeters, Marjan (2003), p. 82.

9.   Enforcement of the EU greenhouse gas emissions trading scheme

Marjan Peeters[1]

1.   INTRODUCTION

One of the remarkable developments in environmental law is the evolution of emissions trading from a theoretical concept towards a main tool of actual climate change policies. The inclusion into the Kyoto Protocol of inter-state emissions trading and project-based crediting of emissions reductions (joint implementation (JI) and clean development mechanisms (CDM)) is, indeed, an important milestone of international environmental law. In addition, the European Union (EU) has established the largest domestic emissions trading scheme worldwide, which actually started on 1 January 2005.[2] It will be interesting to see whether this application of emissions trading will have the expected positive results in practice. Outcome assessments are needed in order to learn about the practical effects of the instrument in different respects, and to facilitate discussions about improving the present scheme.

One core obligation of an emissions trading scheme (ETS) is that each covered installation needs to surrender an amount of emissions rights at least equal to the emissions during a certain well-defined period. In this respect, close attention needs to be paid to the compliance behaviour of firms. It is obvious that compliance with the rules of the emissions trading scheme is a precondition for letting that market work. When some firms would act in breach of the rules, the environmental effectiveness would become less, and the trust within the market would become at risk. Enforcement issues of emissions trading deserve especially close attention as it can be supposed that it might be financially attractive for firms to behave illegally. Furthermore, the fact that enforcement activities towards industries are to be executed by the Member States leads to considerations to harmonize to a certain extent the enforcement provisions at the EU level. Different enforcement strategies (strong and weak strategies) among the Member States could affect the level playing field between industries, and could lead to a distortion of the market for greenhouse gas (GHG) allowances.

In this chapter, it will be discussed how the EU GHG emissions trading programme deals with the fact that compliance by the covered industry must be ensured. In its Green Paper on GHG emissions trading within the EU, the European Commission (the Commission) already recognized the need for a robust enforcement regime:

> The purpose of strict compliance provisions and enforcement is to enhance confidence in the trading system, make it work in an efficient way in accordance with the rules of the internal market and at the same time increase the likelihood of achieving the desired environmental result.[3]

The Green Paper stressed the need for adequate monitoring, tracking of the allowances, and the need of reporting, verification and control. In addition, strong penalties should be established; they should be foreseeable and must significantly exceed the cost of compliance, according to the Commission. Subsequently, several provisions for monitoring, verification and enforcement are included in the Greenhouse Gas Emissions Trading Directive (GHG ET Directive). Those provisions, which will be discussed below, have harmonized the enforcement approaches by Member States to a large extent.

Within this chapter, section two will first outline some general preliminary remarks. Section three will examine which obligations are established for the covered industry under the present greenhouse gas emissions trading scheme (GHG ETS). Section four will then discuss the enforcement package as included within the EU ET Directive, and, finally, section five will end with an outlook forcast.

Besides focusing on (enforcing) compliance by industry, the level of compliance by the Member States in fulfilling their duty to implement the ET Directive is, of course, essential for letting the market work. This chapter will, however, focus on the enforcement issues with regard to industry. Within Chapter four issues of Member State liability when failing to implement sufficiently the ET Directive have already been discussed.

2.   GENERAL REMARKS

2.1   The Need for a Strong Enforcement Package

The rationale for introducing emissions trading is that a price-driven decision-making by potential polluters will be enhanced, facilitating the covered industry (or other sectors of society to be covered) to choose between reducing emissions at a certain cost, or getting emission rights from the market, or, depending on the specific design of a scheme, from the government. Of course, within the first allocation of the allowances, a great deal or all of the available rights may be given for free to the industry (this is often called 'grandfathering'). This is in fact what happened during the first allocation round within the EU ETS. Nevertheless, also in this case a market incentive will be included as firms are in the position to choose between (at least): (1) reducing their emissions and selling their allowances; (2) using their allowances in order to cover their emissions; or (3) increasing emissions and buying extra allowances on the market. The market incentive will differ depending on the specific design of the permit-market. However, in any ETS a certain market incentive is automatically included, and is thus in fact to be seen as a basic feature of emissions trading. The price of the allowances not only creates an incentive to reach environmental protection at low costs, but serves at the same time as an incentive for firms to seek the loop-holes within the rules. It is even possible that some firms would deliberately choose to be in non-compliance, when that would be profitable and when, for instance, the chance to be detected is presumed as being low. Hence, the main and most attractive feature of market-based regulation – establishing price-driven decision-making by industry with regard to their environmental behaviour – is, in fact, also a threat for the level of compliance. It is clear that adequate enforcement provisions must be established and applied in order to back up the permit market. Those provisions need already to be part of the law establishing the permit scheme. One can particularly think of the following provisions:

- *Clear and sound legislation* – it is important that the rules are crystal-clear and hence do not lead to different points of view whether an activity must be qualified as in breach of the law.
- *Monitoring, reporting and control provisions* – in order to verify how much allowances need to be surrendered.
- *Adequate sanctions* – the preventive function of the sanctions is of course important. However, the possibility to restore through sanctioning the over-emissions in case of non-compliance also deserves close attention, in order to ensure the environmental effectiveness of the scheme.

In this contribution, the notion 'enforcement' refers to this whole package of provisions.

2.2 The Need for Specific Enforcement Provisions for Emissions Trading

In general, the monitoring and control issues in the case of emissions trading differ from the classical command and control regulations. The classical way of enforcement in the case of command and control depends on *static* obligations for firms, like binding limit values for their emissions, or the obligatory application of specific techniques. In fact, within the classical command and control permits the obligations for covered firms do not change unless the permit is amended (as in the case of the Integrated Pollution and Prevention Control (IPPC) permit). With emissions trading however, the legal obligation for covered firms is a *dynamic* one: the amount of allowances that need to be surrendered fluctuates depending on the exact amount of emitted gases. This means that the control and enforcement provisions must be adapted to this dynamic obligation. The measurement of the fluctuating emissions is crucial in order to determine the amount of allowances that need to be surrendered by the operator of the installation. It in fact means that a continuous control provision for measuring and/or accounting the emissions is inevitable, ie as has been installed within the acid rain allowance trading programme that runs in the USA.

2.3   Towards Multi-pollutant Emissions Trading?

The present GHG ETS actually only covers CO_2 emissions. This approach runs counter to the idea that environmental law should have a comprehensive approach. The challenge for the future is to examine the possibilities of multi-pollutant emissions trading systems, including the greenhouse gases and even also other air pollutants. One of the main questions is, of course, whether a multi-pollutant trading scheme can indeed be controlled effectively. CO_2 emissions seem rather easy to monitor – so here big problems would hopefully not occur with the design of monitoring requirements within the GHG permit. Expansion of the ETS can be considered in order to enhance cost-effectiveness in climate change policies.

Nevertheless, the monitoring then may become more complicated, as then pollutant equivalents must be taken into account. Hence, it needs to be agreed upon whether emissions trading will still be a preferable regulatory instrument for a multi-pollutant approach. This approach is in fact already included within the Kyoto Protocol. It is, however, not yet clear whether the monitoring issue of multi-pollutant trading can indeed be solved for the national or regional schemes for GHG emissions trading.

3.   LEGAL OBLIGATIONS FOR INDUSTRY WITHIN THE GHG ETS

Before the necessary enforcement provisions can be discussed, we first need to identify what the main possible offences by industry can be within the EU GHG ETS. It appears that the GHG emissions permit, as established by the directive, forms the crucial legal tool for specifying the obligations of industry. Actually two core obligations for industry can be distinguished:

1. *Greenhouse gases may only be emitted by the covered industry when a greenhouse gas permit has been issued.*

   Besides the transferable allowances, to be allocated by the national governments and which may be sold on the emissions trading market, a so-called 'greenhouse gas permit' is included within the GHG ETS Directive. Greenhouse gas emissions (during the first allocation phase running from 2005 until 2008 this concerns only CO_2) may only be emitted by an installation when the operator holds a GHG emissions permit.[4] The permit is meant as a legal document in which the specific GHG activities of the installation(s) are described, and in which main obligations for the operators are included. According to the Directive, the permit may only be issued by the competent authority if it is satisfied that the operator is capable of monitoring and reporting emissions.[5]

2. *To comply with the GHG permit conditions.*

   The GHG permit is, in fact, an important tool for ensuring compliance with the scheme. Within the permit, monitoring and reporting requirements need to be included.[6] Moreover, the permit must contain the obligation for the operator to surrender allowances equal to the total emissions of the installation in each calendar year, within four months following the end of that year.[7]

The operator is obliged to inform the competent authority of any changes planned in the nature or functioning, or an extension, of the installation that might require updating of the GHG emissions permit.[8] Of course, this rule needs also to be monitored and enforced by the competent authority.

The specific requirement mentioned above, that the permitting authority must be satisfied that the operator is capable of monitoring and reporting requirements, will not always be easy to execute in practice. It can be assumed that bad behaviour in the past – especially non-compliance with monitoring and reporting obligations – could be a reason for not having enough trust in the capabilities of the operator. However, the denial of a permit request based on the ground that the operator probably would not be able to comply with the monitoring and reporting emissions needs a sound argument. Such a denial would bring a firm into the unfortunate position that it would not be allowed to emit greenhouse gases anymore, which could mean in fact that the whole unit would no longer be able to operate. Such a high consequence needs a very strong justification.

4.   THE ENFORCEMENT PACKAGE OF THE EU GHG ETS

4.1   Monitoring, Reporting and Verification

4.1.1   Introduction: a harmonized approach

The GHG ET Directive provides for an extensive framework for the monitoring, reporting and – to a lesser extent – the verification provisions to be implemented by the Member States. Nevertheless, as will discussed below, some discretion is given to the Member States and their competent authorities in deciding on the specific monitoring, reporting and verification duties. Such discretion is necessary in order to make the system pragmatic and workable and, moreover, to respect the autonomy and the specific legal cultures of the Member States. On the other hand, it could be the case that this discretion – when not implemented in a sound way – would be a risk for the environmental effectiveness of the European ETS.

Article 14 of the Directive obliges the Commission to adopt 'guidelines' for the monitoring and reporting of emissions by the industry. These guidelines need to be based on the principles for monitoring and reporting as set out in Annex IV of the Directive. The Member States are obliged to ensure that emissions are monitored in accordance with these guidelines, and must ensure that each operator of an installation reports the emissions from that installation during each calendar year to the competent authority after the end of that year, in accordance with the guidelines. Considering these provisions within the Directive, it can be argued that the guidelines of the Commission have a rather strong legal meaning.[9] According to the Commission, Member States are, indeed, not allowed to give derogations to the application of the guidelines unless such derogations are explicitly provided for *in* the guidelines.[10] However, the guidelines as such give some discretion for deciding on the specific duties for industries.

4.1.2   Balancing between accuracy and costs

According to the Directive, the Commission had to adopt the guidelines by 30 September 2003, which, remarkably, was even before the Directive entered into force.[11] In fact, the guidelines were adopted on 29 January 2004.[12] The monitoring and reporting guidelines set out general guidelines (in Annex I) and activity-specific guidelines (in several annexes). The principles (included in Annex IV of the Directive) state that carbon dioxide emissions shall be monitored either by calculation or on the basis of measurement. For the calculation of the emissions a formula is prescribed (Activity data × Emission factor × Oxidation factor).

Of course, the specification of the exact monitoring methodology to be used by the operator is an important element of the enforcement package. The guidelines stress not only the accuracy and completeness of the data, but also the cost-effectiveness of a monitoring methodology: monitoring and reporting of emissions should be done with the highest achievable accuracy, unless this is technically not feasible or will lead to unreasonably high costs.[13] The monitoring and reporting guidelines of the Commission thus try to balance between the problem of technological uncertainties in monitoring emissions and the need for accurate reporting. In the specific annexes, several tiers of approaches are presented, representing different levels of accuracy. An operator has to choose the tier with the highest level of accuracy. The guidelines point out that only if it is shown to the satisfaction of the competent authority that the highest tier is technically not feasible or will lead to unreasonably high costs, the next lower tier may be used, subject to approval by the competent authority.[14] According to the guidelines, lower tiers may be applied for minor sources, which would also need the approval of the competent authority.[15]

The assessment of whether specific costs of a monitoring methodology would be unreasonable can result in a conflict between operators and administrative authorities. For a good understanding of the environmental integrity of the scheme, and also of the (comparable) costs of this directive for industry, it will be interesting to assess what the content of the decisions (with respect to the monitoring methodologies) are within different Member States, and whether those monitoring decisions are rather similar, especially when it would concern comparable industrial installations.

The exact degree of harmonization by these guidelines of monitoring and reporting within the EU depends on the wording and details thereof – and it can be concluded that the guidelines do indeed leave some discretion to the Member States, especially where it concerns the specification of the exact monitoring methodologies.

4.1.3 The permit

The monitoring and reporting guidelines point out that the competent authority shall approve a detailed description of the monitoring methodology prepared by the operator 'before the start of the reporting period, and again after any change to the monitoring methodology applied to an installation'.[16] The directive prescribes, in art. 6(2)(c), that the GHG emissions permit shall contain the monitoring requirements, specifying the monitoring methodology and frequency. This means that, after the initial issuance of the GHG permit, changes towards the monitoring methodologies need to be approved by the permitting authority, and to be included within the updated GHG permit.[17] However, the guidelines suggest that, 'where consistent with the Directive' the monitoring methodology will be specified in general binding rules. According to art. 6(2)(b) of the Directive, the GHG permit must be seen as the central place for these requirements. Also, according to art. 7 of the Directive, the operator shall inform the competent authority of 'any' changes planned in the nature or functioning of the installation, which may require updating of the GHG permit. An update of the monitoring requirements could be qualified as being such a change that would require an update of the GHG permit. Hence, it is questionable whether and how Member States could make use of the general rules, as suggested by the guidelines.

4.1.4   Reporting

In addition to the monitoring requirements, the permit must contain 'reporting requirements', according to art. 6(2)(d) of the Directive. The guidelines of the Commission also refer to reporting. The Member States have to ensure that each operator of an installation reports the emissions from that installation during each calendar year to the competent authority after the end of that year in accordance with these guidelines.[18] According to Annex IV of the Directive, on which the guidelines must be based, the reporting requirements need to refer to data identifying the installation, specific data referring to the calculation of emissions and the measurement of emissions, and some specific data for emissions from combustion. According to Annex IV, in the case of reporting data, with respect to the calculation or measurement of emissions, any 'uncertainty' has to be pointed out, and specific for measurement, 'information on the reliability of measurement methods' should be included. It can be expected that these kinds of points of attention can lead to different opinions between operators and the authorities concerned, and thus will be a (probably to a certain extent unavoidable) source of legal conflict.

4.1.5   Uncertainty about the data, and misstatements

Following the guidelines there is some room for uncertainty about the data, or even misstatements. This can be illustrated by the following terms as included in the guidelines:

> As a general rule the guidelines specify a 'permissible uncertainty', which is expressed as 'the 95% confidence interval around the measured value'.[19]

The monitoring and reporting guidelines introduce the term 'materiality': 'materiality means the professional judgment of the verifier as to whether an individual or aggregation of omissions, misrepresentations or errors that affects the information reported for an installation will reasonably influence the intended users decisions. As a broad guide, a verifier will tend to class a misstatement in the total emissions figure as being material if it leads to aggregate omissions, misrepresentations or errors in the total emissions figure being greater than 5 percent'. The guidelines express that the verifier has to assess the materiality of the reported facts. When the verifier concludes that the emissions report does not contain 'any material misstatement' (and it can be assumed that this assessment includes the margin as specified in the definition of materiality), the report is verified as satisfactory. The operator can then submit the report to the competent authority.

Of course, the question arises as to whether operators would try to gain as much profit from the room for interpretations as been given by the Directive and the guidelines. Principally, one would prefer that not any error or mistake would be accepted, so that no room for misstatements would be included. It is to be recalled that art. 6(2)(e) of the Directive states that there is an obligation to surrender allowances *equal* to the total emissions of the installation in each year, as verified. Annex V states that a high degree of certainty needs to be achieved. Full certainty in every case is probably not feasible in practice. However, the assessment of 'materiality' can obviously lead to legal disputes, and accepting a rather broad margin of errors would be in conflict with the provisions of the Directive.

4.1.6 Verification
The annual reports have to be verified in accordance with criteria set out in Annex V. The Directive does not oblige the Commission to set guidelines for verification, but the published guidelines do, in fact, refer to a certain extent to this.

The verification has to be accomplished by 31 March each year for emissions during the preceding year. A strict sanction has to be applied automatically by the Member States when this deadline is passed: the Member States must ensure that such an operator cannot make further transfers of allowances until a report from that operator has been verified as satisfactory.[20] Here it can be questioned what the term 'transfer' exactly means, as it has not been defined in the Directive. If it only means the selling of allowances, then the ability to buy allowances in order to cover ex post its emissions during the preceding year would still be available to the operator. According to art. 16(3) of the Directive, any operator must surrender sufficient allowances by 30 April of each year to government. It seems reasonable to interpret the term 'transfer' in this respect, so it would only mean that an operator is only prohibited from selling any allowance until he has got a satisfactory verification.

The Directive clearly sees the possibility that non-governmental verifiers will act. Member States may choose whether or not to attribute the verification task to administrative authorities, or to independent private verifiers. In any case, the verifier must carry out its activities in a 'sound and objective professional matter'. A practical question (with possibly legal consequences) is whether there will be enough (qualified) verifiers, at reasonable prices, who are able to verify in time. Member States (and industry and advisers) should aim at organizing adequate verification possibilities. It is not necessarily the case that the costs of verification must be carried by the operators (as this is not indicated in the Directive) but it can be assumed that, to a large extent, the operators will have to pay for the verification. This can depend on the choice of the Member States whether to allow private verifiers, or to execute this task themselves (although it even then is not excluded that a fee will be charged).

Since the reports only have to be presented *after* a calendar year, it can become problematic to control *afterwards* whether the presented report is true. Hence, the verification process seems to be only adequate if it also concerns the credibility of the monitoring *during* the preceding year. Only ex post verification – on the basis of a report – would not be sufficient. A more or less continuous verification – by, for instance, random checks – during the whole period of monitoring is necessary, although probably rather costly. It is clear that the verification process must be accompanied by site inspections in order to investigate whether the presented data or facts are the same as in practice. According to Annex V, the verifier shall indeed be given access to all sites and information in relation to the subject of the verification. In addition, the following criteria have been provided for:

> The verification of the information submitted shall, where appropriate, be carried out on the site of the installation. The verifier shall use spot-check to determine the reliability of the reported data and information.[21]

This provision is an important one, and according to Annex V(11), the verifier must make clear in its verification report how the verification has been executed, including information on the executed site inspections.

One question posed is whether a verifier will indeed be a credible one? Annex V states that the verifier shall be independent of the operator. It can be assumed that some provisions are necessary in order to make the verifiers as credible as possible, like the accreditation of private verifiers by government.

The case could arise that a verifier has stated that the report is satisfactory, but that the government would doubt whether the verifier is – in that specific case – really independent. Doubt can also exist about whether the verification process has been executed in a professional way. It can not be excluded that a situation might occur where a government has evidence that the verifier made an error in its assessment. Hence, in order to comply – as a Member State – with the obligations laid down in the Emissions Trading Directive, additional competences for administrative authorities are required to overrule the findings of the verifier. This also means that administrative authorities need competences for on-site inspections. Of course, it is extremely important that the verification tasks will be carried out sufficiently by the verifiers, but – as a second and additional inspection provision – the government is not dismissed from the duty to control whether the rules of the emissions trading framework will be complied with.

The relationship between the operator and a verifier also deserves attention. For instance, what is the legal position of a firm that has presented in time a report to a verifier, but where subsequently the verifier appears not to be able – for instance, due to internal bad organization – to verify in time? In such a case, the consequences are heavy for the firm, as it will mean an offence of the rule laid down in art. 6(2)(e) of the Directive specifying the obligation to surrender allowances equal to the total of emissions of the installation in each calendar year, as verified, within four months following the end of that year. In addition, the firm will not be allowed to take part in emissions trading (meaning in particular the selling of allowances). It can be recommended to operators, contracting a verifier, to refer to this possible situation in the contract with the verifier, and to claim financial compensation for the damage occurred. The Directive does not give any discretion to the Member States in suspending the requirement that by 31 March of each year the report must have been verified as satisfactory. One could suppose that in cases of force majeur a power to suspend would be reasonable.

In addition, it also needs to be taken into account that the integrity of the civil servants and administrators is not naturally guaranteed. Unfortunately, we have learned that examples of fraud exist within administrative institutions as well. Thus, as it can be financially attractive to help industry with cheating or falsifying data, the integrity of the administration also needs attention.

4.1.7 Conclusion

In summary, it can be concluded that the GHG ET Directive has harmonized the monitoring and reporting provisions to a rather large extent, and has also given some direction towards the verification provisions. A careful approach by the Member States is needed in implementing the monitoring, reporting and verification obligations of the Directive – thereby filling in the room for discretion left by the Directive and the guidelines. It is clear that important administrative action needs to be taken, such as the issuance of the GHG permit (including an assessment of the capability of the operator to monitor and to report), and the specification of monitoring and reporting requirements – thereby following the guidelines of the Commission. In addition, a sound credible verification process must be facilitated. Above all, the monitoring requirements for industry must be controlled effectively.

The issuance and update of the GHG permit is a crucial step, because in this document essential provisions must be included. Legal disputes might arise between the operators and the administration with respect to several decisions and assessments that need to be taken within the monitoring, reporting and verification provisions. It can, for instance, certainly not be excluded that with choosing the methodology dissenting opinions may rise between the operator and the competent authority, and that subsequently legal disputes may thus occur in the permitting process with regard to the monitoring methodology. Also, when the monitoring methodology has been specified, it is to be expected that operators will not always follow this legal obligation. For instance, false data may be used, or errors may be made in calculations or measurements.

Disputes might even arise between the operators and the private verifiers, and diverging assessments between a private verifier – and the ultimate responsible administrative authority are also not to be excluded. The authorities remain ultimately responsible in checking whether the monitoring activities are sound. Unannounced random checks by authorities will be an inevitable part of the enforcement package.

Ultimately, much will depend on the practical control strategies developed and executed by the Member States. It would not be surprising to see Member States completing the given framework for monitoring, reporting and verification with different levels of ambition. A basic question in this respect is whether the Member States will provide for enough financial capacity for executing the necessary control tasks, such as, for instance, making regular on-site inspections by competent civil officers.

4.2 Sanctions

4.2.1 Introduction
Most of the Member States have practically no experience with emissions trading, and thus no experience with establishing sanctions for breaches with the legislative obligations that are part of an ETS. The Directive seems to support the legislators of the Member States to a certain extent, as it specifies several specific sanctions in art. 16. Besides the well-known and common prescription that the penalties provided for must be effective, proportionate and dissuasive, some particular sanctions are prescribed which need to be implemented by the Member States.

4.2.2 Naming and shaming
The first penalty mentioned in art. 16 is a remarkable one:

> Member States shall ensure publication of the names of operators who are in breach of requirements to surrender sufficient allowances under Article 12(3).[22]

This provision is, in fact, not strictly needed for enforcing an ETS. More remarkably, this provision marks a new enforcement approach at least in the field of European environmental legislation. The duty to publish the names of offenders can be called a 'naming and shaming' approach, and has not yet been included in European secondary environmental legislation.

With the Emissions Trading Directive, the Member States will thus be obliged to 'name and shame' offenders. About naming and shaming, especially as a criminal sanction, an intense academic debate has already been held, and there is discussion about the moral aspects and – more practically seen – the effectiveness and usefulness of the approach.[23] It has even been argued that name and shame sanctions would be an unacceptable style of governance.[24] Introducing this apparently questionable enforcement approach within European environmental law would require a careful discussion, which does not seem to have taken place. The explanatory memorandum by the proposal of the Commission does at least not justify why this new approach has been chosen.

In addition, the implementation of the 'naming and shaming' provision embodies interpretative questions. For instance, at what time is it reasonable to publish? Immediately at 1 May each year, when according to art. 12(3) of the Directive every operator had to surrender a number of allowances equal to the total emissions? As the 'naming and shaming' approach is presented as a penalty, would not it be necessary to consider the specific circumstances under which the operator was not able to surrender the allowances, such as the question of whether or not it is the operator or probably the verifier (or even the government) who did not act in time? Also, would not it be preferable to publish the name after a judge has stated that indeed the operator acted culpably or deliberately in breach of art. 12(3)?

However, the advantages of naming and shaming must be taken into account as well. Within the US acid rain allowance trading programme, information technology has been used for making the information transparent. It is suggested that making the information derived from the continuous emissions monitoring system online available to the public would increase compliance.[25] Such positive effects should, of course, be taken into account when assessing the necessity and acceptability of a naming and shaming approach. However, the instrument has now already been prescribed to the Member States, unfortunately without an extensive discussion on the rationale for introducing this, in fact, challenging instrument.

4.2.2　Financial penalty

The second penalty the Member States must implement is a fixed and automatic penalty, which is called the 'excess emissions penalty'.[26]

The penalty is 'fixed', since the Directive prescribes the amount to be paid by the offender. For the first phase 2005–07, the excess emissions penalty will be €40 for each ton of carbon dioxide equivalent emitted for which no allowances have been surrendered by 30 April of a calendar year. From 2008 onwards, the penalty is fixed at €100 per ton of carbon dioxide equivalent.

The proposal for the directive from the Commission included another approach: the excess emissions penalty would be €100 'or either twice the average market price between 1 January and 1 March of that year for allowances valid for emissions during the preceding year'. Within this idea, a connection with the prevailing market price would be provided, which is a more flexible approach.

Of course, the core question is whether or not the penalty would be severe enough to deter possible offenders. The experience with the acid rain emissions trading programme – which has a fixed and indexed penalty price – has shown that the established penalty is far more than twice the market price. In fact, within that programme the penalty is 10 and even (approximately) 20 times higher than the average market price of the allowances.[27] The level of the penalty within the European ETS does not *seem* to be very high but before concluding on this one should know what the market price of the allowances will be, which is probably hard to predict.[28] On the other hand, it has to be considered that, in other literature, it has been argued that a high penalty is not necessarily always effective.[29] Additional sanctions, such as imprisonment, would then be needed.

Even more remarkable is the fact that the Directive prescribes that the penalty in any case should be imposed. The financial penalty is thus 'automatic', because art. 16(3) of the Directive prescribes that this excess emissions penalty has to be paid by 'any operator who does not surrender sufficient allowances by 30 April of each year'. It would be more proportionate when the circumstances that lead to the breach with the rule would be taken into consideration by the enforcement authority. The Directive, however, does not specify what leeway Member States would have in applying the financial sanction to operators. On the other hand, discretion in defining the level of the penalty embodies the danger of sparing the industry. From a legal point of view a careful case-by-case consideration of whether the penalty indeed must be imposed would be preferable instead of having an automatic system of sanctioning. It can be expected that through jurisprudence the order to impose the penalty in an automatic way, which vests in fact a risk liability to be followed by a criminal sanction, will mitigate this provision.

The Directive also does not indicate whether an administrative authority, or a judge, must impose that penalty. It can be assumed that the most practical way will be to give an administrative authority the competence to impose the penalty, whose decision then can be appealed in court.

4.2.3   Compensation of excess allowances

In addition to the excess emissions penalty, the operator is still obliged to surrender an amount of allowances equal to the excess emissions.[30] This enforcement provision is needed in order to uphold the environmental effectiveness of the ETS.

4.2.4   Prohibition to sell allowances

Another 'penalty' has already been discussed above: in case of not having a satisfactorily verified report, further transfers (meaning selling) of allowances are not allowed. It is assumed that operators would be still in the position to buy allowances, in order to fulfil the requirement to surrender an amount of allowances equal to the excess emissions. The prohibition to sell excess allowances will, of course, not be felt as a sanction in cases where there are not enough allowances held compared with the actual emissions.

4.2.5 Extra sanctions, not prescribed by the Directive

When an operator would have breached rules on monitoring and reporting obligations, the situation might occur where no or wrong data are available that would state or indicate the amount of emissions during a calendar year. In this situation the basic obligation of emissions trading (which is the rule that allowances have to be surrendered that are equal to the emissions) can not be applied, because the emissions data are missing. The excess emissions penalty as provided for in the Directive would then of course not be applicable, since with a lack of data it is not known how many allowances must be surrendered and penalized. The Directive does not include a provision for this specific but not unimaginable situation. The Member States thus need to develop specific enforcement competences for this situation. What can be considered by the national legislator that is to make the authorities competent to estimate the amount of emissions in a calendar year concerned, and that subsequently a penalty on the basis of that estimation will be applied. Of course, it can be rather difficult for the enforcing authority to assess the just level of the penalty, and the assessment can be brought to court, but this approach is at least an attempt to deal with this offence and should be part of the Member States' legislation, until a better approach can be found.

It is clear that additional enforcement provisions, like the one discussed above, must be considered by the Member States. For instance, sanctions for the situation where an operator fails to report in time or presents a false report, penalties for acting without an (amended) permit (which could ultimately be the closing of the installation, when proportional), or penalties for breaches of the monitoring requirements should be established.

The question also arises as to whether additional prison sentences must be installed by the Member States. In addition, a provision concerning the removal of illegal gains could be part of an enforcement strategy. The duty to surrender allowances for the excess emissions already constitutes a part of such a strategy.

5. CONCLUSION AND OUTLOOK

In order to reach the goals of European climate change policy, and to comply with the Kyoto commitments and burden-sharing commitments for the EU and its Member States, it will be of crucial relevance that the GHG ETS will function effectively. The remarkably fast adoption of an EU-wide emissions trading directive is clearly not enough; a successful outcome (and other climate change measures) is needed in order to show – in the global arena – that Europe indeed takes its responsibility seriously.[31] First points of attention in this respect are: (1) a sound implementation of the ET Directive by the Member States, including national allocation plans that seriously take into account the need to decrease the CO_2 emissions; and (2) the issuance of GHG permits with adequate control provisions. Then, ideally, 100% compliance by the covered industry with the core obligations of the ETS is needed in order to ensure the environmental effectiveness of the scheme.

It is up to Europe now to show that, in our complicated and decentralized EU-law system, a high rate of compliance will occur, and – when this would be lacking – that non-compliance in most cases will be detected, restored and penalized. This is probably not too difficult in the first allocation phase of the scheme where the emissions reductions asked from the covered industry are not severe. The issue of compliance will be a weightier question in the following phases, where more serious emissions reductions would be included in the national allocation plans.

The first phase of the EU ETS can thus be seen as a learning phase, providing the possibility to gain experience with this instrument. According to art. 21 of the ET Directive, the Member States are obliged to report every year on the application of the Directive. In the report particular attention must be paid to:

- The application of the monitoring and reporting guidelines;
- The verification and issues relating to compliance with the Directive.

This means that, every year, important information will (or should) become available on the compliance issues with the Directive. The first reports needed to be sent in by 30 June 2005, and these reports should already give information on compliance with the duty to have a GHG permit when emitting CO_2, and on other parts of the enforcement package for emissions trading.[32]

Subsequently, on 30 June 2006, the Commission is expected to submit a report to the European Parliament and the Council on the application of this Directive.[33] The report must be accompanied by proposals as appropriate, obviously in order to improve the system. In the list of topics that need to be considered by the Commission in this review process, little has been included on enforcement issues. The Commission is only obliged to consider the level of excess penalties, taking into account, inter alia, inflation. Remarkably, nothing has been said on topics such as monitoring provisions, reporting duties, verification, and alternative approaches to sanctioning, (ie connecting the penalty to the prevailing market price). Nevertheless, on the basis of the yearly reports, and independent assessments by the scientific community or other organizations (and probably on the basis of case- law), it will become clear whether current provisions ought to be revised.

The enforcement package of the EU ETS is widely harmonized, but it is clear that the real actions need to be taken primarily within the Member States. Member States have to fulfil an important role in ensuring the environmental integrity of the ETS. They need to implement the ET Directive and to control and to enforce industry where necessary. The active role of Member States – for instance, with setting up enforcement strategies – is not only important for the environmental effectiveness of the scheme, but also for ensuring as much as possible the competition concerns of the covered industry, often qualified as the concern for a 'level playing field'.

However, one must not overlook that within the transboundary, EU-wide emissions trading market, the Member States are in charge and control their own industry. A close eye on the way in which Member States execute this task is strongly recommended. In fact, Member States will also have an economic incentive not to be too severe towards their own industry. The most important enforcement task is, therefore, probably the task of the Commission controlling the enforcement duties of the Member States, and to start an infringement procedure when necessary.

Many of the provisions belonging to the enforcement package are harmonized by the ET Directive or the binding guidelines, such as the duty to have the GHG permit, the monitoring and reporting provisions, and the sanctions. However, this does not guarantee that an intensive enforcement strategy will be followed by the Member States. Indeed, a further harmonization or even centralization of the more practical enforcement aspects of the ETS would cause an intense debate, and does not seems in line with the principle of subsidiarity. Nevertheless, it can be argued that EU-wide assessments should be made of the content and intensity of the enforcement strategies followed by the Member States in practice. That, at least, gives an impression on how tough the Member States do control their own industries, and how many on-site inspections, for instance, take place. The reports which need to be sent to the Commission by the Member States (according to art. 25 of the Directive) are expected to provide important information with regard to this.[34] It can even be considered to extend the questionnaire with specific questions on the practical aspects of enforcement, in order to gain a deep insight into the compliance behaviour by industry and the intensity of the enforcement strategies of the Member States with respect to emissions trading.

An ultimate, though for the moment not very realistic, provision would be that the inspection tasks would not be established only within the Member States, but that the Commission would have additional inspection competences.[35] Such a far-reaching option of course is an ultimate option, only to be taken when the Member States would appear not to be capable of ensuring the enforcement integrity of the GHG ETS.

NOTES

1. Associate Professor of transboundary environmental law, Maastricht University, Faculty of Law, Metro Institute (e-mail: Marjan.peeters@pubr.unimaas.nl).
2. Directive 2003/87/EC of the European Parliament and of the Council of 13 October 2003, establishing a scheme for greenhouse gas emission allowance trading within the Community and amending Council Directive 96/61/EC, OJ EU L 275/32, 25 October 2003. This Directive is amended by Directive 2004/101/EC of the European Parliament and of the Council of 27 October 2004 (establishing a link between the EU emissions trading scheme and the Kyoto protocol's project mechanisms), OJ EU L 338/18, 13 November 2004.
3. European Commission, *Green Paper on greenhouse gas emissions trading*, 8 March 2000, COM/2000/87, p. 24.
4. Directive 2003/87, art. 4.
5. Directive 2003/87, art. 6(1).
6. Directive 2003/87, art. 6(2)(c) and (d), (and arts 14 and 15).
7. Directive 2003/87, art. 6(2)(e). See also art. 12(3) of the Directive.
8. Directive 2003/87, art. 7.
9. Directive 2003/87, art. 14(2) and 14(3).
10. Answers to Frequently Asked Questions on Commission Decision 2004/156/EC of 29 January 2004 establishing guidelines for the monitoring and reporting of greenhouse gas emissions pursuant to Directive 2003/87/EC (Version: 2 March 2005).
11. According to art. 32, the Directive entered into force on 25 October 2003.
12. Commission Decision of 29 January 2004 establishing guidelines for the monitoring and reporting of greenhouse gas emissions pursuant to Directive 2003/87/EC of the European Parliament and of the Council, Brussels, 29.01.2004, C(2004) 130 final; OJ EU L 59, 26 February 2004.
13. Commission guidelines for monitoring and reporting of greenhouse gas emissions, p. 7.
14. Ibid., p. 12.
15. Ibid., p. 15: 'Minor sources are those emitting 2.5 ktonnes or less per year or that contribute 5% or less to the total annual emissions of an installation, whichever is the highest in terms of absolute emissions'.
16. Commission guidelines for monitoring and reporting of greenhouse gas emissions, p. 8.
17. Ibid., p. 8
18. Directive 2003/87, art. 14(3).
19. Commission guidelines for monitoring and reporting of greenhouse gas emissions, p. 19.
20. Article 15 of the Directive.
21. Annex V, sub 7 of the Directive.
22. Article 16(2) of the Directive.
23. See for instance: 'Shame, Stigma and Crime: Evaluating the Efficacy of Shaming Sanctions in Criminal Law' (anonymous author), (2003) *Harvard Law Review*, 2186–2207.
24. Whitman, J.Q. (1997–98), 'What is wrong with Inflicting Shame Sanctions?', *The Yale Law Journal*, 107, pp. 1055–92.
25. It might increase the possibilities for public pressure and even legal action from non-governmental environmental agencies and/or citizens. Tietenberg, T. (2002), 'The Tradable Permits Approach to Protecting the Commons: What have we learned?', in Ostrom, E., Dietz, Th., Dolšak, N., Stern, P.C., Stonich, S., Weber, E.U. (eds), *The Drama of the Commons*, Washington DC: National Academy Press, pp. 197–231.
26. Directive 2003/87, art. 16(3).
27. Peterson, S. (2003), 'Monitoring, Accounting and Enforcement in Emissions Trading Regimes', *Paper for the OECD Global Forum on Sustainable Development: Concerted Action*

on Tradable Emissions Permits County Forum, OECD Paris, 17–18 March 2003, p. 7: the allowance price varied between $65 and $220 per tonne SO_2.

28. In June 2005, CO_2 allowances were sold in the Netherlands for €22.50. This is more than half the price of the penalty.
29. Shavell, S. (1985), 'Criminal Law and the Optimal Use of Non-Monetary Sanctions as a Deterrent', *Columbia Law Review*, 1232–62.
30. Directive 2003/87, art. 16(3).
31. Famous scholars like Stewart en Wiener have expressed their doubts on implementing and enforcing the international climate change obligations by the EU; they even stated that international GHG obligations would be more effectively implemented and enforced in the USA than in other countries. Stewart, R.B. and Wiener, J.B. (2003), *Reconstructing Climate Policy*, Washington DC: The AEI Press, p. 86.
32. On 4 May 2005 the Commission decided on a questionnaire for these reports. Commission decision of 4 May 2005 establishing a questionnaire for reporting on the application of Directive 2003/87/EC, OJ EU L 126/43, 19 May 2005.
33. Directive 2003/87, art. 30.
34. Member State states the need, for instance, to report the following question: 'Did the competent authority carry out any independent checks on verified reports? If yes, please describe how additional checks were undertaken and/or how many reports were checked'.
35. See for example the European CFC Regulation (Regulation 2037/2000/EC), discussed by Peeters, M. (2003), 'Legal feasibility of emissions trading: learning points from emissions trading for ozone-depleting substances', in Faure, M., Gupta, J., Nentjes, A. (eds), *Climate Change and the Kyoto Protocol*, Edward Elgar, pp. 161–2.

10. A decade of emissions trading in the USA: Experiences and observations for the EU

George (Rock) Pring[1]

1. INTRODUCTION

Asking someone from the United States of America – the land of George W. Bush, oil imports, Enron, SUVs, and climate-change denial – to contribute to a discussion on climate change may seem like 'inviting the Devil to church'. But, as another old saying goes, 'give the Devil his due' – the USA still has positive contributions to make to climate change and international environmental law.[2] Contemplate these ironies:

- The USA was the first industrialized nation to ratify the 1992 Climate Change Convention (even if it may be the last to ratify its Kyoto Protocol).[3]
- The USA was the chief advocate for liberal emissions trading (ET) of greenhouse gases (GHGs) in the Kyoto negotiations of 1997 (then conservatively opposed by no less than the European Union (EU)).[4]
- It pioneered the leading successful ET program in the 1990s (for sulfur dioxide (SO_2)).
- The USA today is a surprising 'experimental laboratory' of ET programs for GHGs – below the national government level – in states, local governments, and the private sector.[5]

The USA's constructive efforts on ET, as well as on climate change, deserve more recognition than they generally receive, clouded as they are by the present national administration's negative environmental image. This chapter focuses on the leading experience the USA has to share – the federal government's SO_2 ET program – and the contribution it may make to the EU's very commendable and ambitious plan for an ET program for the GHG carbon dioxide (CO_2).[6] The chapter proceeds in three steps: section 2 describes the goals and operations of the US SO_2 ET program; section 3 analyzes the reasons for its success, and section 4 assesses what insights this might have for the EU's new CO_2 ET program.

2. THE US SO_2 EMISSIONS TRADING PROGRAM

2.1 Background

In 1970, the USA adopted the world's first modern pollution control law, the Clean Air Act of 1970 (CAA),[7] initiating a progressive program to improve air quality and public health. In doing so, the US Congress rejected market mechanisms and voluntary programs from the outset, relying on a 'command-control' model to accomplish its goals – that is, a system of laws and regulations that both 'command' specific environmental obligations and 'control' their performance through active government enforcement.[8] By 1977, it became clear that the CAA was not achieving its goals (particularly in major metropolitan areas), and it was extensively amended, imposing an even more stringent command-control system. While the US experimented with some market-based pollution trading schemes in those early decades,[9] command-control has remained the dominant approach of the CAA and the other US anti-pollution laws to the present day, although exceptions to it are emerging.

During the 1980s, 'acid rain' (more correctly, 'acid deposition') became an increasingly urgent national concern, as precursor emissions of SO_2 and nitrogen oxides (NO_x) were found to be causing acidic rain, snow, fog, and dry particle deposition in the northeastern US and Canada, with harmful effects on plants, animals, crops, ecosystems, structures, materials, visibility, and public health.[10] Given prevailing winds, the blame primarily centered on the emissions from electric power generating plants in the midwestern states (Ohio, Indiana, and Pennsylvania having the highest emitters) and some eastern states. Fossil fuel burning for electric power generation accounts for some two-thirds of all SO_2 and one quarter of all NO_x in the US.

In 1990, the CAA was again extensively amended, with expanded command-control programs. However, the 1990 amendments are best known for attacking the acid deposition problem, not with more command-control, but with an innovative market-based approach – the 'Acid Rain Program', which established a 'cap-and-trade' program for the SO_2 emissions of those targeted midwestern and eastern fossil-fuel-fired electric utility boilers.[11] The program, while not without its critics,[12] is widely viewed as a great success story for market mechanisms in general and ET in particular – effectively providing significant pollution reduction, lowered costs, reduced opposition and litigation, and a greater stimulus to innovation.[13] The US Environmental Protection Agency (USEPA) is certainly not modest about its accomplishments: the Acid Rain Program is already being viewed around the world as a prototype for tackling emerging environmental issues. The allowance trading system capitalizes on the power of the marketplace to reduce SO_2 emissions in the most cost-effective manner possible. The permitting program allows sources the flexibility to tailor and update their compliance strategy based on their individual circumstances. The continuous emissions monitoring and reporting systems provide an accurate accounting of emissions necessary to make the program work, and the excess emissions penalties provide strong incentives for self-enforcement. Each of these separate components contributes to the effective working of an integrated program that lets market incentives do the work to achieve cost-effective emissions reductions.

The General Accounting Office recently confirmed the benefits of this approach, projecting that the allowance trading system could save as much as [US]$3 billion per year – over 50% – compared with a command and control approach typical of previous environmental protection programs.[14]

The ET program's pollution reduction is impressive. It is designed to reduce electric utility SO_2 emissions more than 50% from 1980 levels by 2010, and by 2005 has already achieved most of that reduction.

2.2   Principles of the ET Program

The operating principle of the SO_2 ET program is the market-based 'cap-and-trade' system. To institute it, the federal government first established a numeric maximum on the total allowable amount of the target emissions (the 'cap'). To improve air quality, the cap is set below existing pollution levels and ratcheted down periodically. The government then issued annual allowances or credits (one allowance = one ton of emissions) free of charge to the existing target sources, based on factors such as size, emission performance standards, and fuel use (typical 'grandparenting', as new sources must buy their allowance credits on the market, generally not free from the government). The allowance credits can be kept for the source's own use or marketed to others (the 'trade'). At the end of each year, each source must surrender enough allowance credits to cover its emissions for that year. A source's actual emissions may not be greater than the number of allowances that it holds or else it will be penalized substantially. However, any unused allowances can be sold, traded, or banked by the original owner for future use – for up to 30 years. The trading can take place in a private free-market setting between sources or in a private-sector commodity futures market or auction.

The theoretical principle supporting ET is the view that an effectively functioning private market is the most efficient tool for organizing a nation's economy.[15] While many distrust markets as a means of achieving environmental and other social goals, others trust markets and feel that 'carefully designed public policies can correct [market] failures, enabling markets to work more effectively and delivering tangible gains in environmental quality at the lowest practical cost'.[16] Since the demand for emission allowances exceeds the supply, they will have a positive value in the market, and by allowing them to be freely marketable or bankable, an economic incentive is created, stimulating sources to reduce their emissions even lower and more quickly than the law requires.[17] Sources reducing their emissions below their allowance level are rewarded financially by being able to sell their unneeded allowances, creating an incentive for sources to develop new and better methods to reduce emissions.[18] On the other hand, sources that cannot meet or choose not to meet their allowance level can acquire allowance credits in the marketplace – without prior government approval.

A simple hypothetical shows how this works.[19] Assume Plants A, B, and C each emit 20000 tons per year (tpy) of pollutant X. Implementing a cap-and-trade program, the government sets an aggregate cap that requires a 50% total reduction in their emissions of X and allocates to each plant 10000 annual allowances (= 10000 tpy). Plant A finds it economical to purchase and install new pollution-control equipment and/or institute raw materials, switch fuels, or make process changes sufficient to reduce its 20000 tpy to its 10000 tpy allowance level. Plant B finds that it can economically reduce its 20000 tpy to 7000, putting it 3000 below its new allowance level, and giving it those 3000 allowances to do with as it will. Plant C, however, can only reduce its 20000 tpy emissions to 13000, leaving it emitting 3000 tpy above its allowance. Plant A has done all it needs to do. Plant B has two options: it can keep ('bank') its 3000 credits for its own use in a future year or sell them now to another source, like Plant C. Plant C also has two options: it can acquire the 3000 extra allowances it needs by purchasing them from Plant B or other below-allowance source or a private party or broker holding allowances, or it can invest in more pollution-reducing equipment, materials, and/or processes as A did to reduce its emissions to its allowance level.

In theory, each source acts in an optimal fashion, in its own economic self-interest, while achieving society's environmental goal. The government's traditional command-control role is limited to setting the cap, distributing allowances, taking action against non-compliance, and potentially distributing or selling additional credits it reserves from distribution. Emissions trading is popular because it reduces the cost of achieving regulatory goals, provides the regulated community more self-governance and flexibility in meeting program goals, and generates compliance and political support thereby.[20]

However, pollution credit trading has its sceptics.[21] Some view it as 'buying a license to pollute', 'paying for paper', or an 'outright sham' since it allows some sources (like Source C in the hypothetical above) to keep operating without reducing pollution levels, fails to capture and preserve clean air gains below the allowable level (such as Source B's 3 000 tpy below-allowance achievement), and could have a much lower (cleaner) cap.[22] All agree, ET will inevitably result is cleaner air in some areas at the expense of bad air 'hot spots' for neighbors of sources that would rather pay up than clean up.[23] Therefore, it is essential that existing ambient standards (based on human health or other public-welfare goals) continue to be enforced – in essence to act as a veto on any ET transaction that would violate them and provide a 'cap on the cap'.

Now in its eleventh year, the successful SO_2 ET program, coupled with President George W. Bush's strong preference for non-command-control programs,[24] is inspiring experimentation with a number of ET and other market and economic-based approaches to pollution control, so much so that in the US today federal, state, and local governments and the private sector are now using hundreds of market-trading and other economic-based mechanisms to solve environmental problems, in lieu of command-control regulation.[25] Among these pollution-fighting programs, the federal government's SO_2 ET program remains the undisputed heavyweight.

2.3   Operation of the ET Program

The US SO_2 ET program has many names – 'Acid Rain Program', 'Title IV' (since it comprises the fourth chapter or the 400-numbered sections of the CAA), 'cap-and-trade', 'emissions trading program', 'pollution credit trading', and the USEPA officially preferred term, 'allowance trading system'. All mean essentially the same thing, and for brevity the term 'ET' will be used here. This section provides an overview of the operational details of the US SO_2 ET program, while section 3 will provide an in-depth look at the six key features of the program that make it a success.

In Title IV of the CAA amendments of 1990, Congress set a goal of reducing targeted utility SO_2 emissions by fully 'ten million tons [below] 1980 levels',[26] ordered the reduction to take place in a two-phase progressive tightening of emission restrictions, and authorized the innovative market-based 'cap-and-trade' market-based program to achieve this.[27] The 1980 SO_2 emission level from all US utilities was 17.5 million tons (mmt). From the program's beginning year of 1995, annual caps are set that decline to a level of 8.95 mmt by the year 2010, so that the actual final cap turned out to be somewhat less than Congress's goal of a 10 mmt reduction, but still impressive: the final 8.95 mmt 2010 cap is a reduction of 8.55 mmt or almost 50% of 1980 levels.[28]

Phase I of the program (the period from 1995–99) targeted the largest-capacity and highest SO_2-emitting 263 units[29] at 110 largely coal-burning electric utility facilities in 21 states, principally in the midwestern but also in the eastern US. This generally swept into the program all sources with a generating capacity above 100 megawatts (MW) and an average 1985 emission rate greater than 2.5 pounds of SO_2 per million British thermal units (Btu) of heat input (#SO_2/mmBtu). In addition, another 182 units came under Phase I of the program, in substitution or compensation for other targeted units, which brought the total of Phase I units to 445. In Phase I, USEPA issued each unit allowances (one allowance equaling one ton of SO_2) based on a fixed emission rate multiplied by their historic fossil fuel consumption.[30] The initial year's allocations were set to require most units to begin reductions immediately. Impressively, these most highly polluting sources had reduced their SO_2 emissions over 50% below their 1980 levels by the end of Phase I in 1999.

Phase II of the program (2000–10) expanded the targeted sources to include almost all units with a generating capacity above 25 MW, a total of over 2 000 units and including all new and virtually all existing fossil-fuel-fired electricity generating facilities in the midwest and east. It also 'ratcheted down' the cap further to a permanent ceiling of 8.95 mmt. In Phase II, USEPA is issuing each unit reduced allowances at roughly one-half the amount of Phase I.[31] As a result, total utility emissions nationwide have thus far been reduced more than 30% below 1980 levels.

The allowance system enables companies to select their own compliance strategy. For example, sources may choose to use cleaner burning fuel, adopt energy-efficient measures, add pollution control equipment (such as sulfur scrubbers), or they may opt to buy allowances from other sources. For each ton of SO_2 emitted in a given year, one allowance is retired and may no longer be used. Sulfur dioxide allowances are recorded by the USEPA in a central computerized database called the Allowance Tracking System (ATS). Sources that find themselves short of allowances may then buy them from other sources that have reduced emissions below their allocated level. However, regardless of the number of allowances a source holds, it may not emit at levels that would violate federal or state limits set under Title I of the CAA to protect public health – a very critical check on the system.

The USEPA itself holds some small 'reserves' of allowance credits. It has discretion to distribute these reserves either as incentives (rewarding sources utilizing certain advanced technologies, conservation measures, or renewable energy generation), or it can sell some of them in the annual auction.

Allowances can be marketed privately or in the yearly auction, which is held by the nation's largest private-sector commodities brokerage, the Chicago Board of Trade (on a no-fee basis). Both USEPA and private entities may sell allowances in the auction, and anyone may participate as a purchaser (including environmental groups or individuals who simply wish to 'retire' the allowances they buy in order to clean up the air further!). The allowances marketed are of two 'vintages': 'spot' allowances which may be used in the year purchased and '7 year advance' allowances which are eligible for use after 7 years. Unused allowances of a given vintage year may also be banked or carried forward for future years. The USEPA typically puts into the auction around 250 000 allowances (less than 3% of the total annual allowances) and a few more may be put in by private parties (only 11 private-party allowances in 2004; none in 2005).

Substantial prices are paid for allowances at the auction. In 2004, the spot allowances sold for high, low, and average prices of US$300, 107, and 273, respectively; in 2005, they jumped to US$750, 300, and 703.[32] The total proceeds at each auction are substantial: more than US$50 000 000 in 2004 and almost $125 000 000 in 2005). Large-volume purchasers include major sources, energy companies, and commodities brokers (often purchasing 10 000–75 000 allowances at a time). Smaller purchasers include allowance 'retirement' groups, like university and law school environmental organizations (typically purchasing between 1–10). Since the auction is only 'the tip of the iceberg' of the total market in ET, it is obvious that the SO_2 credits trade actively and why they are considered the most liquid of US environmental markets.

Command-control enforcement is still involved in the ET program. The USEPA maintains an ATS, and sources must record their allowances and transfers in their ATS account and validate them if they wish to use them for compliance. At the end of each compliance year,[33] sources must surrender from their ATS account the number of allowances that match their actual tons of SO_2 emissions, as reported by continuous emissions monitoring (CEM).[34] If a unit has insufficient allowances to match its emissions, USEPA enforces a dual penalty against the offender: (1) a monetary sanction plus (2) the surrender of additional allowances for the following year as 'excess emissions offsets'.

3.   THE SIX KEY FEATURES FOR A SUCCESSFUL ET PROGRAM

The US experience demonstrates that there are six features that are essential for a sound, effective ET program, regardless of the pollutant targeted. To be successful, a program must incorporate (1) an emissions cap, (2) allowances or credits, (3) reliable record-keeping, (4) freedom of choice, (5) a trading market, and (6) enforcement. While each of these has been discussed above, it is worth making some further observations on what they contribute and why they are essential.

3.1   The Emissions Cap

The emissions cap fixes an aggregate limit on the total amount of the pollutant that can be emitted from all regulated sources and should be set so that current emission levels will be substantially lowered. An ultimate cap goal should be set (like the US's goal of achieving a 50% reduction in SO_2 by the end of 15 years of program operation), and ideally intermediate declining caps should be established to 'stair-step' down to the ultimate cap. Naturally, the cap or caps should all be clearly established at the outset, for transparency and planning/investment purposes.

At the outset, it is worth noting the biggest operational difference between the US SO_2 ET program and the EU CO_2 ET program. The former is one unitary decision-making system, operated by the central government, with one set of rules, and the individual states included in the program have no role in determining caps, distributing allowances, record-keeping rules, enforcement strategies, etc. This lack of subsidiarity makes the US ET program a much simpler system to operate from the start.

There are obviously a number of conditions precedent to setting an effective cap. The class of regulated sources must be appropriately comprehensive. The US SO_2 program has been criticized for limiting its regulated group (1) to one industry of the many that emit SO_2, (2) to one geographic region (midwest and east), and (3) initially at least, to the largest-volume polluters only.[35] Put another way, is the program designed to reach a significant percentage of the pollution problem? If not, it may be subject to attack on US legal principles of arbitrariness and equal protection or the similar EU fundamental principles of proportionality and equal treatment.[36] The US SO_2 program certainly covered a very substantial percentage of the acid rain problem, if not of SO_2 nationwide, but the EU CO_2 Emissions Trading Directive may represent only 45% of Community-wide CO_2 emissions.[37] Another question is what industries have the greatest and least potential for reducing the pollutant. The electricity sector in the US obviously had great SO_2 reduction potential (just look at the reductions achieved!), but the German steel industry or EU cement manufacturers may be at a technical control level that does not allow for great reductions of CO_2 without reducing production.[38] Cap-setting can also raise issues of taking of property and freedom to pursue an economic activity under both US and EU fundamental rights law.[39] Generally, the most prudent way to avoid or at least minimize these problems is to set a long-term cap goal (like the US 8.95 mmt in 2010) and provide a declining series of intermediate caps to reach that ultimate goal over a reasonable number of years.

The biggest problem with caps is not the number, the phasing periods, or the scope of sources to be regulated. The biggest hurdle is having the extensive, reliable ambient and emissions data necessary to establish the baseline(s) from which the cap or caps are set. Fortunately, the US had over 20 years of SO_2 data collected under the previous CAA regulatory scheme on which it could rely. Collection of CO_2 data is a newer endeavour and may be substantially less comprehensive and reliable.

3.2   The Allowances

Allowances or credits are the subunits into which the cap is divided. When allocated among sources, they provide both the limit on the amount of the pollutant a source may emit as well as the tradable commodity without which the market-based system cannot function. Their allocation raises some of the thorniest barriers to success. The cap controls the maximum number of allowances available, of course, but it may be unwise to distribute all of them in any one year. Some (minor) portion should be retained by the regulatory agency as 'reserves' for certain purposes (incentives, auction, etc. as in the US program).

Here again, a key to success is having a very reliable inventory of the emissions of the sources (as well as any necessary additional data, such as fuel usage, production capacity, etc.). The US had this data for SO_2 as a result of the pre-existing permit program, but it and the EU do not have comparably reliable data with respect to GHGs currently.

Initial allocations of allowances may need to be generous (near existing levels of emissions) to avoid the same fundamental rights issues mentioned above with regard to caps, so a certain amount of 'grandparenting' may need to be considered. The same is true of the phasing or 'stair-stepping', how fast the allowances can be reduced to reach the ultimate cap goal. The US ET allowances began in the initial 4-year phase at 2.5 pounds/mmBtu, dropping to 1.2 in the second 11-year phase. They may need to be generous in another way to avoid legal/political complications – the US issued all initial allowances free of charge, thus avoiding the legal arguments of taking of property or interference with economic advantage. Issues also arise about how to treat new sources coming on line. The US took the draconian (but successful) approach of allocating no free allowances to new sources (construction initiated from 1996 on); instead, new sources are required to purchase their allowances in the marketplace and take that into account as a cost of doing business from the planning stage on. Furthermore, the regulatory agency should be entrusted with some discretionary power to allocate additional allowances and to make exceptions, substitutions, and other negotiated 'deals' with sources that have special advantages or disadvantages.

Another very important feature contributed greatly to the eventual acceptance of the US SO_2 ET program by the most sceptical interest groups – including public health authorities, environmentalists, and people living near emitting sources. As mentioned earlier, it has a 'cap on the cap'. No matter how many allowances a source holds, it may not emit at a level that violates the National Ambient Air Quality Standards (NAAQS). The NAAQS are numeric standards limiting concentrations of key pollutants in the receptor (airshed) that are set at the level deemed necessary to protect public health (and to a lesser extent other public welfare values). The NAAQS are set under the basic CAA program and are independent of the ET program. Thus, a source that would rather 'pay up instead of clean up' cannot do so to the extent of creating unhealthy air for its neighbours. An ambient air 'override' works well for those pollutants that directly affect public health or some other easily measured public welfare factor (crops, visibility, etc.). However, the effects of CO_2 are global and not measurable in such simple, direct human-centred terms, so it will be challenging to envision how a reassuring 'cap on the CO_2 cap' could be developed.

3.3 Record-keeping

Accurate, reliable, and consistent monitoring, reporting, and recording of emissions is crucial for a successful ET program. The USEPA requires CEM in the Acid Rain Program and has established detailed CEM requirements for exhaust gases covering SO_2, volumetric flow, NO_x, diluent gas, opacity, and even for monitoring or estimating CO_2.[40] The CEM rule also contains requirements for equipment performance specifications, certification procedures, and record-keeping and reporting. As crucial as good information-flow is, it is equally important that the procedures be kept as simple as possible to make ET successful.

Centralized reporting, validation, and recording is also essential for the ET program to have credibility and to create marketable credits. The USEPA serves this role through its ATS which records in individual sources' accounts all allowances issued, deductions for compliance purposes, transfers of allowances, and net allowances held, as well as USEPA's own allowance reserves. To maximize data and minimize compliance costs, the ATS is not a true registry – parties are not required to report to the ATS until the end of the year when allowances are used to meet requirements. The use of a single, centralized ATS system, as above-noted, is more efficient for a centralized ET market than one with multiple, state-level tracking systems.

3.4 Freedom of Choice

No market system is entirely free of government/legal controls, but a successful ET system must strike a balance that minimizes government command-control requirements to the essential-only level and maximizes private parties' freedom of choice in all other decisions. The US SO_2 ET system allows sources the freedom to choose between reducing emissions to meet their allowance level or acquiring additional allowances in the market. Parties that are not sources (from commodity brokers to environmentalists) also have the freedom to participate as buyers and sellers in the SO_2 market. This flexibility produces significant cost savings because it eliminates the expense in time and resources of obtaining government approvals, as with a standard permit system. It also enlarges the supply and demand sides in the marketplace and allows the economy to grow.

Another freedom of choice feature is the 'Opt-in Program', which allows sources that are not required to participate to have the option of voluntarily entering the program and receive their own SO_2 allowances.[41] With SO_2, as with CO_2, there will be many sources outside the program that present additional emission reduction opportunities. With an 'open door' policy, a program offers an economic incentive to other sources to reduce their emissions in order to have unused allowances which they can sell on the market.

3.5　Trading Markets

The ideal ET system will also minimize the government's role in the actual trading of allowances and maximize their free marketability as commodities. The US system does this by allowing trading to take place either in (1) direct sales in the open market (for example, a non-regulated, non-public, voluntary transfer between two utility sources), and (2) in the annual auctions held by the Chicago Board of Trade, as described above. One of the keys to a successful market system is to encourage trading not only in current-year allowances but 'futures' as well. The US auction sells both the spot (current year) allowances as well as the 7-year advance allowances (those sold in 2005 becoming usable in 2012, although they can be freely traded at any time before that). The availability of advance allowances improves the effectiveness of the ET market (as does any futures trading) by helping to provide stability in planning for capital investment and a hedge against inflation.

An important factor to consider in a carbon market is the validity of parties' allowances. To have a functioning market, participants must be able to rely on the legitimacy of the allowances being traded – in other words, to have value, allowances must be actual, quantifiable, surplus (ie beyond the requirements), and enforceable.[42] The USEPA's ATS system performs this validating function in two ways. First, sources must submit end-of-the-year compliance or 'reconciliation' reports certifying as to the quantity and legitimacy of their allowances.[43] Secondly, USEPA uses its computerized Emission Tracking System to cross-check sources' emission reports for accuracy.[44]

Another device that is critical to success and can be thought of as 'trading' is the ability of sources to bank allowances not used in one year for use in future years. If a source banks 2005 allowances for its future use in 2009, it is in effect trading with itself – trading between its 2005 and 2009 balance sheets. This flexibility adds enormously to the attractiveness of the program for advance planning purposes.

3.6 Enforcement

Command-control enforcement is vital to the success of an ET program, otherwise there is no real incentive for compliance. Here again a balance is needed between government power and freedom of the market. Emissions trading is maximized by having a system that is as simple as possible, has minimal limitations, minimal transaction costs, minimal uncertainties, etc. However, even the most minimal compliance-enforcement system places additional limits, costs, and uncertainties on sources.

Overall the US ET program is viewed as successful in balancing these factors. It is estimated that the Acid Rain Program has reduced compliance costs by more than 50% below the cost to sources of a traditional command-and-control system with the same outcomes.[45] Command-control penalties have been set at a high enough level to assure compliance with minimal need for enforcement. The following noncompliance penalties exist:

- fines ($2 900/ton of SO_2 in excess of allowances)
- deduction of allowances from next year's allocation
- possible civil and criminal penalties.[46]

The result has been an impressive 99.9% compliance rate thus far.[47]

4. CONCLUSION: OBSERVATIONS FOR THE EU

The creation of a successful market in tradable emission allowances represents one of the most complex, challenging, but ultimately rewarding environmental protection measures countries can undertake. The USA's success with ET systems like the Acid Rain Program and other national, state, and local government ET programs is proof these market-based systems can and do work effectively – in both environmental and economic senses – and can even work better, cheaper, and faster than the traditional command-control permit regulatory systems under the right conditions.

That said, what are 'the right conditions'? Given the US experience, what are the basic features an ET system must have in order for it to 'transplant' and work effectively in other countries? The proceeding section has discussed six operational features that are essential: caps, allowances, record-keeping, freedom of choice, a trading market, and enforcement. Can these essential ET features be as readily achieved for CO_2 as for SO_2 or are the two pollutants so very different as to make the system non-transferable?

The two systems have striking dissimilarities with serious implications for ET:

- Knowledge: We have been regulating SO_2 – and therefore inventorying, studying, and understanding it – much longer than we have CO_2. Quite simply, we have much less reliable or comprehensive inventories of CO_2 emissions on which to base an ET program. This has serious implications for establishing a baseline year against which to measure the cap. We have good records for a 1980 baseline year for SO_2; however, CO_2 records for 1980 are problematic, and one of the major criticisms voiced against the Climate Change Treaty's adoption of that base year.

- Environmental effects: The acidic effects of SO_2 were localized, tangible, dramatic, well-documented, and well-known by the public thanks to the media during the 1980s. Carbon dioxide is a different story. Its effects are global as opposed to local, generally long-range as opposed to tangible and dramatic, controversial as opposed to well-accepted, and much less well understood by the public. These differences in the risk or harm factor will make any CO_2 program more vulnerable to opposition.

- Regulated entities: The great majority of SO_2 in the US is produced by one industry, making it relatively easy to have a sharply focused and managed program. This also minimizes potential objections from the regulated sources that other significant sources of SO_2 are being ignored. Carbon dioxide, on the other hand, is produced by a very large spectrum of very different sources (energy, minerals and metals, pulp, transportation, households, etc.), giving rise to objections of unequal treatment if more than half of the CO_2 emitters are exempted from in the program.

- Diverse processes: SO_2 is typically produced at only one point in its production process and one mechanism (boiler furnaces), making it more easy to monitor and control. Carbon dioxide, on the other hand, is produced at many different stages and through many different mechanisms (from processing raw materials, to manufacturing, to transportation, to consumption, etc.).[48]

- Sheer numbers: The US SO_2 ET program involves a few thousand units. In contrast, the EU CO_2 ET program encompasses at least an order of magnitude more entities.[49]

- Decentralized control: SO_2 ET in the US is a single, centralized program under exclusively national government control; the states do not have a subsidiary role in the US Acid Rain Program, although they are allowed to innovate and govern other ET programs if not in conflict. The EU's subsidiarity approach means that 25 different states will be making their own programmatic decisions about all of these operational factors that bear on success. Enforcement can also be expected to vary considerably under a decentralized system.

- Reporting: An ET system cannot be successful without rapid, regular, reliable reporting of emissions. CEM was both technologically and financially feasible for SO_2 in the US. It is not so readily feasible for CO_2.

● Certification: USEPA's transparent ATS system fairly validates the legitimacy of SO_2 allowances being traded. Certifying the legitimacy of CO_2 credits is an infant science at best and raises all the specters of 'hot air' or 'allowance inflation' problems that as a practical matter plague the Kyoto Protocol credit arrangements for joint implementation and clean development mechanisms.

These points should not be taken as negating the possibility of a successful application of ET to CO_2, but as cautionary only and indicative of the challenges ahead. The good news is that the US is experiencing very promising success with a variety of CO_2 ET programs – not at the national government level yet, of course – but at the state and local government level and in voluntary programs in the private sector, as a growing number of studies are demonstrating.[50]

Hopefully, the success of both SO_2 and CO_2 ET programs in the USA should provide valuable insights and inspiration for the very ambitious and commendable CO_2 ET program in the EU. The differences in the undertakings are substantial, but, on balance, ET is ET – the similarities outweigh the differences, and the differences can be recognized and rationalized. Success to the EU's pioneering venture now and in the future.

NOTES

1. Professor of Law, Sturm College of Law, University of Denver, Denver, Colorado, USA; email: rpring@law.du.edu. Special thanks are due to my Research Assistant Annie M. Tunheim, Sturm College of Law JD 2004, and to Diane Burkhardt, Faculty Liaison, Sturm College of Law Library, for their research and other contributions to this chapter.
2. See generally Nanda, V.P. and Pring, G. (2003), *International Environmental Law and Policy for the 21st Century*, New York: Transnational Publishers.
3. Pring, G. (2002), 'The United States Perspective,' in Cameron, P.D. and Zillman, D. (eds), *Kyoto: From Principles to Practice*, **185**, Deventer: Kluwer.
4. Ibid., pp. 216–18.
5. There is a wealth of literature on the USA's numerous 'sub-federal' GHG ET programs, for example, Pring, G. and Feger, R.A. (2006), 'Alternatives to Conventional Regulation in United States Environmental Law' in Barton, B., et al. (eds), *Regulating Energy and Natural Resources*, Oxford University Press; Pew Center on Global Climate Change, *Climate Change Activities in the United States: 2004 Update*, available at http://www.pewclimate.org/what_s_being_done/us_activities_2004.cfm; Ellerman, A.D., et al. (2003), *Emissions Trading in the US: Experiences, Lessons, and Considerations for Greenhouse Gases* (Pew Center on Global Climate Change), available at http://www.pewclimate.org/docUploads/emissions%5Ftrading%2Epdf; Bryner, G.C. (2004), 'Carbon Markets: Reducing Greenhouse Gas Emissions Through Emissions Trading', *Tulane Environmental Law Journal*, **17**(267); US Environmental Protection Agency, *The United States Experience with Economic Incentives for Protecting the Environment* (EPA-240-R-01-001 January 2001), available at http://yosemite.epa.gov/ee/

epa/eed.nsf/webpages/USExperienceWithEconomicIncentives.html; Anderson, R.C. and Lohof, A.Q. (1997), *The United States Experience with Economic Incentives in Environmental Pollution Control Policy*, Environmental Law Institute Report for USEPA, http://yosemite. epa.gov/ee/epa/eermfile.nsf/vwAN/EE-0216a-1.pdf/$File/EE-0216a-1.pdf; Riggs, J.A. (ed.) (2004), *A Climate Policy Framework: Balancing Policy and Politics*, Aspen Institute 2004; Aulisi, A., et al. (2005), 'Greenhouse Gas Emissions Trading in US States: Observations and Lessons from the OTC NO_x Budget Program', World Resources Institute White Paper, available at http://pubs.wri.org/pubs_description.cfm?PubID=3954.

6. The other chapters of this book set out an admirably detailed analysis of the new EU ET program. See also Peeters, M. (2003), 'Emissions Trading as a New Dimension to European Environmental Law: The Political Agreement of the European Council on Greenhouse Gas Allowance Trading', *European Environmental Law Review*, **12**(82); Posser, H. and Altenschmidt, S. (2005), 'European Union Emissions Trading Directives', *Journal of Energy & Natural Resources Law*, **23**(60).

7. 42 USC §§ 9601 *et seq.* A copy of the CAA is available at http://www.epa.gov/oar/oaq_caa. html. See Brownell, F.W. (2005), 'Clean Air Act', in *Environmental Law Handbook*, **227**(18), for a detailed explanation of the CAA; the *Environmental Law Handbook* is perhaps the best single-volume summary and analysis of all of the major US environmental laws.

8. See Pring and Feger, n. 5, as well as the excellent introductory chapters of that book.

9. Pring and Feger, text id., at nn. 43–45; US General Accounting Office (1982), 'A Market Approach to Air Pollution Control Could Reduce Compliance Costs Without Jeopardizing Clean Air Goals, GAO/PAD-82-15.

10. Possibly the best detailed reference on the history, goals, operations, and accomplishments of the Acid Rain Program is the US Environmental Protection Agency's website: 'Clean Air Markets …,' available at http://www.epa.gov/airmarkets/acidrain and http://www.epa.gov/air-markets/arp. This section of the chapter draws extensively on those USEPA materials, and, for brevity, without extensive additional footnoting to the website.

11. A copy of the Title IV Acid Rain Program statute provisions and its extensive regulations can be found at http://www.epa.gov/airmarkets/arp/regs/index.html. The 1990 amendments also established a parallel NO_x Program to deal with its contribution to the acid deposition problem. However, at the federal level, it does not utilize a 'cap-and-trade' system like that for SO_2, but instead relies on a somewhat flexible command-control system. See http://www.epa.gov/air-markets/arp/nox/index.html. See Aulisi, et al., n. 5.

12. Eg, Driesen, D. (1998), 'Free Lunch or Cheap Fix? The Emissions Trading Idea and the Climate Change Convention', *Boston College Environmental Affairs Law Review*, **26**(1).

13. Pring and Feger, n. 5 (see text at n. 46).

14. See http://www.epa.gov/airmarkets/arp/overview.html, and click on 'A Model Program'. (The General Accounting Office (renamed the Government Accountability Office in 2004), mentioned in the quote, is an independent inspection agency reporting to the US Congress.) For more objective confirmation that the program is successful, see the 'ex post evaluation' of it in Ellerman, A.D. (2004), 'The U.S. Cap-and-Trade Programme', in Organization for Economic Co-Operation and Development, *Tradeable Permits: Policy Evaluation, Design and Reform*, 71.

15. REP America, *REP America's Policy on Market-based Environmental Policies: A Conservative Approach to Solving Environmental Problems* (2004), available at http://www.rep.org/policy/market-based.html.

16. Ibid.

17. Stewart, R.S. (2000), 'Economic Incentives for Environmental Protection: Opportunities and Obstacles' in Revesz R.L. et al. (eds), *Environmental Law, The Economy, and Sustainable Development*, 171–75.

18. Examples include (1) installing new pollution control equipment, (2) switching fuels, (3) employing energy-efficiency measures, (4) utilizing renewable energy sources, etc.

19. Hypothetical from Pring and Feger, n. 5, at text following n. 51.

20. Bryner, n. 5, at 331–38 (citing US General Accounting Office, 'Overview and Issues on Emissions Allowance Trading Programs' (GAO/T-RCED-97-183, 1997), available at http://www. gao.gov/archive/1997/rc97183t.pdf). GAO estimated that the emission reductions would cost

as much as US$4.5 billion/year by 2002 if a traditional command-control regulatory approach was used, but only US$1.4 billion/year by 2002 if maximum trading was allowed, for a saving over 50%.

21. Eg, Driesen, n. 12.
22. GAO 'Overview,' n. 20, at 1 estimates that the SO_2 emissions could be reduced by approximately 2 mmt below the cap set in the CAA.
23. Bryner, n. 5, at pp. 330–31.
24. President Bush has stated that '[M]y government has set two priorities: we must clean our air, and we must address the issue of global climate change. [E]conomic growth is key to environmental progress, because it is growth that provides the resources for investment in clean technologies. This new approach will harness the power of markets, the creativity of entrepreneurs, and draw upon the best scientific research. And it will make possible a new partnership with the developing world to meet our common environmental and economic goals'. President George W. Bush, 'President Announces Clear Skies and Global Climate Change Initiatives', Address Before the National Oceanic and Atmospheric Administration (14 February 2002), see http://www.whitehouse.gov/news/releases/2002/02/20020214-5. html.
25. See authorities, n. 5.
26. CAA § 401(b), 42 USC § 7651(b).
27. CAA § 403 et seq., 42 USC § 7651b *et seq.* Again, see generally, USEPA, 'Clean Air Markets' website, available at http://www.epa.gov/airmarkets/arp/overview.html for the most detailed set of materials and data on the ET program. As mentioned above, n. 11, the Acid Rain Program also includes a system for reducing NO_x emissions from the same sources, but the federal NO_x program is not a market-based cap-and-trade system.
28. Technically, the statute says the ultimate cap is to be '8.90 million tons' (403(a)), and USEPA's regulatory selection of 8.95 presumably represents congressionally allowed bonus credits.
29. A 'unit' is defined in the CAA as a 'fossil-fuel combustion device', so each separate furnace at a given source is a unit.
30. The very complex formulas for computation of allowances are established in the Act itself, in Table A of CAA § 404 and in USEPA's extensive regulations. In general, allowing for many exceptions and variations, allowances in Phase I were set based on an emission rate of 2.5 pounds of SO_2 per one million British thermal units (Btu) of heat input multiplied by the unit's baseline mmBtu (the average fossil fuel consumed in the baseline period of 1985–87. Alternative or additional allowances were allocated for some units to ease them into the program.
31. Again, allowing for complexities, alternatives, and other variables (see n. 29), allowances in Phase II are based on an emission rate that is dropped to 1.2 pounds of SO_2/mmBtu multiplied by the unit's 1985–87 baseline fuel consumption.
32. Seven-year advance allowances sell for considerably less. Details of the auction transactions are available at http://www.epa.gov/airmarkets/auctions/2005/05summary.html; http://www. epa.gov/airmarkets/auctions/2004/04summary.html; and so on.
33. The year end is 1 March (or 29 February in a leap year).
34. Some 36% of all units must use CEM, but this covers 96% of the Program's total SO_2 emissions. Kruger, J. (2004), 'Compliance with Cap and Trade Programs: The US Experience,' PowerPoint presentation at INECE Conference, Worcester College, Oxford, UK, 5, http://www.inece.org/emissions/kruger.pdf.
35. The US defense to these criticisms is that it was focusing on the issue of acidification in the eastern US and Canada and therefore logically focused on the chief causes of that problem. Still, fossil fuel burning by the electricity generating sector accounts for only two-thirds of all SO_2 and one quarter of all NO_x nationwide, and the ET program only addresses a portion of that (albeit a substantial portion).
36. Posser and Altenschmidt, n. 6, pp. 68, 70.
37. Ibid., p. 63.
38. Ibid., p. 65.
39. Ibid., p. 67.
40. See details of CEM at http://www.epa.gov/airmarkets/monitoring/factsheet.html.

41. For more details on the Opt-in Program, see http://www.epa.gov/airmarkets/arp/optin/index.html.
42. Bryner, G.C. (1999), *New Tools for Improving Government Regulation: An Assessment of Emissions Trading and Other Market-Based Regulatory Tools*, http://www.colorado.edu/law/centers/nrlc/publications/NewToolsForImproving.pdf, pp. 11–13.
43. Further see, http://www.epa.gov/airmarkets/arp/reconcil/index.html.
44. Further see, http://www.epa.gov/airmarkets/reporting/etsfactsheet.html.
45. See, eg Ellerman, et al., n. 5, p. iv.
46. Kruger, n. 34, p. 9.
47. Ibid., p. 10. In the program's first 9 years, only 15 excess emissions penalties were imposed (but they were substantial, ranging from US$2 682–1 580 000) and only eight enforcement cases were filed (average fine US$50,000).
48. Bryner, n. 5, pp. 293–94.
49. EU Emissions Trading, Vertis Environmental Finance, available at http://www.vertisfinance.com/page.php?page=38&l=1.
50. Authorities, n. 5.

11. Climate change taxes, emissions trading, and international trade law

Geert van Calster[1]

1. INTRODUCTION: OF TREMS AND MEAS

At the moment of conception of the United Nations Framework Convention on Climate Change (UNFCC), the integration of a future emissions trading regime with international trade law was, arguably, not the main preoccupation of the negotiators. This was all the more so given that the outlook for the agreement at that time was not one which would have focused heavily on trade-sensitive instruments if only because, as a framework convention, the UNFCC does not provide for much in terms of policy detail. Rather, trade concerns started arising once the Kyoto Protocol to the UNFCC was signed. Even that Protocol, readers will remember, was not initially designed to harbour many 'new' approaches such as emissions trading or flexible instruments – these were championed by the United States (US) that eventually did not adhere to the Protocol, and were only embraced by the European Union (EU) after some time.

Once, of course, one starts having major regional groupings such as the EU, implement multilateral environmental agreements (MEAs) by way of a trade instrument, the interest of the World Trade Organization (WTO) is assured.

Sadly, perhaps, little progress has been made in formal discussions in the WTO, ever since the issue of Trade-Related Environmental Measures (TREMs) was brought to its attention. On the other hand, there is reason for cautious optimism. First, that little progress has been made in *formal* discussions is not necessarily a reason for concern. What is more important are the practical ways in which the WTO takes account of environmental concerns in its trade practice. Furthermore, while the WTO still suffers from the phantom of the General Agreement on Tariffs and Trade (GATT) in the ways in which it is perceived as dealing with environmental concerns, WTO practice actually reveals no immediate bias against TREMs.[2] Finally, all TREMS-related dispute settlement at the WTO and its predecessor, the GATT, concerned unilateral measures, not MEAs. This author, for one, has argued in the past and remains convinced that, on the environment issue, too much time and political capital at the WTO has gone into talking about the integration of MEAs into the multilateral trading system.

While academically interesting, trade disputes in practice arise not out of MEAs but out of unilateral TREMs.[3]

To a degree, that is exactly what we are seeing with respect to emissions trading. Trade concerns have been raised not so much at the room which the Kyoto Protocol leaves for flexible and trading instruments, but rather at the implementation of this room for manoeuvre by individual parties to the Protocol.

This contribution offers a crash course into the difficulties which emission trading schemes and their satellites may encounter with the rules of international trade law. However, the specific focus of the analysis is on climate change levies, or taxes.

2. CALMING OF NERVES

It is important, first, not to repeat the chilling effect which occurred in particular during the 1990s. This was a period in which the GATT and later on the WTO (but in particular the GATT) were reputed to block a wide range of regulatory measures on the basis that these allegedly violated international trade law. The very potential for GATT-incompatibility reportedly led to States' representatives in multilateral environmental negotiations to soften the range of suggested measures, so as not to risk GATT condemnation.

Thankfully, this chilling effect – evidence of which is at any rate sketchy and anecdotal rather than firm – would now seem to have subsided. A variety of studies have shown that the GATT and the WTO are not in the business of sacrificing regulatory interests such as environmental protection, and health and safety, on the altar of a free trade Armageddon. Furthermore, where tension has been identified (in particular because of faulty design rather than substance of the trade measures of MEAs), these have often pre-emptively been remedied through the more or less permanent contacts which the WTO Secretariat has established with MEA Secretariats. Finally, although we shall not go into detail on this issue in this chapter, there is a forceful argument that under international law of treaties, WTO Members who are also a Member to the Kyoto Protocol (some even argue that membership of the UNFCC suffices), forego their rights of complaint under the WTO Agreement.[4]

There are remnants of this GATT/WTO nervosa in the *modus operandi* of a number of the EU's trading partners, in particular the United States (US). Indeed, it has become more or less standard in recent years for these to waive the WTO flag at any more or less environmentally proactive plans of the EU, including the Directives on Waste Electric and Electronic Equipment (WEEE), and on the Restriction of Hazardous Substances (RoHS) – which in the end did not lead to formal complaints in Geneva. On the 'flip-side', a US-led complaint against the pre-existing regulatory regime of the EU with respect to GMOs is, of course, currently making its way through the WTO Dispute Settlement system.

3. FLASHPOINTS

Notwithstanding the need not to exaggerate the threat, there are obviously some flashpoints in any climate change regulatory regime, which merit attention under international trade law rules.

3.1 Which Agreement?

A first issue which will need to be settled, should it ever come to WTO dispute settlement, is which of many potentially relevant WTO Agreements, may apply.[5] As readers will be aware, since the Uruguay Round, the GATT, which covers goods, has become something of a *lex generalis* with many other agreements acquiring the status of *lex specialis*. Moreover, a range of other agreements apply to services and investment measures. While these agreements share a skeleton structure and even common legal principles such as transparency, there are important differences in the depth of trade liberalization contained in each of them and, consequentially, in the room for manoeuvre left to WTO Members in protecting regulatory interest.

In particular with respect to the Kyoto Protocol's flexibility mechanisms:[6] are allocated allowance units, emission reduction units and credit emission reductions, 'goods' or 'services'? It has been argued that since all these instruments in the end boil down to carbon, they are commodities. Others emphasize their financial nature, as well as their investment character, thus making the General Agreement on Trade in Services – GATS, or even TRIMS (the Agreement on Trade-Related Investment Measures) more likely to be applicable.[7]

GATS, much like TRIMS, is very much unfinished business. It does harbour some core international trade law principles, such as Most Favoured Nation (MFN), as well as, as noted, transparency. It also includes an Article XX-type general exceptions regime. This, interestingly, includes a potential exception for the protection of human and animal/plant health, but not for exhaustible natural resources.

However, GATS is not so much characterized by its general principles, but rather more by its sketchy nature, including its schedules of commitments. A Member is permitted to maintain a measure inconsistent with the general MFN requirement, if it has established an exception for this inconsistency.

If any aspects of the EU's emission trading regime were to be judged *vis-à-vis* the GATS rather than GATT, this would be an interesting prospect, precisely because the GATS Agreement is so sketchy and untested. In trade and environ-ment disputes, the WTO is used to pitching well-established GATT principles, refined through case-law, and hence more or less clear, against relatively unclear provisions of international environmental law. If the GATS were pitched against the EU's, Kyoto-related climate change regime, the relative vagueness would occur on both sides.

3.2   MEA or Unilateral TREM?

The element of vagueness brings one to one of the threats of the Kyoto Protocol as a whole. World Trade Organization negotiators, within the organization's Committee on Trade and Environment, have for some time now attempted for-mally to accommodate at least a number of well-known MEAs within the WTO framework. One of the reasons they have failed so far – other than through the Members' lack of political courage, perhaps – is that MEAs are typically frame-work conventions, leaving further clarification to a combination of decisions of the meetings of the Parties, and domestic implementation measures.

The EU can certainly be seen as a pioneer in the implementation of the regime (including a start well before the Kyoto Protocol actually entered into force). It is as yet unclear to what extent its example will be followed by other Parties to the Protocol. For instance, partially because of subsidiarity considerations, the EU has let its Member States devise crucial details of the emission trading scheme. The more divergent the implementation by Parties to the Protocol, the more cumbersome a future meeting with WTO rules may become.

3.3   The Allocation of Emission Permits

In the process of allocating *tradable* emission permits, there are some trade pitfalls which one needs to keep in mind and which we shall look into below. In fact, trade tension may originate with the very allocation *per se* of emission rights – even if not tradable. First, by way of a general restriction on business activities, issuing permits for the emission of selected pollutants, may act as a measure prohibited under GATT's Article III:4:

> The products of the territory of any contracting party imported into the territory of any other contracting party shall be accorded treatment no less favourable than that accorded to like products of national origin in respect of all laws, regulations and re-quirements affecting their internal sale, offering for sale, purchase, transportation, distribution or use. The provisions of this paragraph shall not prevent the application of differential internal transportation charges which are based exclusively on the economic operation of the means of transport and not on the nationality of the product.

Just as any other feature of a State's environmental regime, so too could allocation be employed to grant an unfair advantage to one's domestic industry. In the
case of the EU, scrutiny therefore was first directed towards the selected industries which in the first phase of the tradable permits regime at least, are subjected
to limited emission rights. It became swiftly apparent that the EU's choice of
industries was not geared in any way towards granting an advantage to its domestic pet industries. That would have been the case, for instance were the EU
to have limited the application of the restrictions to industries mostly not held
by national undertakings, or were distribution or sales allowances to be used to
improve the EC's or a Member State's security of supply (by promoting the sale
of energy products from EC or national sources).[8] Moreover, a further concern
could be raised where emission rights are overwhelmingly granted to incum-
bents, and not enough capacity (pollution capacity if one likes) allocated to new
entrants,[9] especially if these new entrants are more likely to come from abroad.
That, too, would not seem likely in the EU. However, it remains to be seen how
the various national allocation plans (NAPs) with respect to newcomers will
work out in future.

3.4 The Allocation of Tradable Emission Rights

The application of EU law with respect to State aid, falls outside the scope of
this chapter. In general, the guidelines for State aid for environmental protection,
have by now put into place a fairly well-established and environmentally generous regime.[10] The European Commission (the Commission) is currently
reviewing it, so as to make it better geared towards environmental innovation – a
move which should make the regime better equipped to deal with the challenges
and opportunities posed by small and medium sized enterprises.

On the WTO side, relevant guidelines under the Agreement on Subsidies and
Countervailing Measures (SCM) are a lot less generous, but also a lot less
amendable, thus making them less well geared towards 'new' environmental
concerns such as climate change.[11]

4. THE CARBON LEVY

A prime instrument in combating climate change, is the introduction of a 'carbon levy', in one of various forms.[12] In its widely applied format, a carbon levy
simply works as a tax on the consumption of energy – whereby States in varying
measure link the height of the tax to the carbon content of the energy consumed.
Indeed most do not link the taxation of energy products at all to carbon content.
In its most radical form, not yet applied but mooted by a variety of politicians,
the carbon tax would levy a tax simply on the origin of a product from a State,

not Party to the Kyoto Protocol, with the trigger being the absence of action, in that State, to combat climate change.

Imposing a tax of any nature, typically involves a phenomenon named 'Border Tax Adjustment' (BTA). Border Tax Adjustments essentially allay competitiveness concerns of the State imposing a tax. On the import side, BTA refers to imposing internal taxes on imports to the same extent as they would be charged on internally produced goods at the same stage of production.[13] On the export side, BTA refers to rebating taxes imposed on domestically produced goods, so as to put them on an equal footing with foreign competitors. In other words, stated simply, BTA refers to the equalizing tax on imports on the one hand, and the equalizing rebate for exported goods on the other.[14] The OECD has defined BTA as:

> any fiscal measures which put into effect, in whole or in part, the destination principle (i.e. which enable exported products to be relieved of some or all of the tax charged in the exporting country in respect of similar domestic products sold to consumers on the home market and which enable imported products sold to consumers to be charged with some or all of the tax charged in the importing country in respect of similar domestic products).[15]

This is a 1960s quote, hence it is clear that BTA is not a new phenomenon.

As noted, the more conventional form of addressing the climate change impact of imported products is by imposing relevant energy taxation upon them. This is not an ideal instrument: not all States' energy taxes are necessarily geared towards addressing the products' climate change impact. Hence, in most cases, energy taxation does not directly (through taxing relevant greenhouse gases) but indirectly (simply through taxing energy use) tackle climate change.

Whether BTA in the case of energy taxation is compatible with the rules of the WTO, involves a combination of the GATT and SCM Agreement. There are two anchor points for BTA in the GATT and WTO Agreement: GATT Article II:2.(a) *juncto* GATT Article III (National treatment on internal taxation and regulation), and GATT Article XVI (Subsidies) *juncto* the WTO Agreement on Subsidies and Countervailing Measures (SCM).

GATT Article II aims to ensure that Members abide by their schedules of concessions which they agree to in the multilateral rounds. Article II:2.(a) provides that Members may impose at any time on the importation of any product, a charge equivalent to an internal tax imposed consistently with the provisions of paragraph 2 of Article III in respect of the like domestic product or in respect of an article from which the imported product has been manufactured or produced in whole or in part. Article II applies to bound products (ie those subject to tariff concessions), while Article III applies both to bound and unbound products.

GATT Article XVI lays down the rules with respect to Members' subsidies. Importantly, the Note *Ad* Article XVI states that '(t)he exemption of an exported product from duties or taxes borne by the like product when destined for domestic consumption, or the remission of such duties or taxes in amounts not in excess of those which have accrued, shall not be deemed to be a subsidy'. It is this provision which allows for BTA on the export side. The concept of 'like products' of course plays a predominant role, as does the mirror provision on the import side (Articles II–III). In GATT Articles II–III, one compares products of domestic origin, with those of foreign origin; under Article XVI, exported domestic products are compared with those that stay within the domestic market.

The WTO's SCM Agreement has included an important provision with respect to the inputs in the production processes. Annex II to the Agreement, and footnote 61 to it, pave the way for BTA with respect to energy products at least. They do not give *carte blanche* however for BTA with respect to all PPMs (see below).

4.1 BTA, Energy Taxes, Climate Change Levies and their (il)legality under GATT 1947

The BTA debate in the GATT and now in the WTO, continues to centre around a 1970 Report of a Working Party on BTA. The Working Party examined the tax adjustment practices of GATT contracting parties, their trade effects and relevant provisions of the GATT, in particular the changeover in certain countries from cascade tax systems to the value-added tax.[16] The Report concluded that there was a convergence of views amongst the delegates that taxes directly levied on products (indirect taxes) were eligible for tax adjustment. It cited as examples of such taxes, specific excise duties, sales taxes and cascade taxes, and the tax on value added. The Group also noted that there was agreement among its members that certain taxes that were not directly levied on products (direct taxes) were not eligible for BTA. These include social security charges, and payroll taxes.[17] The Group's conclusions mean that GATT's BTA provisions incorporate the destination principle with respect to indirect taxes, and the origin principle with respect to direct taxes.[18]

Other than the elements cited above, where the Group noted consent, consent could not be found on a number of remaining issues. Importantly, the report identified two categories of taxes where it was generally felt that there was no clear view on eligibility for adjustment, but where the scarcity of complaints was not such so as to justify examination. These two categories were:

(a)   'Taxes occultes'. These are defined by the OECD as 'consumption taxes on capital equipment, auxiliary materials and services used in the transportation and production of other taxable goods'.[19] The Group identified taxes on advertising, energy, machinery, and transport, as among the more important taxes which may be included in this group. It observed that adjustment was not normally made for these taxes, except in countries with a cascade tax;

(b)   A number of other taxes, such as property taxes, stamp duties and registration duties. These are normally not made subject to adjustment.

It should be underlined that the Group left the discussion on these two categories open; its conclusions merely refer to too wide a divergence in the practice of GATT contracting parties, for it to come to a unanimous conclusion.

Since Article II:2(a) refers to 'like products', the Group of course examined the meaning of this concept. Broadly speaking, the Group's findings resulted in the acceptance of product-based distinctions only, as an acceptable criterion for distinguishing between products. Distinctions made purely on the basis of Production Processes and Methods (PPMs) would in this respect be unacceptable. Importantly, the Group concluded with respect to this element, that there was quite some uncertainty.[20]

Whatever its merits, the distinction between direct and indirect taxes has been sustained ever since the 1970 Report, and so has the corresponding application of the destination principle in respect of indirect taxes, and the application of the origin principle for direct taxes. In our analysis, we have chosen the methodology that was used by New Zealand in a presentation before the Committee on Trade and Environment (CTE).[21] This approach assesses the gains, from an environmental point of view, and the GATT compatibility, of environmental taxes and charges in four different situations, depending on the environmental 'externality' involved. Externalities are either positive or negative. A negative externality results when the activity of one person or a business imposes a cost on someone else.[22] Examples of environmental externalities (which, realistically, are mostly assumed to be negative) are abundant:

> In economic theory, an environmental tax levied on a polluter in a way that equals the approximate damage caused by the pollution will internalise the long-term costs of climate change into the price system and thus balance the private costs of carbon dioxide emissions with the environmental and social costs of global warming.[23]

Climate change levies typically involve global/transboundary production externalities.

4.1.1   Global/transboundary production externalities

Tackling *production* externalities through fiscal techniques, including the use of BTA, is problematic from a GATT point of view.

(a)   On the import side   Environmental considerations might justify the use of BTA on imports. Any reduction in the use of the PPMs which result in the externalities involved, would serve to protect the State imposing the tax on imported products. Non-taxation of the imported products produced using the targeted PPM, might encourage domestic producers to move their place of production.

An important complication is that the taxes at issue use PPMs as a tax base. The conclusions of the 1970 working group on BTA clearly indicate that PPMs must not be used as a base for distinguishing between products. Only where the tax is levied on the use of certain inputs, which are physically incorporated in the imported product, would this be possible. Most importantly, WTO Panels and the Appellate Body stick to the firm line that distinctions between products must not be based on anything else but physical end-characteristics of the products involved: non-product based distinctions between products cannot be accepted, even if the distinction does not seek to protect domestic production.

(b)   On the export side   With respect to exports, BTA would be both environmentally unsound and GATT incompatible. For the proportion of output which is exported, the price effect will vanish and so will the output-reducing effect. The note *Ad* GATT Article XVI, and point (e) of Annex I of the WTO Subsidies Code exclude BTA eligibility.

In the long term, and with respect to particular industries, not rebating the tax on exports could have unsolicited effects. In sectors where environmental costs count for a high proportion of production costs, the taxation of domestic production and the absence of rebate for exports, would encourage these companies to move to those countries which do not employ similar taxes. In other words, the sales in foreign markets of the country which does not rebate the tax on exports would be replaced by expanded foreign production (using the globally harmful PPM). Not rebating the tax would then only serve to relocate production. The damage to the global and/or transboundary environment would not have been remedied. Arguably, precisely because the industry concerned would relocate to States with less stringent environmental standards, the damage to the environment might even be higher. However, were these considerations to lead to tax rebate, the State concerned may violate its WTO obligations.

4.2   The Same Ambivalence under the WTO Agreement?

4.2.1   The WTO Agreement and energy inputs in the production process
The WTO Agreement has resulted in a more positive stance with respect to BTA for energy inputs in the production process, at least formally on the export side.

Energy has a dual nature, as it is both a product and a fundamental part of most production processes.[24] Therefore, it does not easily fit into the direct–indirect distinction. It is this uncertainty which has led to the current provision. Annex I to the SCM Agreement provides for an 'illustrative list of export subsidies'. Paragraph a(h) allows for Members to remit taxes on exports in respect of prior stage cumulative indirect taxes on goods or services, if they are 'levied on inputs that are consumed in the production of the exported product (making normal allowance for waste)'.

This provision has to be read together with Annex II to the SCM Agreement, which is entitled 'Guidelines on consumption of inputs in the production process'. It provides that 'indirect tax rebate schemes can allow for exemption, remission or deferral of prior-stage cumulative indirect taxes levied on inputs that are consumed in the production of the exported product (making normal allowance for waste)'.[25] Footnote 61 adds the clarification that 'imports consumed in the production process are inputs physically incorporated, energy, fuels and oil used in the production process and catalysts which are consumed in the course of their use to obtain the exported product'.

Footnote 59 of the SCM Agreement defines 'Indirect taxes' as 'sales, excise, turnover, value added, franchise, stamp, transfer, inventory and equipment taxes, border taxes and all taxes other than direct taxes and import charges' (the latter two are also specifically defined). 'Prior stage indirect taxes' are those indirect taxes which are levied on goods or serves used directly or indirectly in making the product. 'Cumulative indirect taxes' are multi-staged taxes levied where there is no mechanism for subsequent crediting of the tax if the goods or services subject to tax at one stage of production are used in a succeeding stage of production.

Whereas footnote 61 now includes inputs that 'are consumed in the course of their use to obtain the exported product', the Tokyo Round Subsidies Code used the term 'goods that are physically incorporated in the exported product'.

Extent of the 'environmental window' in the SCM Agreement   The Annex clearly allows for BTA on the export side for energy consumption during the manufacturing of the exported product. Caution is called for. Even though this provision is a breakthrough, it most definitely does not represent a general opening for BTA with respect to environmental taxation, such as in the context of climate change.

First, the Annex does not clarify how the rebate is to be calculated. This should not hamper the implementation of such BTA, since it is incumbent upon the Member claiming that another Member is unjustifiably granting subsidies, to prove that subsidies have been granted (see Article 4 of the SCM). Secondly, the possibility of BTA for environmental purposes is limited to taxation of energy inputs, thus leaving aside a vast proportion of taxation of certain PPMs. This includes PPMs which have a negative impact on climate change.

Most importantly, the environmental window of the SCM is, at least formally, limited to rebate of taxes at the moment of exportation. However, such rebate is often not in the interest of the environmental goals at stake: and certainly not for the issue of climate change. Could the acceptance of such BTA be extended to the import side?

Economic considerations obviously would argue for it. If a State is not in a position to impose taxes on energy inputs on imported products, the competitive position of its domestic output might suffer. This might result in an impaired position for the domestic production concerned, possibly leading to the disappearance of it. Strict environmentalism might in fact applaud such events, as the disappearance of national production would protect the local environment (through the disappearance of energy consumption). Realistically, however, no government would be convinced to allow such development.[26] It is more likely that States, in the absence of room for BTA on the import side, will be tempted to skip altogether the introduction of the tax concerned. In the long term, therefore, only BTA on the import side as well, would seem fit to guarantee the maintenance of the tax at issue.

Apart from considering the ecological soundness of BTA on the import side, one is still left with the question as to whether the SCM Agreement allows one to conclude that States can impose similar taxes on products entering their territory. Does the mere similarity of environmental considerations on the export and import side justify the extension of the regime to imports as well?[27] The Uruguay Round's (UR) set-up would seem to argue against this. The principles which it lays down are the result of protracted negotiations, and, for the sake of certainty, ought to be interpreted literally. As the SCM Annex clearly mentions export arrangements only, one has to conclude *a contrario* that it is not meant to cover BTA on the import side. One can only look at the SCM's intervention as a limited concession to energy-related inputs, and not as a radical overhaul of the GATT's approach of PPM-related regulatory intervention.[28]

Just as one should not read into the SCM Agreement what it does not say, one must not deny what it does expressly provide for.[29] The Annexes and footnote 61 allow for BTA on the export side for energy inputs which are consumed in the production process.[30] Even though, as noted above, BTA on the export side may not be effectively introduced in the absence of a similar provision on the import side, it does not correspond to the clear wording of the Agreement to deny such rebate with respect to carbon taxes specifically. The United States, however, disagrees – claiming that there was agreement at the time of adoption of the WTO Agreement, that carbon taxes specifically would not be covered by the provision.[31] The discussion evolves around the concept of 'cumulative indirect tax'. A carbon and energy tax is usually set up as a single-stage tax, and not as a multi-stage tax.[32]

5. CONCLUSION

The concession in the 1996 GATT allowing remission of energy taxes has been called both appropriate and unsurprising.[33]

Appropriate, as the main reason of course for introducing it, was the growing concern of governments that their (plans for) energy taxation, including provision for BTA, would be found GATT incompatible. Furthermore, 'unsurprising', as the context of GATT 1947 itself may be, in fact, hints to a more PPM-friendly attitude than generally thought. Analysis of the preparatory works of the GATT by Demaret and Stewardson shows that it was originally intended that taxes on all inputs to an imported product, whether raw materials or processes, would be eligible for adjustment.[34] The language of Article III:2 could in principle be read to allow for adjustment, with respect to the import of taxes on any input to any stage of the production or distribution process.[35]

The rather clear view of the delegates to the 1970 Working Group – that only *indirect* taxes are refundable – leaves no doubt that at that moment in time, this was the stance of the international trade community. The clue to the legislative history however is important, because it would seem to indicate that the GATT's position is not absolutely reflected in the very text of the agreement, but rather in the Parties' interpretation of it. One could also argue that the 1970 Working Party Report has in some ways acted as a self-fulfilling prophecy. Parties have looked upon the Report as somewhat of a bible of GATT's BTA rules, and of the GATT's 'like products' analysis, leading to a dogma that non-incorporated inputs are not eligible for adjustment.

To this author, it would seem that the aforementioned provision of the UR SCM has in fact deteriorated the prospect of allowing for BTA with respect to both import and export, for a large range of environmental taxes. As noted, the non-allowance of BTA for such taxes prior to the UR was, in fact, more the result of States' practices and of their interpretation of the relevant GATT requirements, than the irreversible conclusion of GATT's language. However, the express provision in the UR SCM with respect to BTA on the export side only, and for certain environmental taxes only, in this author's view excludes room for overall BTA of all environmental taxes, including those on inputs not physically incorporated in the product, without express amendment of the Agreement.

In any event, it is clear that energy taxes of any kind, and their BTA, involve design, thought, and some degree of technicality. This rules out so-called 'Kyoto' or 'climate change' levies, which simply levy a charge on products

originating in States which do not adhere to the Kyoto Protocol.[36] The mantra
of such proposals goes along the lines that if a State does not adhere to the Protocol and 'hence' does not contribute to the fight against climate change, it
subsidizes its industry. This, in turn, would justify 'countervailing measures'
by those States which do adhere to the Protocol.

While appealing in its simplicity, this measure in fact attracts so many 'ifs'
and 'buts' as to be unworkable. Obviously, the very adherence to one or other
international Treaty cannot suffice to impose sanctions of any kind. That is of
such common legal sense as not to warrant much analysis – it is, moreover, an
issue which has been studied extensively in the general literature on the issue
of MEAs. What one could consider is whether public international law could
grant a title for action against those States which indeed take no action whatso-
ever to mitigate climate change, an issue of international concern which affects
us all and hence arguably gives States a territorial link to undertake action.
However, even if international law were to give such claim of jurisdiction, difficult considerations remain. For instance, one would have to put into place a
regime which would not impose the climate levy on States which, even without
adhering to the Protocol, put into place measures which equally effectively as
the Kyoto Protocol, tackle climate change. This equation involves two elements.
First, the very effectiveness of the Kyoto Protocol would have to be assessed;
that could be a tricky exercise in itself. Subsequently, the State at the receiving
end of the tax, is certain to put forward a number of measures which it will claim
to be just as effective as the Protocol – energy savings measures, for instance,
such as under the United States 'no regrets' doctrine. Furthermore, that such a
tax would be directed at the United States alone is inconceivable.[37] Were it to
be directed at a number of emerging economies, however, more question marks
and legal arguments will be raised: emerging economies, for instance, may even
cite the principle of common but differentiated responsibilities, so as not to have
to be subjected to a similar tax as the United States.[38]

Generally, the token of arguments listed above in themselves are of such a
nature, as to exclude any serious suggestion that a simple 'Kyoto' tax would
ever be introduced.

NOTES

1. Co-director, Institute of Energy and Environmental law, K.U. Leuven; Member of the Brussels
   Bar, DLA Piper. Email: gavc@law.kuleuven.be; www.law.kuleuven.be/imer/master.
2. See Van Calster, G. (2000), *International and EC Trade Law – The Environmental Challenge*,
   London: Cameron May.
3. Note, in this respect, that the Kyoto Protocol, unlike the usual suspects when it comes to
   MEA/WTO tension (the Basel Convention on movements of hazardous waste, the Montreal
   Protocol on ozone-depleting substances, the CITES Convention on the trade in plant and ani-
   mal species), does not in itself contain any trade sanctions vis-à-vis States which are not a

Party to the Protocol. See in this respect also Bradnee Chambers, W., 'International trade law and the Kyoto Protocol: Potentian incompatibilities', in *Global Climate Governance: Inter-linkages Between the Kyoto Protocol and other Multilateral Regimes*, undated, United Nations University, available at www.geic.or.jp/climgov, p. 52ff; Brewer, T. (2002), 'The Kyoto Protocol and the WTO: Institutional evolution and adaptation', *CEPS Policy Brief*, No. 28, December 2002, available at www.ceps.be; and Sampson, G., 'WTO rules and climate change: The need for policy coherence', in *Global Climate Governance: Inter-linkages Between the Kyoto Protocol and other Multilateral Regimes*, undated, United Nations University, www.geic.or.jp/climgov, p. 29.

4. Abundant literature regarding the issue of foregoing rights under an earlier Treaty by adhering to a later; in particular with respect to the WTO/Kyoto issue: Petsonk, A. (1999), 'The Kyoto Protocol and the WTO: Integrating greenhouse gas emissions allowance trading into the global marketplace', *Duke Environmental Law and Policy Forum*, **194**(185); Sampson, G., n. 3 above, p. 31.

5. See also eg Green, A. (2005), 'Climate change, regulatory policy and the WTO', *Journal of International Economic Law*, **143**, p. 151ff.

6. Other climate change measures may trigger yet different WTO Agreements. Most authors justifiably point to the Agreement on Technical Barriers to Trade (TBT) for the issue of product standards, for instance with respect to energy efficiency; and to the Agreement on Government Procurement (AGP) for the procurement of, say, energy-efficient machinery by Members' authorities. See eg *Climate and Trade Rules – Harmony or Conflict?*, Swedish National Board of Trade, 2004, available via www.kommers.se. On procurement, see Van Calster, G. (2002), 'Green procurement and the WTO – A different shade of grey', *Review of European Community and International Environmental Law – RECIEL*, 298–305.

7. It has been suggested that coverage by the GATS rather than the GATT Agreement would be beneficial for the regime: under GATT, limiting the trade in the goods (permits) concerned to Annex I countries, would be a restriction to trade which would be difficult to justify under GATT rules. By contrast, under GATS' coverage of financial services, it would – by and large – suffice for the market to be open to foreign service providers (banks and the like). While opening the trade in emission rights itself to non-Annex I countries would be a direct threat to the very regime, allowing foreign financial service providers into the market is far less so. See Cosbey, A. (1999), 'The Kyoto Protocol and the WTO', seminar note (reflecting the views of seminar participants), see www.riia.org.

8. Example quoted by Werksman, J. (1999), 'WTO issues raised by the design of an EC emissions trading system', (contracted work for the European Commission), available at http://www.field.org.uk/, p. 14.

9. A particular problem in the case of 'grandfathering' ie allowing emmision rights on the basis of historic emissions.

10. See eg Van Calster, G. (2000), 'Greening the EC's State Aid and Tax regimes', *European Competition Law Review*, 294–314.

11. Detailed analysis i.a. in Assuncao, L. and Zhang, Z. (2004), 'Domestic climate policies and the WTO', *The World Economy*, **359**, p. 361ff; Petsonk, A., n. 4 above, 204ff.

12. For some practical insights, see Ismer, R. and Neuhoff, K. (2004), 'Border tax adjustments: a feasible way to address nonparticipation in emission trading', *Cambridge Working Papers in Economics*, http://ideas.repec.org/p/cam/camdae/0409.html.

13. McGovern, E. (1995), *International Trade Regulation*, Exeter: Globefield Press, (looseleaf), 11.32–5.

14. Cameron, J. and Chaytor, B. (1995), 'Taxes for Environmental Purposes: The scope for border tax adjustment under WTO rules', *WWF International Discussion Paper*, p. 5.

15. Referred to in Report of the Working Party of 2 December 1970 on Border Tax Adjustments, GATT doc. L/3464, reproduced in General Agreement on Tariffs and Trade, Basic Instruments and Selected Documents, Geneva, GATT Secretariat, 1972, (pp. 97–108) at 4.

16. World Trade Organization, Guide to GATT Law and Practice, Geneva, 1995, 2 vols, vol. 1, p. 145. Also available online at http://www.worldtradelaw.net/reports/gattpanels/bordertax.pdf.

17. Report of the Working Party of 2 December 1970, n. 15 above, at p. 14. The Group noted that

for payroll taxes and for social security charges on employers, some countries did provide for adjustment. The Group's subdivision of direct and indirect taxes was broadly confirmed by the 1979 Subsidies Code, which some GATT contracting parties adhered to. See Demaret, P. and Stewanderson, R. (1994), 'Border Tax Adjustments under GATT and EC Law and General Implications for Environmental Taxes', *Journal of World Trade*, **4**, pp. 5–66, p. 13.

18. Demaret and Stewanderson, ibid, p. 8; Fauchald, O.K. (1998), *Environmental Taxes and Trade Discrimination*, London: Kluwer Law International, p. 165.
19. Report of the Working Party of 2 December 1970, n. 15 above, p. 15.
20. Ibid, p. 18, in fine.
21. TE 010, 11 October 1994, at 4 et seq. Available through the WTO online documentation center.
22. Harrison, J.L. (1995), *Law and Economics*, St. Paul (Minn.): West Publishing Company, p. 42.
23. Biermann, F. and Brohm, R. (2003), 'Implementing the Kyoto Protocol without the United States: The satrategic role of energy adjustments at the border', Global Governance Paper No. 5, see www.glogov.org.
24. See Schlagenhoff, M. (1995), 'Trade Measures Based on Environmental Processes and Production Methods', *Journal of World Trade*, **6**, p. 143.
25. The room offered for the 'normal allowance for waste' is designed to cover BTA practice to assign a surplus rebate for the taxes on those inputs which are not reflected in the final product, as they were 'wasted' in the production process. The Annex includes provisions to ensure that this allowance does not exceed an acceptable level, thus avoiding that the waste proviso turns into a straightforward export subsidy.
26. Along the same lines: Swedish National Board of Trade, n. 6 above, p. 64.
27. Demaret and Stewardson, n. 17 above, p. 31.
28. This limited reading is supported by the origins of the provisions, which indicate that they were subject to a 'Gentlemen's agreement', excluding its use with respect to carbon taxes in particular: see Demaret and Stewardson, n. 17 above, p. 30. As this agreement is not recorded anywhere, one can however not rely on it for the purpose of this analysis.
29. Indeed Biermann, F. and Brohm, R., n. 23 above, p. 23, observe that the US are alone in their objection against the prima facie meaning of the wording.
30. Cameron, J. and Chaytor, B., n. 14 above, p. 11.
31. See n. 28.
32. See Schlagenhoff, M., n. 24 above.
33. Westin, R.A. (1997), *Environmental Tax Initiatives and Multilateral Trade Agreements: Dangerous Collisions*, Den Haag: Kluwer Law International, p. 77.
34. Demaret and Stewardson, n. 17 above, pp. 18–19.
35. Ibid.
36. As proposed for instance by a group of European Parliamentarians, in a piece published in *The European Voice* on 17 February 2005 – the piece may also be consulted via http://www.corbey.nl/nieuwsbericht.php?id=229.
37. Suggested however by the group of European Parliamentarians, n. 35 above.
38. Along those lines, see also Sampson, G., n. 3 above, p. 34.

PART III

Energy and climate change measures

PART II.

Energy and climate change measures

12. EU energy policy and legislation under pressure since the UNFCCC and the Kyoto Protocol?

Véronique Bruggeman and Bram Delvaux[1]

1. INTRODUCTION

To meet our socio-economic development goals, people and countries must have access to reliable and affordable energy and energy services. However, the production, distribution and consumption of a substantial part of present energy resources (namely fossil fuels) contribute to a significant increase in greenhouse gases. As a result, a great deal of new legislation establishing a link between environmental protection and energy production, distribution and use has been created.

Alongside the international efforts to limit emissions of greenhouse gases and their possible adverse effects on the global climate system with the United Nations Framework Convention on Climate Change (UNFCC) in 1992 and the Kyoto Protocol in 1997,[2] the European Union (EU) provides this goal as well by identifying climate change as one of its key priorities. The very beginnings of the European Community climate change policy saw the light in 1989, the year in which the European Commission (the Commission) submitted to the Council a communication on climate change, which led to a non-binding Council resolution to stabilize CO_2 emissions by the year 2000 at their 1990 level.[3] This initiative has been followed by various ambitious programmes and a bunch of legislation on the reduction of greenhouse gases, of which a significant number include energy-related measures. The latter will be addressed and discussed critically in this chapter. First, however, the international and European climate change background will be sketched shortly as a memorandum.

2. THE UNFCCC AND THE KYOTO PROTOCOL

The main objective of the Kyoto Protocol[4] is to limit the emissions of greenhouse gases[5] and, unlike the UNFCCC, the Protocol established, for the industrialized or 'Annex I' countries, legally binding reduction targets for greenhouse gases within fixed timetables (art. 3.1).[6] In addition to these fixed levels, the Kyoto Protocol also introduced three international flexible mechanisms, namely: international emissions trading, joint implementation (JI) and clean development mechanism (CDM).

To achieve the Protocol's targets the Contracting Parties need to implement national climate change policies and measures. To this end, art. 2 of the Kyoto Protocol lays down a list of policies and measures, which might help to mitigate climate change and to promote sustainable development. As this list is indicative, the Protocol gives the national governments absolute discretion to implement any particular policy. The indicative list includes enhancing energy efficiency, promoting sustainable agriculture, promoting renewable energy, carbon sequestration and other environmentally friendly technologies.[7]

Energy-related measures are thus prominently present as one of the ways in which climate change can be mitigated. This can not be a surprise, knowing that about three-quarters of the world's energy is provided by fossil fuels such as coal, oil and gas (which have fuelled industrial development and continue to fuel the global economy) and that burning these fossil fuels emits greenhouse gases, which are recognized as being responsible for climate change. Therefore, there is a need for drastic measures in the energy and transport sector in order to achieve greenhouse gas emissions reduction and general enhancement of the environment. Hereto national governments should be encouraged to focus on reducing energy and carbon intensity, whereby the solution exists in energy efficiency improvements and a greater reliance on environmentally sound energy.[8]

3. THE EU AND ITS COMBAT AGAINST CLIMATE CHANGE

3.1 Background

Climate change and climate change measures encompass a large variety of policy sectors, such as environment policy, energy policy, transport policy, agricultural policy and so on.[9] As a result, climate change is not expressly mentioned in the list of Community policies and activities of art. 3 of the EC Treaty or other Treaty provisions, so that the legal basis for measures combating climate change will depend on the nature of the measure, eg arts 93 (taxation), 95 (internal market) or 175 (environment) of the EC Treaty. As there is no general EU competence for energy policy, art. 3 (1.u.) only allows the EU to take measures in the area of energy, energy-related measures addressing climate change (e.g. Decision on energy savings; Directive on the promotion of renewable energy sources; Directive on the energy performance of buildings) are most of the time based upon art. 175 of the EC Treaty and thus treated as environmental protection issues.[10]

3.2 The European Climate Change Programme

In June 2000, the Commission established the European Climate Change Programme (ECCP) with the purpose of identifying the most environmentally beneficial and cost-effective policies and measures enabling the EU to meet its commitments under the Kyoto Protocol.[11] The ECCP focuses on action in the areas of energy, transport, industry and research sectors, and on the issue of emissions trading in the framework of the three flexible mechanisms.[12] For these areas, six technical working groups were established, which identified in the first ECCP report of June 2001 42 cost-effective emission reduction options with a total combined potential to cut CO_2 emissions between 664 and 765 million tonnes CO_2 equivalent against a cost lower than €20/tonne CO_2 equivalent. In comparison with the 8% Kyoto reduction target, this means that the technical potential of the cost-effective measures identified by the first report of the ECCP is twice the size of the required emissions reduction to fulfil the commitments under the Kyoto Protocol. However, achieving this ambitious reduction depends on a number of factors, such as the timeframe within which measures are implemented, the accuracy of data, the public acceptance and so on.[13]

After the publication of the first ECCP report the Commission adopted a package of three legislative initiatives: (1) a proposal to ratify the Kyoto Protocol; (2) a proposal for a Directive establishing an internal EU scheme for greenhouse gas emissions trading; and (3) a communication outlining the priority actions for reducing greenhouse gas emissions (hereafter the Communication with the EU priority actions).[14] Given the subject of this chapter we will only focus on this latter Communication and more, in particular, on its energy-related measures enabling the EU to meet its commitments under the Kyoto Protocol.[15] The Communication with the EU priority actions highlighted a package of 12 priority measures, which represent an emission reduction potential between 122 and 178 million tonnes CO_2 equivalent. These priority measures are grouped into four sections: cross-cutting, energy, transport and industry. The energy section included, for example, legislation on minimum efficiency requirements for end-use equipment, energy demand management, combined heat and power, as well as non-legislative proposals on increased energy-efficient public procurement, a public awareness campaign and a parallel campaign for take-off, which will be discussed in more detail further on in this chapter.[16]

At the 2001 Gothenburg Council, the Heads of State and governments indicated that combating climate change is a major priority of the EU's Sustainable

Development Strategy. To make this commitment clear, the EU adopted a Kyoto ratification decision, and together with its Member States the EC jointly submitted the ratification instrument to the United Nations.[17] The European Community, had from then onwards, the target to reduce its emissions of greenhouse gases by 8% to be reached during the 2008–12 period compared with its 1990 emissions. This reduction corresponds, in absolute terms, to a reduction of 336 Mt CO_2 equivalent in 2010 with respect to 1990. Furthermore, the EU also developed and adopted various other policies and measures to reduce its emissions of greenhouse gases.[18]

Unfortunately, the ECCP has almost been abandoned since 2002, except for the emissions trading scheme. A second critique that the ECCP has to deal with is that the programme has a piecemeal approach – due to the lack of competence of the EU in all the sectors where measures need to be undertaken. Besides, the latest report of the European Environment Agency, however, concluded that Europe is moving further away from the compliance with its Kyoto reduction commitments. The report shows that with the existing policies and measures, Europe will not fulfil its commitments by the end of the first commitment period.[19]

4.   THE EU'S ENERGY-RELATED MEASURES TO COMBAT CLIMATE CHANGE

With regard to the energy-related measures to combat climate change a distinction can be made between measures relating to the supply of energy (generation, transmission and distribution) on the one hand and measures dealing with energy demand (consumption) on the other hand. The latter include energy efficiency measures in the household, tertiary and industry sector.

4.1   Measures Dealing with the Supply of Energy

4.1.1   Directive on the promotion of electricity from renewable energy sources

Since the Commission, in 1997, adopted the Communication 'Energy for the Future: Renewable Sources of Energy',[20] the EU has recognized the need to promote renewable energy sources as a priority measure in various documents, given the important contribution that renewable energy can make towards improving security and diversification of energy supply. It also reduces the strain which is being placed by conventional energy on the environment.[21] The Commission therefore has set out a strategy and target to double the share of renewables in the gross energy consumption from 6% in 1997 up to 12% in 2010.[22] According to a Communication from the Commission of 26 May 2005 (on the share of renewable energy in the EU), the analysis of national reports shows that the policies and measures currently in place will probably achieve a renewable energy share between 18–19% of the electricity market in 2010. If this is the case, in 2010 then the share of renewable energy in energy consumption as a whole will reach no more than 9%.[23] The fall short of expectations can be clarified by the administrative barriers such as long and complex authorization procedures and also that, some Member States still lack a legal framework based on objective, transparent and non-discriminatory criteria for the access to the grid. Therefore, further steps need to be taken in many Member States in order to accelerate growth in the use of renewable energy and thereby ensure that the EU's targets will be met.

The objective of Directive 2001/77/EC on the promotion of electricity from renewable energy sources in the internal electricity market[24] is to create a framework which will facilitate the future development of renewable electricity (RES-E) within the EU and its access to the internal electricity market.[25] Under the Directive, Member States have the obligation to set national indicative targets for the encouragement of a greater consumption of electricity produced from renewable energy sources (art. 3). Indications for these targets are being set out in an Annex to the Directive 2001/77/EC and if they are all met, the consumption of renewables will rise in 2010 to 22%, compared with 14% today. However, the targets are indicative and not binding as a result of the political process, which makes their legal nature unclear. If Member States would not comply with these targets, it is uncertain whether the Community has (besides political pressure) any instrument to make sure that Member States are obliged to comply. Hopefully, this does not impede Member States to reach those indicative targets with the eye on meeting their Kyoto commitments.

Article 4 of the Renewables Directive 2001/77/EC provides that the Commission shall evaluate the mechanisms of direct or indirect support to producers of electricity used by Member States and their compatibility with the legal regime on state aid. The Directive abstains from proposing a harmonized Community-wide support scheme for electricity from renewable energy sources. However, it obliges the Commission to make, if necessary, within 4 years a proposal for such a harmonized support system, on the basis of a Commission's report assessing the various support schemes in favour of electricity production from renewable as well as from conventional energy sources.[26] Besides this, Directive 2001/77/EC tackles a number of technical issues that are fundamental to the further development of renewable electricity. It thus requires Member States to introduce accurate and reliable certification of green electricity, to ensure guaranteed access for green electricity and to ensure that the calculation of costs for connecting new producers of green electricity to the electricity grid is transparent and non-discriminatory.[27] Meanwhile, also at the European level, efforts have been made to introduce a European Renewable Electricity Certificate Trading (RECerT) system. This project was conceived in 1999 by an experienced team of EU energy companies, researchers and consultants.[28] In the literature, the question is now being asked if it would not be better to introduce a Tradable Green Certificate system at the European level in order to reduce overall costs.[29]

4.1.2 Directive on the promotion of the use of biofuels

Since 1985, the European Commission and the Council have been fostering the development of renewable energies and in particular biofuels, more precisely through Directive 85/536/EEC relating to crude oil savings through the use of substitute fuel components.[30] The importance of biofuels to improve partly the security of energy supply has explicitly been stressed and the Directive authorized the incorporation into petrol of up to 5% of ethanol by volume and up to 15% of ethyl-tertiary-butyl-ether (ETBE) by volume. In 1993[31] and 1997,[32] the ALTENER programmes – to promote the use of renewable energies in the European Community – emphasized the need to increase the use of biofuels and introduced also the objective for biofuels of 5% market share.[33]

The production of biofuels is relatively expensive; however the additional costs are justified by the benefits of biofuels across several policy fields. They would provide alternative supplies for the oil consuming sectors, and in particular the transport sector. At present biofuels are the only technically viable means of using renewable energy to replace oil as a transport fuel, which accounts for more than 30% of the final energy consumption in the Community. Taking into account the advantages of using biofuels in terms of climate change, security of supply and its good employment balance, the increase of their share on the European market became a priority of the Community.[34]

In November 2001, the Commission submitted a communication and two draft Directives concerning alternative fuels.[35] When considering these drafts the European Parliament adopted a series of amendments, and therefore obliged the Commission to produce an amended proposal. The purpose of this proposed Directive is to ensure and to promote a minimum percentage of biofuels for transport in the EU (starting with 2% in 2005) to be marketed and distributed. It is clear that biofuels will only be able to achieve market penetration if they are widely available and competitive. To realize this competitiveness and to promote biofuels, the proposal foresees Member States stimulating the technological development of biofuel production and of the companies involved in its production by using financial instruments for research, the environment and regional development. Furthermore, the proposal also authorizes the Commission to consider in its evaluation report the possibility of introducing some fiscal incentives, on the basis of environmental criteria.[36]

Finally, in May 2003, at the end of a long evolution, the Council and the European Parliament adopted Directive 2003/30/EC on the promotion of the use of biofuels or other fuels for transport which aims at promoting the use of biofuels by largely ensuring a minimum proportion of biofuels and other renewable fuels for transport.[37] This Directive states in art. 3 that Member States: 'should ensure that a minimum proportion of biofuels and other renewable fuels is placed on their markets, and to that effect, shall set national indicative targets'. For these targets the Diretive sets reference values: 2% by the end of 2005 and 5.75% by the end of 2010. Furthermore, Member States are obliged to report to the Commission each year on the measures taken to promote biofuels and on the share of biofuels placed on the market in the previous year. As with the Directive on the promotion of electricity from renewable energy sources, the targets in the Biofuels Directive have an indicative nature and are not binding, which makes their legal nature questionable.

4.1.3 Directive on combined heat and power

Cogeneration, or combined heat and power (CHP), is a technique allowing the production of heat and electricity in a single process. Although the given benefits of CHP[38] with regard to saving primary energy, avoiding network losses and reducing emissions (in particular greenhouse gases) the use of cogeneration as a measure to save energy is under-used in the European Community. It was, therefore, necessary to establish a directive to facilitate the installation and operation of such plants, while being in line with the Green Paper on Security of Supply, the Kyoto Protocol, the Renewables Directive and the Energy Efficiency in Buildings Directive. Thus, Directive 2004/8/EC was born.[39]

In the short-term, this CHP Directive should serve as a level playing field for existing cogeneration installations and for new plants, while in the medium to long-term, the necessary framework for high efficiency cogeneration needs to ensure the reduction of CO_2 emissions. High efficiency cogeneration is, in this Directive, defined by the energy savings obtained by combined production instead of the separate production of heat and electricity. Energy savings of more than 10% qualify for the term 'high efficiency cogeneration' (Annex III). Member States must ensure that all forms of electricity produced from high efficiency cogeneration can be covered by guarantees of origin, which are clearly distinguished from exchangeable certificates. To maximize the energy savings and to avoid energy savings being lost, great attention must be paid to the functioning conditions of cogeneration units. Moreover, Member States may operate different mechanisms of support for cogeneration at the national level, including investment aid, tax exemptions or reductions, green certificates and direct price support schemes.

4.1.4 Integrated Pollution Prevention and Control Directive

As Council Directive 96/61/EC concerning Integrated Pollution Prevention and Control[40] (hereafter IPPC Directive) does not tackle climate change as such, it suffices here to say that the energy efficiency requirements of the IPPC Directive fully apply to the sectors and installations which are not covered under the EU emissions trading scheme – although this depends on the national implementation of the Emissions Trading Directive 2003/87/EC.[41] Moreover, the IPPC Directive applies to energy efficiency in production processes and not in products itself.

4.2   Measures Dealing with Energy Demand[42]

Due to the ongoing surge in oil prices and the vast threat of climate change, energy saving through energy efficiency must be seen as a new fundamental priority for the European Community.[43] According to the European Commission, energy saving is allegedly the quickest and most (cost-) effective manner for reducing greenhouse gas emissions and will therefore help Member States in meeting their Kyoto commitments.[44] In the longer term, it will also constitute a major contribution to further emissions reductions, as part of a future post 2012 regime within the UNFCCC. Finally, energy efficiency is one of the key methods to cap EU energy demand and to promote the security of energy supply for the EU.

A spectrum of policy tools is at hand to increase the efficiency of energy use. These tools range from raising the energy price to more specific actions, such as detailed information (in the form of labelling), efficiency regulations, or subsidies. The best policy includes elements that both 'push' the market (ie mandatory efficiency requirements for manufacturers and real prices for consumers) and 'pull' the market, encouraging consumers and/or manufacturers. In the following, the most important legislative measures on energy efficiency and on the energy-efficient use of certain particular appliances will be described and commented briefly.

4.2.1   Directive on the energy performance of buildings
The residential and tertiary sector, of which the major part is buildings, represents approximately 40% of the final energy consumption in the European Community (and is expanding), and the potential energy saving accounts for more than 20%. These are enough reasons for the EU to focus on energy efficiency legislation in this area, starting with key Directive 93/76/EC, which requires Member States to limit CO_2 emissions and which includes insulation measures and heating requirements.[45] The Action Plan to Improve Energy Efficiency in the European Community,[46] the SAVE Programme for the Promotion of Energy Efficiency,[47] as well as the Green Paper on Security of Supply emphasized the potential of this Directive and called for continuation, expansion and more concreteness of the legal rules, the latter being realized through Directive 2002/91/EC.[48]

This Directive 2002/91/EC focuses on a general methodology of calculation of the integrated energy performance of buildings, the application of minimum requirements on the energy performance of new buildings and of large existing buildings subject to renovation, energy certification of buildings, and a regular inspection of boilers and air-conditioning systems inside the buildings (art. 1). The minimum standards mentioned should be drawn up by the Member States themselves and this by the end of 2006.

The crucial role the Energy Efficiency in Buildings Directive can play in the public sector must be emphasized, as public buildings displaying prominently and permanently a building certificate relating to energy efficiency benchmarks can give an excellent example to a large number of citizens (art. 7.3). However, the way to reach such a building certificate can be bumpy: the department responsible of buying equipment and investing in the building needs to work closely together with the department responsible for purchasing energy. Co-ordination in the top management concerning these matters is thus indispensable.

It is worth mentioning here also that the 2000/55/EC Directive on energy efficiency for lighting because one of the most effective actions the Community can take to reduce energy consumption in (public) buildings is the production of performance standards for fluorescent lamp ballasts.[49] Moreover, the EU also finances a 'Green Light Programme' concerning lighting in commercial buildings.

4.2.2 Directive on energy end-use efficiency and energy services
As there is relatively limited scope for further influence on energy supply and distribution conditions in the short to medium term, either by building new capacity or by improving transmission and distribution,[50] the European Commission proposed, on 10 of December 2003, a plan for a Directive on energy end-use efficiency and energy services, which would create a European market for energy services and remove existing market barriers for the efficient end use of energy.[51] This Directive shall apply to all providers of energy efficiency improvement measures (but Member States may exclude small distributors and small retail energy sales companies) and to all final customers.

The initial proposal of the Commission aimed at an annual amount of energy to be saved that is equal to 1% of the amount of energy distributed and/or sold to final customers, and a 1.5% savings target for the public sector.[52] These – in principal indicative – targets were given a uniform 6-year period for achieving them, while the national target would be subdivided into two 3-year targets of energy savings. The European Parliament, however, requested a binding 3% target for the period 2006–09, 4% for 2009–12 and 4.5% for the period 2012–15. Moreover, the MEPs rejected requiring energy distributors or energy sales operators to offer their customers free audits to evaluate their energy saving needs. A later, new, version of the proposal by the European Council featured a national indicative savings target of 6% for the sixth year of the application of the Directive.[53] During the first reading, the European Parliament even proposed an improvement of national energy efficiency by over 11% within a decade, being a mandatory target.

At the end of June 2005 the EU Council of Ministers and the European Parliament differed considerably in their views on the targets in the draft Energy End-use Efficiency Directive: whereas Member States wish to set a 6% savings target of their energy consumption within 8 years as of the entry into force of the Directive, the European Parliament supports a 15% savings proposal by 2015. The Member States are also much more in favour of a voluntary approach with indicative targets only, while the European Commission's original proposal suggests that the targets should be binding. The EU Council, moreover, disapproves the possibility of setting different targets for public bodies. Finally, the European Council reached political agreement on 29 June 2005 on a watered down directive with a non-binding indicative target to increase efficiency by 6% over 8 years, with no separate public sector target, and with allowance on the use of voluntary agreements as instruments which can contribute to the achievement of the indicative targets. Final agreement on this Directive should be reached by December 2005.

4.2.3   Green Paper on energy efficiency

The very recent Green Paper on Energy Efficiency or Doing More With Less[54] represents the top document in the battle of the European Commission to increase energy efficiency. The paper seeks to identify the bottlenecks preventing cost-effective efficiencies from being captured, presenting key actions such as the establishment of Annual Energy Efficiency Action Plans at national level, providing better information to the citizens, improving taxation with the emphasis on the polluter pays principle, targeting state aid where justified, using public procurement to kick-start new energy efficiency technologies, using new or improved financing instruments, going further regarding buildings and, finally, using the CARS 21 Commission initiative to speed up the development of a new generation of more fuel-efficient vehicles. This enumeration solely acts as an idea-basis for discussion and is not exhaustive.

More precisely, half the cuts required to curb energy consumption need to come from 'rigorous implementation of adopted measures' (legislation on buildings, domestic appliances or energy services), the other 10% of the energy reduction from additional and proactive actions. The transport and buildings sectors are singled out as offering the greatest potential reductions. The Green Paper will further be analyzed in a new 'European Sustainable Energy Forum', following the successful Florence and Madrid Forums, which will bring the Commission, Member States, European Parliament, national energy regulators, representatives of European industry and NGOs around the discussion table. In

2006, at the end of the consultation process, the European Commission will come forward with a comprehensive Action Plan which will identify measures to be put forward.

Furthermore, on 6 September 2005, the European Commission presented a five-point plan, as a strong reaction on the current surge in oil prices. One of the main points of this plan consists in the acceleration of the European Action Plan on Energy Efficiency, identifying different ways energy savings can be achieved, in increased pressure for the full and rapid implementation of the new Buildings Directive, in a strong push for an agreement on the Energy Services Directive in the December Energy Council, in the promotion of more effective international action on energy efficiency, but through its bilateral contacts and the IEA, in the promotion of an international conference on energy efficiency in November, and in the establishment of a Sustainable Energy Forum on 13–14 October 2005.[55] Meanwhile, a group of European Parliamentarians called for an 'Energy Intelligent Europe' that would make of Europe 'the most energy efficient society in the world by 2020', saving to over 380 million tonnes of oil equivalent (mtoe) by 2020, compared with 369 mtoe in the Green Paper.

4.2.4   The various Directives on efficiency in energy using products

Household appliances   The labelling of household appliances was the first step in applying the policy of energy efficiency to specific products. The provision of accurate, relevant and comparable information on the specific energy consumption of household appliances may influence the public's choice in favour of those appliances which consume less energy, thus indirectly encouraging the efficient use of these appliances. However, although the fact that this measure will play a more important role – as the energy prices will keep on increasing – one cannot lose sight of the reality that it can be doubted if this information will give sufficient stimulus to potential buyers. To this end, and to avoid the creation of barriers to intra-Community trade, a uniform label for all appliances of the same type has been made obligatory with the establishment of Directive 92/75/EC on the labelling of household appliances.[56]

Suppliers have the duty to provide a fiche and a label which details the consumption of electric energy, other forms of energy and other important resources, while it is the task of the dealer to attach this product information onto the household appliances. The energy label thus indicates the energy-efficient class of the appliance. The provision of this standardized information on labels must be insured for all the appliances concerned through the rejection of voluntary schemes. As a continuation of this Household Labelling Directive, various 'daughter' Directives for specific product groups have been drawn up, ie for household electric refrigerators, freezers and their combinations,[57] washing machines,[58] dish washers,[59] ovens, tumble dryers, household lamps, and so on.

Office appliances   A large proportion of electricity consumption in the tertiary sector can be attributed to office equipment (such as personal computers, monitors, fax machines, scanners, etc.). In this area, and following the Community's climate change commitments, energy efficiency has a significant role to play. Thereto, a very important voluntary labelling system referring to the qualified energy-efficient product types has been elaborated, known as the 'Energy Star'. This energy label has been established in the USA and adopted some years later by the European Community,[60] through the signature in December 2000 of an agreement with the Government of the USA on the cooperation and co-ordination of energy-efficient labelling programmes for office equipment, which was a logical step as the Energy Star was already the *de facto* standard for office equipment manufacturers. This Agreement has been incorporated into Council Decision of 14 May 2001[61] and the error on the legal basis of this Decision has been rectified in Council Decision of 8 April 2003, where art. 133 EC is seen as the correct legal basis for the Energy Star Agreement.[62]

The labelling programme will act as a worldwide efficiency standard for office information and communication technology equipment. The logo will enable consumers to identify energy-efficient office equipment and should therefore result in the maximization of electricity savings and environmental benefits, while stimulating the supply of and demand for energy-efficient appliances. Hereto, the most effective measure for reducing electrical consumption of office equipment is to reduce the standby consumption and to support switching off when not needed. In October 2003, a Working Plan on the development of the Energy Star programme was delivered, announcing the revision of all Energy Star office equipment over the next years.[63] Herein Member States are encouraged to incorporate the Energy Star requirements in policies for 'Green Public Procurement', considering energy efficiency requirements at the time of public purchasing.

Some critical thoughts   Energy Star is arguably the largest voluntary energy efficiency programme in the world. Both critical and cheering voices can be heard, although both agree that, in order to remain effective, both technical and administrative problems have to be addressed. More and more environmental groups and even members of the EU's Energy Star Board call the labelling programme 'ineffective' and 'outdated', so that the technical challenges include defining sets of new products, updating the technical specifications, creating test procedures, and selecting the appropriate Energy Star product specifications bearing in mind the technological developments.[64] On the administrative level, manufacturers will be less eager to participate in the programme as the requirements become more demanding. However, overall, a clear position on the Energy Star label has to be elaborated, since there exist also other energy efficiency labels for office equipment, such as the Energy Label or the Swedish

TCO Label. It would be a good idea to position Energy Star as a label with modest requirements in the medium term, in comparison with the Energy Label which provides stricter criteria but at the same time is not very widespread. Finally, one should draw attention to the fact that the Energy Star label is created as an energy-saving 'sleep mode', which does not express necessarily whether the product in its whole is energy efficient.

4.2.5 Directive on eco-design for energy-using appliances

Besides putting energy efficiency requirements to particular groups of products, it was considered necessary to focus also on energy-using products (EuPs) in general, as they account for a large proportion of the consumption of natural resources and energy in the European Community. Action should be taken during the design phase of EuPs, since it appears that the pollution caused during a product's life cycle is determined at that stage. Following the new Directive 2005/32/EC,[65] manufacturers are now encouraged to produce products with an environmentally-friendly design throughout their entire life-cycle, thus implementing integrated product policy and bearing in mind the goals and priorities of the Sixth Environmental Action Programme. The Directive does not introduce directly binding requirements for specific products, but does define conditions and criteria for setting, through subsequent implementing measures, requirements regarding environmentally relevant product characteristics (such as energy consumption) and allows them to be improved quickly and efficiently.

Although these implementing measures are foreseen, priority should be given to alternative courses of action such as self-regulation by industry where such action is likely to deliver the policy objectives faster or in a less costly manner than mandatory requirements. Legislative measures may be still needed where market forces fail to evolve in the right direction or at an acceptable speed.

5. CONCLUSION

In the light of the foregoing, the conclusion can be made that the EU is indeed placing a greater emphasis on energy-related measures addressing climate change and therefore has created a comprehensive EU regulatory framework. However, the latest reports show that the policies and measures currently in place will probably achieve a renewable energy share between 18–19% of the electricity market in 2010, which is under the assumed 22%. Therefore, further steps need to be taken in many Member States to accelerate growth in the use of renewable energy and thereby ensure that the EU's targets will be met. With regard to the EU's energy efficiency policy there is still an urgent need to revise it. The large potential for improvements in energy efficiency and conservation must be realized through immediate action, both by speeding up technology diffusion and introducing incremental changes in standard practices.[66] Immediate priority should be given to existing buildings and housing, as they constitute a great yet untapped potential for decreasing energy consumption.

NOTES

1. Véronique Bruggeman, Junior Researcher at the University of Maastricht and Lawyer DLA Piper Rudnick Gray Cary UK LLP (Brussels Bar); E-mail: V.Bruggeman@PR.unimaas.nl and Bram Delvaux, Researcher at Institute for Environmental and Energy Law (IEEL) – Collegium Falconis at the K.U. Leuven and Counsellor at the private office of Vice-minister-president of the Flemish Government, Fientje Moerman; E-mail: bram.delvaux@law.kuleuven.be.
2. See for an extensive overview of international and European developments concerning climate change, the contribution of Pallemaerts, M. and Williams, R. elsewhere in this book.
3. See Krämer, L. (2003, 5th edn), *EC Environmental Law*, London: Sweet & Maxwell, p. 299; Council Resolution of 21 June 1989 on the greenhouse effect and the Community, OJ C 183, 20 July 1989, pp. 4–5.
4. For a detailed analysis of the Kyoto Protocol, see Leal Arcas, R. (2002), 'Is the Kyoto Protocol an adequate environmental agreement to resolve the climate change problem', *EELR*, 282–94.
5. The Kyoto Protocol emission targets cover six of the main greenhouse gases: Carbon dioxide (CO_2), Methane (CH_4), Nitrous oxide (N_2O), Hydrofluorocarbons (HFCs), Perfluorocarbons (PFCs) and Sulphur hexafluoride (SF6).
6. The Annex I countries must cut emissions to at least 5% below the 1990 levels by 2008–12. See Wang, X. and Wiser, G. (2002), 'The Implementation and Compliance Regimes under the Climate Change Convention and its Kyoto Protocol', *RECIEL*, p. 186.
7. Climate Change Secretariat (2002), *Guide to the Climate Change Convention and its Kyoto Protocol*, p. 24; See http://r0.unctad.org/ghg/sitecurrent/download_c/publications.html, Assunçao, L. and Zhang, Z.X. *Domestic Climate Policies and the WTO*, p. 1.
8. Climate Change Secretariat, *loc.cit.*, pp. 24–38. See also Communication from the Commission on Energy for the future: Renewable Sources of Energy, White Paper for a Community Strategy and Action Plan of 26 November 1997 COM (97)599 final, pp. 4–5 and 19.
9. See the contribution of Krämer, L. elsewhere in this book.
10. See Krämer, L., *loc. cit.*, p. 300.
11. Communication from the Commission to the Council and the European Parliament on EU policies and measures to reduce greenhouse gas emissions: Towards a European Climate Change Programme (ECCP) of 8 March 2000, COM/2000/88 final.
12. See http://europa.eu.int/comm/environment/climat/eccpreport.htm – The European Climate Change Programme long report of June 2001, pp. 5–6.
13. See http://europa.eu.int/comm/environment/climat/eccp.htm, Second ECCP Progress Report, 'Can we meet our Kyoto targets', April 2003, pp. i–iv. Catalayud, A. (2003), 'Atmospheric Pollution', *YEEL*, pp. 355–57.
14. Communication from the Commission on the implementation of the first phase of the European Climate Change Programme of 23 October 2001, COM /2001/580 final.
15. For a more detailed analysis of the other legislative initiatives see Catalayud, A., *loc. cit.*, pp. 358 and 359.
16. *Ibid*, at p. 359 and Second ECCP Progress Report, 'Can we meet our Kyoto targets', *loc. cit.*, p. 6.
17. Council Decision 2002/358/EC concerning the conclusion, on behalf of the European Community, of the Kyoto Protocol to the United Nations Framework Convention on Climate Change and the Joint Implementation fulfilment of commitments there under, OJ L 130/1, 15 May 2002.
18. See Lefevere, J. (2003), 'Greenhouse Gas Emission Allowance Trading in the EU: a Background', *YEEL*, pp. 149–52 and 173–77.

19. 'Greenhouse gas emission trends and projections in Europe 2004. Progress by the EU and its Member States towards achieving their Kyoto Protocol targets', EEA Report, No. 5/2004, p. 40.
20. Communication of the Commission on Energy for the Future: Renewable Sources of Energy, White Paper for a Community Strategy and Action Plan of 26 November 1997, COM/1997/599 final, pp. 1–55.
21. See Council Resolution of 16 September 1986 concerning new Community energy policy objectives for 1995 and convergence of the policies of the policies of the Member States, OJ 25 September 1986, C 241/1, where the Council listed the promotion of renewable energy sources among its objectives and stated: '(f) to maintain the development of new and renewable energy sources, including conventional hydroelectricity, in particular by continuing with the effort made and by placing greater emphasis on arrangements for disseminating results and reproducing successful projects. The output from new and renewable energy sources in place of conventional fuels should be substantially increased, thereby enabling them to make a significant contribution to the total energy balance'.
22. European Commission Green Paper towards a European strategy for the security of the energy supply of 29 November 2000, COM/2000/769 final, p. 48.
23. Communication from the Commission to the Council and the European Parliament on the share of renewable energy in the EU, Commission Report in accordance with art. 3 of Directive 2001/77/EC, evaluation of the effect of legislative instruments and other Community policies on the development of the contribution of renewable energy sources in the EU and proposals for concrete actions of 26 May 2004, COM/2004/366 final, pp. 21 and 22.
24. Directive 2001/77/EC of the European Parliament and of the Council of 27 September 2001 on the Promotion of Electricity Produced from Renewable Energy Sources in the Internal Electricity Market, OJ L 283, 27 October 2001.
25. Gual, M. and del Río, P. (2004), 'The promotion of green electricity in Europe: present and future', *European Environment*, 14, pp. 219–34.
26. See art. 4 of Directive 2001/77/EC on the promotion of renewable energy sources, loc. cit.; Communication from the Commission to the Council, the European Parliament, the Economic and Social Committee and the Committee of the Regions on the implementation of the Community Strategy and Action Plan on Renewable Energy Sources, COM/2001/69 final.
27. Articles 5–7 of Directive 2001/77/EC on the promotion of renewable energy sources, loc. cit.; Communication from the Commission on the implementation of the Community Strategy and Action Plan on Renewable Energy Sources, loc. cit.; http://europa.eu.int/comm/dgs/energy_transport/home/aids/doc/energy_inventory_en.pdf, Commission staff working paper on the inventory of public aid granted to different energy sources, Brussels, SEC(2002)1275, p. 46; Delvaux, B. (2003), 'EC State Aid Regime on Renewables', *EELR*, pp. 103–12.
28. The European Renewable Electricity Certificate Trading Project, Final Technical document, No. NNE5/1999/00051, see http://recert.energyprojects.net.
29. Del Río, P. (2005), 'A European-wide harmonized tradable green certificate scheme for renewable electricity: is it really so beneficial?', *Energy Policy*, 33, pp. 1239–50.
30. Council Directive 85/536/EEC on crude-oil savings through the use of substitute fuel components in petrol, OJ L 34/20, 15 December 1985.
31. Council Decision 93/500/EEC of 13 September 1993 concerning the promotion of renewable energy sources in the Community (ALTENER programme), OJ L 235/41, 18 September 1993.
32. Council Decision 98/352/EC of 18 May 1998 concerning a multiannual programme for the promotion of renewable energy sources in the Community (ALTENER II programme), OJ L 159/53, 3 June 1998.
33. Communication from the Commission to the European Parliament, the Council, the Economic and Social Committee and the Committee of the Regions on alternative fuels for road transportation and on a set of measures to promote the use of biofuels, of 7 November 2001 COM/2001/547, p. 38.
34. Communication from the Commission to the Council and the European Parliament on the share of renewable energy in the EU, Commission Report in accordance with art. 3 of Directive 2001/77/EC, evaluation of the effect of legislative instruments and other Community

policies on the development of the contribution of renewable energy sources in the EU and proposals for concrete actions of 26 May 2004, COM/2004/366 final, pp. 28 and 29.

35. Proposal for a Directive of the European Parliament and of the Council on the promotion of the use of biofuels for transport, COM/2001/0547 final – COD/2001/0265, OJ C 103E, 2002, pp. 205–7.

36. Articles 3 and 4 of the amended proposal for a Directive of the European Parliament and of the Council on the promotion of the use of biofuels for transport presented by the Commission pursuant to art. 250(2) of the EC Treaty, COM/2002/508 final – COD 2001/0265, OJ C 331E, 2002.

37. Directive 2003/30/EEC of 8 May 2003 of the European Parliament and of the Council on the promotion of the use of biofuels or other renewable fuels for transport, OJ L 123/42 17 May 2003. For a detailed analysis of the Directive promoting the use of biofuels see Delvaux, B. (2004), 'Promoting Biofuels in Energy Supply: the European Legal Framework', *EELR*, pp. 66–78.

38. Electricity/heat cogeneration installations can achieve energy efficiency levels of around 90%. The process is more ecological, since during combustion natural gas releases less carbon dioxide (CO_2) and nitrogen oxide (NO_X) than oil or coal. The development of cogeneration could avoid the emission of 127 million tonnes of CO_2 in the EU in 2010 and 258 million tonnes in 2020.

39. Directive 2004/8/EC of the European Parliament and of the Council of 11 February 2004 on the promotion of cogeneration based on a useful heat demand in the internal energy market and amending Directive 92/42/EEC, OJ L52, 21 February 2004, pp. 50–60.

40. Council Directive 96/91/EC of 24 September 1996 concerning integrated pollution prevention and control, OJ L257, 10 October 1996, 26–40.

41. See hereto the contribution in this book dealing with IPPC and emissions trading.

42. For an earlier version of this chapter, see Bruggeman, V. (2004), 'Energy Efficiency as a Criterion for Regulation in the European Community', *EELR*, 13(5), pp. 142–6.

43. Recent studies indicate that an effective energy efficiency policy could allegedly save at least 20% of its 2005 energy consumption by 2020, equivalent to €60 billion per year, or the present combined energy consumption of Germany and Finland!

44. 'Europe could save 20% of its energy by 2020', *Rapid Press Releases*, 22 June 2005, following the adoption of the Green Paper on Energy Efficiency.

45. Council Directive 93/76/EEC of 13 September 1993 to limit carbon dioxide emissions by improving energy efficiency (SAVE), OJ L237, 22 September 1993, p. 28.

46. Committee and Committee of the Regions: Action Plan to Improve Energy Efficiency in the European Community, COM/2000/247final, 26 April 2000.

47. Council Decision No. 647/2000/EC of the European Parliament and of the Council of 28 February 2000 adopting a multiannual programme for the promotion of energy efficiency (SAVE) (1998 to 2002), OJ L79, 30 March 2001.

48. Directive 2002/91/EC of the European Parliament and of the Council of 16 December 2002 on the energy performance of buildings, OJ L1, 4 January 2003, pp. 65–73.

49. European Parliament and Council Directive 2000/55/EC of the 18 September 2000, on energy efficiency requirements for ballasts for fluorescent lighting, OJ L279, 1 November 2000.

50. Green Paper, 'Towards a European Strategy for Energy Supply', COM/2000/769.

51. Proposal for a Directive of the European Parliament and of the Council on energy end-use efficiency and energy services, COM/2003/739 final, 10 December 2003.

52. *Ibid.*

53. Proposal for a Directive of the European Parliament and of the Council on energy end-use efficiency and energy services, 19 May 2005, Council of the European Union, interinstitutional file 2003/0300 (COD).

54. Green Paper on Energy Efficiency or Doing More With Less, 10 June 2005, COM/2005/265 final.

55. 'Five-point plan to react to the surge in oil prices', 6 September 2005, MEMO/05/302.

56. Council Directive 92/75/EEC of 22 September 1992 on the indication of labelling and standard product information of the consumption of energy and other resources by household appliances, OJ L297, 13 October 1992, pp. 16–19.

57. Commission Directive 2003/66/EC of 3 July 2003 amending Directive 94/2/EC implementing Council Directive 92/75/EEC with regard to energy labelling of household electric refrigerators, freezers and their combinations, OJ L170, 9 July 2003, 10–14. As of 2004, consumers can easily identify those refrigerators and freezers which consume the least energy through the labels A++ and A+.

58. Commission Directive 95/12/EC of 23 May 1995 implementing Council Directive 92/75/EEC with regard to energy labelling of household washing machines, OJ L136, 21 June 1995, pp. 1–27, as amended by Directive 96/89/EC, OJ L388, 28 December 1996.

59. Commission Directive 1999/9 of 26 February 1999 amending Directive 97/17/EC implementing Council Directive 92/75/EEC with regard to energy labelling of household dishwashers, OJ L56, 4 March 1999, p. 46.

60. A Commission Decision of 11 March 2003 established a European Community Energy Star Board (ECESB) to manage and co-ordinate the introduction of the Energy Star in Europe: Commission Decision 2003/168/EC of 11 March 2003 establishing the European Community Energy Star Board, OJ L 67, 12 March 2003, 22–24 and Commission Decision 2003/367/EC of 15 May 2003 establishing the Rules of Procedure of the European Community Energy Star Board, OJ L125, 21 May 2003, pp. 9–11.

61. Council Decision 2001/469/EC of 14 May 2001 concerning the conclusion on behalf of the European Community of the Agreement between the Government of the United States of America and the European Community on the co-ordination of energy-efficient labelling programmes for office equipment, OJ L172, 26 June 2001, 1–2. The detailed arrangements for implementing this Agreement at Community level have been put forward in Regulation (EC) No. 2422/2001 of the European Parliament and of the Council of 6 November 2001 on a Community energy-efficient labelling programme for office equipment, OJ L 332, 15 December 2001, pp. 1–6.

62. See Case C-281/01 *Commission of the European Communities* v. *Council of the European Union*. According to the European Court of Justice, the Energy Star Agreement aims at energy-efficient labelling of office equipment, which facilitates trade and therefore constitutes a commercial-policy measure, so that the Agreement should have been based, following the 'centre of gravity' theory, on art. 300(3) EC (para. 43).

63. Energy Star – European Community – Energy Efficiency labelling programme for office equipment – Working Plan, 9 October 2003.

64. Meier, A., 'The future of Energy Star and other voluntary energy efficiency programmes', IEA/EET Working Paper, March 2003.

65. Directive 2005/32/EC of the European Parliament and of the Council of 6 July 2005 establishing a framework for the setting of ecodesign requirements for energy-using products and amending Council Directive 92/42/EEC and Directives 96/57/EC and 2000/55/EC of the European Parliament and of the Council, OJ L191, 22 July 2005, pp. 29–58.

66. Multi-stakeholder task force on climate change convened by the Centre for European Policy Studies (Ceps) in Brussels – Report of 2 June 2005.

13. Energy taxation within the EU

Manfred Rosenstock[1]

1. ENVIRONMENTAL POLICY IN THE EU

Environmental policy at European Community (Community) level has been developed since the 1970s. Its main instrument has traditionally been technical regulation.[2] In recent years, however, the European Commission (the Commission) has proposed to complement this approach by expanding the use of market-based instruments (MBI),[3] such as environmental taxes, emissions trading and incentives. The Council and the European Parliament have supported this approach in strategic documents, such as the 6th Environmental Action Programme, where an intensified use of MBI is also demanded.[4]

Given the importance of the climate change issue and the key role of energy in our economy, Community activities to pursue the increased use of MBI have focused on energy-related areas. Prime examples are the European Emissions Trading Scheme for greenhouse gases (ETS),[5] but also the directive on the taxation of energy products.[6] It should, however, not be forgotten that some provisions on the use of economic instruments can also be found in Community environmental legislation relating to non-energy areas (cf. below, section 4 of this chapter).

The argument for an increased use of economic instruments rests on the fact that they are, in the majority of cases, the most efficient means to curb pollution and encourage energy efficiency as they ensure the internalization of external costs in the prices of goods and services. At the same time, they provide industry with a flexible means of response to environmental objectives, as they only provide financial incentives but do not prescribe the level of emissions of an individual enterprise or the instruments that are to be used to achieve such objectives. A final important point is that, by putting a price on the remaining pollution, an incentive is maintained to reduce it further. This encourages polluters to seek technological innovations that will reduce emissions further.[7] This aspect is usually referred to as dynamic efficiency, and can be further encouraged through e.g. differentiated rates for various fuel types.

An additional aspect which concerns the use of environmental taxes is that revenues from such taxes allow a reduction of other, more distorting taxes, e.g. on labour, with the objective to further employment. This way, it might be possible to earn a so-called double dividend – an improvement in both environmental quality and employment – in the context of an environmental tax reform (ETR).[8]

2.   IMPORTANCE OF ENVIRONMENTAL TAXES IN THE EU

A few data on the development of the revenue from environmental taxes allow to put the legislative developments in context: Member States have over time made increasing use of environmental taxes. In 2002, revenue from such taxes was €238bn, which is equivalent to 6.5% of the total revenue from taxes and social security contributions, or 2.7% of the GDP of the 15 Member States of the EU before the enlargement in 2004 (EU-15). The highest shares of environmental taxes in total tax revenue could be observed in Denmark, Ireland, the Netherlands and Portugal where the share exceeded 8%, while the revenue from such taxes is least important in Austria, Belgium, France and Sweden where the share was below 6%.[9] In those new Member States, that are OECD members (Czech Republic, Hungary, Poland Slovakia), and where comparable data are thus available, the share is in the range of 5.8–7.6% of total tax revenue.[10] The relative importance of environmental taxes grew slowly but steadily between 1980 and the mid-1990s from 5.8% to 6.5%. Since then, the share has remained roughly stable.

Concerning their tax base, these taxes can be broken down into three broad categories (see also Table 13.1 below):

- energy taxes, which are responsible for about three-quarters of total revenue from such taxes and charges;
- taxes on transport, which are not energy-related. These include mainly vehicle registration and circulation taxes, and make up for about 20% of all environmental tax revenue;

Table 13.1   Environmental taxes in the EU by type (% of total taxes)

	1980	2002
Energy taxes	4.23	5.0
Transport taxes	1.43	1.3
Pollution taxes	0.18	0.1
Environmental taxes	5.84	6.5

- taxes on pollution, such as emissions, pesticides or packaging waste, and on resources, such as water. This category covers a large number of taxes with usually small revenues, so that their share in total revenue is about 2.5%.

Over time, the highest increase in revenue among these groups could be observed for taxes on energy and on pollution.

3.   THE DIRECTIVE ON ENERGY PRODUCTS TAXATION

As mentioned above, the primary importance of energy taxes, both in environmental and in competition and internal market terms, made the Community develop a European Union (EU) framework here first.

In 1992, the EU had achieved a partial and incomplete harmonization in the field of the taxation of energy products, with common rules and the introduction of minimum tax rates limited to mineral oil products only. Given the increasing risk of distortions of trade within the internal market, in particular in the context of the ongoing energy markets liberalization, and of relocations of consumption (in particular in border areas) through growing divergences of applied rates of the tax, the European Council and Parliament asked the Commission in 1996 to develop a new comprehensive approach to the taxation of energy products.

This request also had to be seen in the context of international developments in relation to climate change. The commitments for the reduction of CO_2 emissions undertaken by the EU in the context of the Rio conference of 1992 and in the negotiations on the Kyoto Protocol constituted a case for the introduction of additional instruments on those sectors that would not participate in a greenhouse gas ETS.[11]

3.1   Concept and Content

The Commission duly submitted its proposal for a directive on energy products taxation in 1997.[12] Negotiations in the Council were long and difficult, not least due to the fact that in matters of taxation, unanimity is required for a directive to be adopted. Finally, a political agreement on a modified proposal was reached in the Economic and Financial Affairs Council (ECOFIN) in March 2003.[13] After a re-consultation of the European Parliament on this modified text, the Council adopted it definitively in October 2003. The directive came into force on 1 January 2004 and replaced the mineral oil tax directives.

The directive does not introduce a new tax, but rather provides a framework of rules to restructure and harmonize national tax systems within the context of the single market. It is based on two key ideas:

- The scope of the community framework is widened beyond mineral oil products to include the competing sources of energy, such as coal, lignite and natural gas, whenever they are used for heating purposes and as motor fuels. They are, however, not taxed when they are used for purposes such as chemical reduction in the steel industry. Electricity is also to be taxed.
- As an incentive for a more rational energy use, minimum tax rates on mineral oil products were increased beyond the rates agreed upon in 1992 while for other energy products, Community minimum rates were introduced for the first time. The energy tax character of the levy is underlined by the fact that both the tax rates for natural gas and for coal were set at an identical level per unit of energy created. The table below shows the changes in Community minimum tax rates through the directive.

Table 13.2   Community minimum tax rates on energy products

	Previous rates since 1992	Commission proposal, rates for 1998	New minimum rates since 2004
Motor fuels in (€/1000 l)			
Leaded petrol	337	417	421
Unleaded petrol	287	417	359
Diesel	245	310	302*
Other products (household use)			
Electricity (€/Mwh)	0	1	1
Natural Gas (€/Gj)	0	0.2	0.3
Coal (€/Gj)	0	0.2	0.3

Note: * € 330 from 2010 onwards. l = litres; Mwh = Megawatt hours; Gj = Gigajoules.

While the Commission proposal had foreseen gradual increases of minimum rates in three steps, the adopted directive introduces an increase in only one step in 2004; the exception to this is a second step increase in 2010 for diesel used as motor fuel.

One of the criticisms that was originally raised against the proposal was that it would have a negative impact on the competitiveness of European industry, in particular the energy intensive part of it. This criticism was voiced most forcefully by the cohesion countries,[14] who have higher growth rates of the industrial sector than other Member States.

In order to analyze whether this concern was well-founded, the Commission undertook a number of studies when it presented its proposal in 1997. The studies showed, however, that the impact of the proposal on Gross Domestic Product (GDP) would instead be marginally positive, while its implementation would at the same time lead to a net creation of jobs.[15] None of the cohesion countries would have suffered a reduction of GDP. The effect of the directive on consumer prices, which had been another concern, was found to be very limited while emissions of CO_2 and other pollutants would have fallen by 2005. These results were later broadly confirmed by a study about the effects in Spain undertaken by the Spanish Ministry of Economic Affairs in 2002.

3.2 Detailed Provisions of the Directive

The concept and intentions of the directive can best be understood by looking in some more detail on its key provisions.

3.2.1 Tax rates

The increase in tax rates for mineral oil products roughly reflects accumulated inflation since 1992 (25%), so that the erosion of Community minimum rates for mineral oil products that had taken place in real terms since 1992, has been reversed. Also, the introduction, for the first time, of minimum rates on other energy products, is to be welcomed from both an economic and an environmental point of view. On the one hand, this development reduces distortions between different energy products as these products are substitutes for each other; while on the other hand, it leads to progress towards the internalization of external costs, and thus most likely to a more efficient energy use and a reduction in emissions.

Given the prevailing concerns about competitiveness and also inflationary impacts, some Member States obtained transitional periods to adjust to higher tax rates for specific products.[16] Upon enlargement in May 2004, a similar procedure was employed for the new Member States. They were also granted transitional periods for specific products through the adoption of a separate directive.[17] Given their lower income levels and the fact that many energy products constitute basic consumption goods, the prices of those products represent a significant social policy issue for them, which provided another argument for a gradual phasing-in of higher tax rates.

As the directive foresees no further increases in minimum rates (except for diesel fuel) so that minimum rates in real terms will erode again, the directive delivers fewer incentives for energy efficiency and emission reductions than might have been hoped for from an environmental point of view. Its environmental effects in the old EU–15 will, therefore, most likely be limited, while more substantial effects can be expected in the new Member States.

3.2.2 Taxation of electricity

The directive introduces an output tax, ie a tax on the quantity consumed, instead of an input tax on the energy products used in its generation. This approach facilitates the application of the destination principle of taxation and the taxation of electricity produced with inputs outside the scope of this directive, such as nuclear fuel. Furthermore, it allows Member States to maintain a differentiation of tax rates between various types of users, in particular between industry and households. This enables them to reduce the tax burden for energy-intensive industries in order to limit any sectoral competitiveness impact of the tax.

From an environmental policy point of view, an output tax has the drawback that a differentiation of tax rates in line with the environmental quality of the fuel used to produce electricity is not possible. The directive gives Member States the option, however, to introduce an additional input tax on environmentally undesirable fuels.

Furthermore, Member States can grant tax reductions or even total exonerations for electricity produced with renewable energy, so as to enable them to support these forms of energy. For high-efficiency combined heat and power (CHP) plants, it is possible to grant a partial or total exemption for fuels used, while the heat output from CHP plants is not taxed at all.[18]

3.2.3 Energy-intensive industries

To address their concerns that the higher tax rates could harm their international competitiveness, the directive allows for tax reductions for energy-intensive firms. For the first time, the directive provides a definition of what constitutes an energy intensive firm. They are defined as firms whose purchases of energy products exceed 3% of their production value or, alternatively, whose payments of energy taxes exceed 0.5% of their value-added.

While they can obtain reductions from standard national tax rates, the tax rates applied to them must still be higher than Community minima. This does not apply, however, to firms which have entered into voluntary agreements with authorities or into tradable permit schemes that lead to an increase in their energy efficiency or a reduction of their emissions that is broadly equivalent to what would have been achieved by an application of minimum rates. In such cases, Member States can reduce the tax rates applied to them to 50% of community minima for non-energy intensive firms, and down to zero for energy intensive companies.

3.2.4 Aviation fuel

Traditionally, aviation fuel used by commercial aircraft was completely exempted from excise taxes via bilateral air service agreements (ASA) between countries.[19]

This situation is unsatisfactory from an environmental policy point of view, as CO_2 and NO_x emissions from aircraft cause a substantial and increasing share of the external costs of air pollution and climate change. The Energy Products Taxation Directive, while maintaining this principle as long as international treaties have not been changed, allows Member States to tax national flights and – on the basis of bilateral agreements between Member States – intra-EU flights. It remains to be seen whether Member States will use this new option. The Netherlands has been the first Member State to introduce the taxation of kerosene on domestic flights. At the same time, the Commission is trying to renegotiate bilateral ASAs to allow the taxation of fuel purchases by commercial aircraft at EU airports.

Two alternative instruments to address, in particular, the climate change costs of aviation are also under discussion. They are, on the one hand, a system of en-route charging, where charges would be applied on aircraft at an airport in proportion to the aircraft type and the distances it has flown, and on the other hand, a participation of the aviation sector in the ETS.[20]

3.2.5   Issues outside the directive

As mentioned above, the directive covers the use of energy products for heating purposes and as motor fuels. When they are used for other purposes, eg for chemical processes, the provisions of the directive thus do not apply. This concerns in particular, the following cases:

- The so-called dual-use processes, such as the steel industry, where energy products on the one hand provide the energy for the process of steelmaking from iron ore and on the other hand, through a series of chemical reactions, are part of this process. The same logic is applied to metallurgical and electrolytic processes as well as to other processes of chemical reduction.
- The directive widens this exemption to mineralogical processes, such as the cement, chalk and glass industries. These processes were previously considered as heating processes and were thus taxed, when mineral oil products were employed.

The exclusion of these processes from the directive means that neither the minimum tax rates nor the rules about the tax base apply, but nevertheless leaves Member States the freedom to tax energy products used in these processes or to exempt them. It should be kept in mind that both steelmaking and mineralogical processes are covered by the ETS.

4. COMMUNITY RULES ON ENVIRONMENT-RELATED TAXES IN OTHER AREAS

Many Member States apply environment-related taxes to other products. Tables 13.3 and 13.4 in the Appendix to this chapter provide an overview of the taxes applied in different Member States and candidate countries. Nevertheless, no detailed provisions at Community level were introduced for any of those. Where there are rules, they are limited to a few general provisions. These concern in particular the following areas.

4.1 Packaging waste directive

This directive ensures in particular the collection and recycling of packaging waste and fixes collection and recycling quotas for different types of waste. It foresees, however, that the Council can adopt economic instruments at Community level to help reach these objectives.[21] The directive also clarifies that, as long as no such decision has been taken – and so far none has been taken – Member States can adopt national measures to implement the polluter-pays principle. These measures have, however, to respect their EC Treaty obligations, in particular those arising from the internal-market rules and the principle of non-discrimination.

4.2 Batteries directive

A similar approach has been followed in the Commission's proposal for a new batteries directive.[22] Also here, it is specified that Member States can adopt economic instruments to promote the collection of spent batteries and to promote the use of batteries containing less polluting substances. The proposal has passed its first reading in the European Parliament and the Council adopted a common position in July 2005.[23]

4.3 Water framework directive

The key objective of this directive[24] is to establish a framework for the protection of inland surface waters, transitional waters, coastal waters and groundwater. For this purpose, Member States shall introduce water-pricing policies by 2010 that will ensure that users bear the costs – including the external environmental costs – of water consumption and thus implement the polluter-pays principle. In this way, Member States shall provide adequate incentives to consumers to use water efficiently. All the different groups of users, such as industry, households and agriculture, shall make their contribution to cost coverage. The directive also includes an obligation for Member States to

report in their river-basin management plans on the implementation of these
measures.

5.   STATE AID ISSUES RELATED TO ENERGY TAXATION

The reimbursement of revenues from environmental levies to firms constitutes
state aid in the sense of Art. 87 of the EC Treaty. This also holds for exemptions
or reductions from an environmental tax, if they are based on the decisions by
individual Member States.

As is the case for state aid in general, Member States have to notify the
measures they intend to apply to the Commission and the latter has to decide
on their compatibility with the EC Treaty, in order to avoid distortions of competition in the internal market. The obligation to notify applies to the aid
schemes, in this case the legal provisions for tax reductions, not to the individual
payments to single enterprises within such schemes.

The fact that a measure constitutes State aid does not necessarily mean that
the Commission will prohibit it. The provisions of art. 87 of the EC Treaty list
in a general way types of state aid which can be compatible with the internal
market. The Community Guidelines on State Aid for Environmental Protection[25]
establish more specific conditions under which environmental aid may be authorized. They contain, inter alia, a number of precise criteria for judging the
compatibility of exemptions from environmentally related taxes. In particular,
these provisions allow for the fact that reductions may prove necessary because
of temporary risks of loss of international competitiveness for energy-intensive
sectors.

As any state aid will lead to a distortion of competition, derogations from
environmental taxes can only be approved if this distortion is balanced by real
environmental benefits. Such benefits can usually be achieved, when firms or
associations of firms sign voluntary agreements with the authorities where they
undertake to achieve environmental protection objectives that go beyond business as usual. The participation of companies in an ETS, which fixes overall
quantitative emission limits is judged to be equivalent to this. In all cases, the
Commission has to assess whether the environmental benefits do indeed
occur.

Furthermore, in line with the Commission's general position in State aid
matters, the guidelines specify that any derogation has to be limited in time so
that the justification of the aid can be regularly re-assessed. For reductions in
environmental taxes, the maximum duration is fixed at 10 years, again subject
to the condition that companies participate in voluntary agreements or emission
trading schemes. The Commission has implemented these rules in a number of
decisions on specific cases.[26]

Finally, as concerns the area covered by the energy products taxation directive, the guidelines contain detailed provisions on the types of support and specific conditions that Member States can grant to promote the use of renewable energies.

6. ONGOING AND FORTHCOMING PROJECTS

An issue that has already figured prominently in the discussions about the Energy Products Tax Directive, is the taxation of fuel used by the road haulage sector, the so-called professional diesel. Major differences in diesel taxation between Member States gave rise to concerns about competitive distortions between operators in different Member States. Furthermore, these differences led truck operators to change their routes in order to be able to fill their tanks in countries with lower tax rates, thus covering longer distances and increasing their emissions. This problem led the European Council of Barcelona, in 2002, to call on the Commission to make a proposal that would eliminate the problem. The Commission submitted a proposal in the same year[27] that had the following key features:

- The separation of the tax rates for professional and other diesel fuels. Professional diesel was essentially defined as diesel fuel used by trucks with a weight of at least 16 tonnes. For such diesel, a central tax rate would be introduced with fluctuation margins around it that would be narrowed over time. This process would lead to a unified tax rate for such diesel by 2010.
- The indexation of this rate in line with inflation. This element would solve the problem of both the mineral oil tax and energy tax directives that minimum tax rates lose their real value over time, due to inflation.
- The increase of the minimum rate for non-professional diesel to that of petrol. Both these rates would then also be indexed over time. From an environmental point of view, such a step would be positive as a litre of diesel has a comparable – or for some pollutants, such as CO_2, NO_x or particulate matter – even higher environmental impact than petrol. Currently, diesel fuel is taxed at a significantly lower rate than petrol in all Member States except the United Kingdom (UK), mainly as a measure to support the road haulage sector. With the proposed decoupling of the tax rates for both user groups, this political argument would fall away.

The negotiations on the proposal in the Council were very difficult and the European Parliament issued a negative opinion on it, due in particular to doubts about the importance of the 'gasoline tourism' and a preference for a degree of tax competition for fuel taxes over a complete harmonization.[28] As a result, the Commission has recently withdrawn the proposal. In the meantime, the Energy Products Tax Directive, which was adopted in the year following the professional diesel proposal, already allows Member States to differentiate the tax rates on professional and ordinary diesel, as long as they do not lower their current rates on diesel. A revised proposal by the Commission, taking into account the comments made by the Council and the European Parliament, is expected.

If such a proposal was to be adopted, Member States would need to reflect on the treatment of diesel vehicles under circulation taxes, where currently most Member States compensate for the lower taxation of diesel fuel compared with petrol by charging diesel-powered cars at a higher rate.

Beyond questions of energy taxation for transport fuel in the narrow sense, discussions are ongoing in Member States on possibilities to address problems of local air pollution and congestion caused by road transport via a system of infrastructure charges. In the UK, urban congestion charges have been introduced in London and Durham, and also in several other Member States (as well as in third countries, such as Switzerland) charges on the use of specific types of infrastructure by heavy duty vehicles have been implemented.[29]

At the EU level, infrastructure charging for heavy goods vehicles on major roads is currently regulated by the so-called Eurovignette Directive.[30] The directive does not require Member States to charge for motorway use, but stipulates conditions in case they do.[31] It allows both charges over certain time periods and tolls for certain stretches of road, including distance-dependent road pricing. In 2003, the Commission proposed an amendment to this directive.[32] One of the key motivations for this proposal was the desire to raise finance for the further development of the Trans-European transport networks (TEN-T). Therefore, it included a requirement that the revenues from the motorway charges should be spent for developing the transport system. The cost base for the toll or charge levels is mainly related to the infrastructure costs and does not include environmental externalities, with the exception of the cost of noise protection measures. However, environmental considerations can be built into the tolls as they are allowed to vary – within certain maximum boundaries – as a function of the damage to the roads and the emission class. The scope of the directive is limited to the TEN-T and other streets where traffic might be deviated to, but it leaves Member States free to charge for the use of other roads. Member States may introduce mark-ups on standard charges in environmentally sensitive zones, subject to Commission approval and with the condition that the related revenues be used for investment in alternative infrastructures (typically rail) in the same zone. In June 2005, the Council reached a common position on this proposal. Contrary to the Commission's proposal that the revenue would have to be used for road infrastructure investments, it would leave Member States complete freedom on the use of revenues.[33] It is notable that references to external costs introduced by the European Parliament in its first reading were removed by the Council in its common position.[34]

The current environmental state aid guidelines expire by the end of 2007. In its recently adopted State aid action plan,[35] the Commission announced that it would start the review process of these guidelines later in 2005. The review will extend to all provisions of the guidelines in the light of the experience gained since 2001. This would, for example, include measures to encourage eco-innovations and eco-efficiency and could include the exemption of certain environmentally-beneficial aid measures from the obligation to notify. A public consultation on the experience with the guidelines has been undertaken in mid-2005 as part of the review process.

7. CONCLUSIONS

As seen above, Member States have intensified the use of environmentally-related taxes over time, in particular in the area of energy taxation. Apart from questions of distributive justice, concerns about the competitiveness impact of such taxes on specific industries has often slowed progress towards their increased use and has led to generous exemptions that are problematic from the perspective of the polluter-pays principle. The Energy Products Taxation Directive has allowed some progress to overcome these obstacles. Further proposals by the Commission are on the table of the Council.

The environmental state aid guidelines, combined with the relevant provisions of the directive, have also ensured that any derogations from taxes are designed in such a way as to minimize both distortions of competition and a loss of incentives for environmental protection. The review of the guidelines will provide an opportunity to asses whether further refinements of the rules are necessary.

It could also be observed that recently, the types of economic instruments used by Member States have become more varied – in particular through the growing use of tradable permit systems[36] – and are increasingly used in combination (policy packages) to optimize their effect.[37] Further analysis is required to see whether this requires adaptations to different sets of Community rules.

NOTES

1. Dr Manfred Rosenstock, European Commission, Brussels, Directorate-General Environment, Sustainable Development and Economic Analysis unit. E-mail: Manfred.Rosenstock@ec. europa.eu. The views expressed in this chapter represent exclusively those of the author and may not in any circumstances be regarded as stating an official position of the European Commission.

2. Often referred to as command-and-control regulation.
3. Also known as economic instruments.
4. Cf. art. 3 of the 6th EAP, OJ L 242/1 of 10 September 2002.
5. Directive 2003/87/EC of 13 October 2003, OJ L 275 of 25 October 2003, p. 32.
6. Directive 2003/96/EC of 27 October 2003, OJ L 283 of 31 October 2003, p. 51.
7. Cf. Sterner, T. (2003), *Policy Instruments for Environmental and Natural Resource Management*, Washington DC: Resources of the future, pp. 136–49.
8. The existence of such a double dividend is, however, disputed. For a discussion of the issue, see OECD (2001), *Environmentally Related Taxes in OECD Countries – Issues and Strategies*, Paris.
9. Cf. Eurostat (2004), *Structures of the Taxation Systems in the European Union 1995–2002*, Luxembourg.
10. Cf. OECD/EEA database on instruments used for environmental policy and natural resources management, see http://www2.oecd.org/ecoinst/queries/index.htm.
11. The EU ETS covers energy activities, the production and processing of ferrous metals, the mineral industry and the pulp and paper industry. Cf. Directive 2003/87/EC, op. cit, n. 5, p. 42.
12. COM/1997/30 of 12 March 1997.
13. For more details on the negotiation process, cf. Boeshertz, D. and Rosenstock, M. (2003), 'Energy Taxation in the EU: a case study of the implementation of environmental taxes at a supranational level', pp. 151–62, in Milne, J., Deketelaere, K., Kreiser, L., Ashiabor, H. (eds), *Critical Issues in Environmental Taxation*, **I**, Richmond, pp. 151–62.
14. At the time, these were Ireland, Greece, Spain and Portugal.
15. Cf. Boeshertz, D. (2003), 'The EU Energy Taxation Directive of 2003 – competition and state aid issues in the political negotiations', in Schneider, F. and Steinmüller, H. (eds), *Forum Econ013*, Linz., pp. 9–10.
16. Member States did not necessarily make use of these provisions. While Belgium was allowed a transitional period until 2007 for the adjustment of its tax on diesel fuel to the new minimum rates, it had already increased the tax beyond this rate by the beginning of 2005.
17. Directive 2004/74/EC of 29 April 2004, OJ L 157/87 of 30 April 2004.
18. Finally, Member States can reduce taxes on biofuels to compensate for their higher production costs.
19. These agreements are based on the Chicago Convention of 1944, which precludes, however, only the taxation of aircraft fuel on board of aircraft. The full exemption from all fuel taxes, even for fuel purchased at airports, was probably introduced in order to promote the growth of civil aviation in the period after the Second World War.
20. The Commission has recently adopted a communication on the use of economic instruments to reduce the climate-change impact of aviation. COM/2005/459 of 27 September 2005.
21. Directive 94/62/EC of 20 December 1994, in particular art. 15, OJ L 365/10 of 31 December 1994.
22. COM/2003/723 of 21 November 2003.
23. 'Ministers approve EU batteries directive', *Environment Daily* 1921, 20 July 2005.
24. Directive 2000/60/EC of 23 October 2000, OJ L 327/1 of 22 December 2000.
25. OJ C 37 of 3 February 2001. For a detailed discussion of the provisions of the guidelines, see also Boeshertz, D., op. cit. n. 13, pp. 23–29.
26. For details, cf. the Commission's annual *Reports on Competition Policy*.
27. COM/2002/410 of 24 July 2002.
28. Report on the proposal for a Council directive amending Directive 92/81/EEC and Directive 92/82/EEC […], Rapporteur: Piia-Noora Kauppi, A5-0383/2003 of 5 November 2003.
29. For more information on these specific instruments, see European Environmental Agency (2006), *Using the Market for Cost-Effective Environmental Policy – Market-Based Instruments in Europe*, Copenhagen.
30. Directive 1999/62/EC.
31. I am grateful to my colleague Günter Hoermandinger for further clarifications and explanations on this issue.
32. COM/2003/448 of 23 July 2003.

33. Cf. 'Ministerial breakthrough on Eurovignette revision', *Environment Daily* 1865, 22 April 2005.
34. At the Council when the common position was reached, the Commission declared that it would deepen its analysis of external costs, particularly in view of the second reading of the proposal.
35. COM/2005/107 of 7 June 2005.
36. E.g. the CO_2 emission trading scheme in the UK or the NO_x trading scheme introduced this year in the Netherlands.
37. For a detailed analysis of Member States' experiences, cf. European Environmental Agency (2006), op. cit. n. 29.

APPENDIX

Table 13.3 Types of Environmental Taxes in Member Countries of the EEA

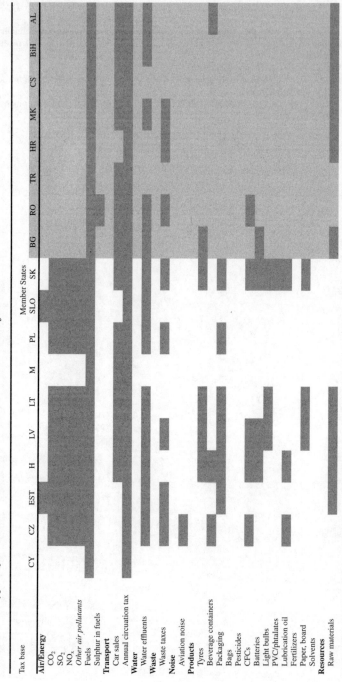

Source: EEA (2006), Using the Market for Cost-Effective Environmental Policy – Market-Based Instruments in Europe, Copenhagen, pp. 26–7.

Table 13.4   Types of Environmental Taxes in Member Countries of the EEA

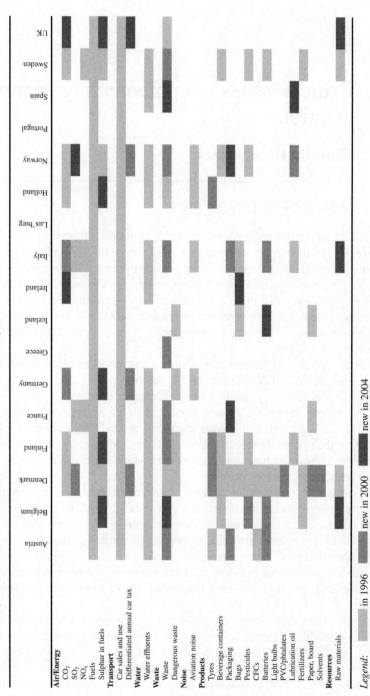

Legend: in 1996   new in 2000   new in 2004

Sources: EEA (2006), *Using the Market for Cost-Effective Environmental Policy – Market-Based Instruments in Europe*, Copenhagen, pp. 26–7.

14. Critical issues in implementing energy taxation

Claudia Dias Soares[1]

1. FRAMING THE DISCUSSION

European Union (EU) energy policy has among its priorities supporting the achievement of the targets set within the EU Climate Change Policy. On the other hand the EU Green Paper on security of energy supply of November 2000 reflects a transition from a 'dirigiste', intergovernmental and mercantilistic view of security of supply to a more market-based one.[2] Thus, in order to fulfil its environmental goals in the new regulatory context, the EU will have to fine-tune its intervention, as this has to be a market-facilitating one rather than a market-replacing one.

To design a 'soft' and 'effective' regulatory intervention, fiscal and economic instruments are critical. Therefore, the EU has been taking decisive steps to promote energy efficiency and investment in cleaner energy with resource to such instruments. Scanning the new legal framework at this point might be helpful to access possible unexploited opportunities and policy failures in order to improve the system. In this chapter we will develop a critical analysis over the present regime, pinpointing some aspects which we consider worthy of special attention by the EU regulator.[3]

We will start by targeting our analysis to the constitutional and institutional levels. Apart from increased restraints posed in an enlarged EU by the unanimity rule in tax matters, there is another mounting problem. This is the absence of an explicit energy chapter in the EC Treaty, which translates into a lack of explicit authority for EC treaty-making in the energy field. The maintenance of this situation might bring problems with the emergence of more detailed and comprehensive specific EU energy law aimed at an Internal Energy Market (IEM).[4]

The new Treaty Establishing a Constitution for Europe has addressed the issue. However, this is not a guarantee that the problem will be overcome soon. On one hand, following the decision taken by the European Council meeting on 16–17 June 2005, the treaty ratification process was suspended in Member States *sine die* to restart. On the other hand, the explicit reference to a 'Union policy on energy' found in art. III-256 of the Treaty is insufficient. In spite of clarifying further who is the competent authority in this subject, whether the Member States or the Community, the Treaty provision does not ease the way for the insertion of environmental concerns within this policy.

The EU enlargement increased the disparity in energy prices charged within the European market. Despite the efforts made so far, there are still important differences among Member States' environmental and tax regimes with a direct effect on energy prices charged by national producers.[5] The problem has been brought up by some Member States, who invoked unacceptable state aid regimes in other Member States and non-EU countries that harm their industry competitiveness.

With the liberalization of energy markets in Europe the referred problem can build up. Trade restrictions in the IEM, which have their roots in the regulation posed by the energy directives require special attention. Furthermore, it should be questioned whether the present legal framework is adequate to promote renewable energy sources and overcome the negative environmental effects associated to the European Court of Justice (ECJ) decision in the *Outokumpu Oy* case.[6]

EU energy directives promote energy efficiency and the use of some energy sources against others. Directive 2003/96/EC[7] bares special importance in this issue. It sets a favourable context for cogeneration, development of renewable energy, rail and river transports. There are, hence, several aspects in this set of rules that might raise concern. First, it can be questioned if the EU energy tax regime is in agreement with the Polluter Pays Principle. Moreover, market distortions due to such promotional measures occur. These distortions should not advantage or disadvantage energy production and trade beyond what is required for environmental reasons. We shall address these topics below through the analysis of some specific cases.

Some of the criticisms drawn in this work to the EU energy tax regime ended up being recognized by the EU Commission in its Green Paper on Energy Efficiency.[8] In this document the Commission pointed out the need to analyse further opportunities with a view to strengthening the positive impact of taxation of policies in favour of greater energy efficiency. Among the aspects emphasized was the need to concentrate efforts as regards excise taxes on a few essential political areas (for example, harmonize rates where substantial problems involving distortion of competition arise and use differentiated tax measures in order to promote renewable energy).

Furthermore, the Green Paper pointed out the importance of bringing excise rates on energy products and electricity consumed in production activities closer together, but at the higher end of the scale, and introducing automatic indexing of all excise rates in order to avoid erosion by inflation. Other aspects considered worthy of attention were the tax treatment of transport, both as regards excises and VAT, the conditions for application of border trade adjustments, the tax treatment of inputs for heat production, in particular for large housing developments, and the rationalization of tax exemptions and exceptions.

2. INSTITUTIONAL COMPETENCES FOR ENVIRONMENTAL AND ENERGY POLICY INTERVENTIONS

2.1 Inserting Environmental Concerns in the EU Energy Policy

An explicit reference to a common energy policy (Union policy on energy) was done for the first time in the Treaty Establishing a Constitution for Europe (art. III-256). The introduction of such provision has been repeatedly pointed out as necessary to support Community's intervention in energy related issues.[9] In the absence of such legal provision, the European Union (EU) acts based on art. 308 of the EC Treaty, since Member States have competence under the ordinary law whereas the EU must enjoy only the competences allocated to it. These competences have to be defined pursuant to the principles of subsidiarity[10] and proportionality.

The Community's actions in the field of energy policy, for the most part, have pursued the aim of establishing a common market. This single-minded approach might be due to the fact that there has been no primary law provision on substantive objectives such as the supply of energy services in conformity with the principle of sustainability. With the new treaty expressly mentioning environmental objectives within energy policy, the capacity of the Community to introduce sustainable development considerations into this policy design can thus be enhanced and the burden on art. 308 of the EC Treaty eased.

The two main concerns supporting the new provision on a common energy policy are the establishment of an internal market and the protection of the environment. In order to fulfil these objectives, the EU shall, on one hand, ensure the functioning of the energy market and security of energy supply in the EU, and, on the other, promote energy efficiency and saving, as well as the development of new and renewable forms of energy (art. III-256).

According to art. III-256, European laws or framework laws on energy policy shall not affect a Member State's right to determine the conditions for exploiting its energy resources, its choice between different energy sources and the general structure of its energy supply. However, this right can be restricted based on art. III-234/2(c), and only on this provision, which belongs to the section of the EC Treaty dedicated to EU policy on the environment. However, when the option provided by art. III-234/2(c) is used, the European measures significantly affecting a Member State's choice have to be adopted unanimously by the European Council (the Council).

The regime found in the Treaty on European Union (TEU) is identical to the one just described: '[A]ll the rules or conditions relating to the taxation of Members or former Members shall require unanimity within the Council' (art. 190/5 TEU), namely provisions of a fiscal nature taken by the Community in order to achieve the objectives of its environmental policy referred to in art. 174 TEU (art. 175/2(a) TEU). Furthermore, measures significantly affecting a Member State's choice between different energy sources and the general structure of its energy supply shall also be adopted by the Council acting unanimously on a proposal from the Commission and after consulting the European Parliament and the Economic and Social Committee (art. 175/2(c) TEU).

Therefore, it was not given any step to remove the long-time obstacle to the implementation of environmental taxation at a supranational level ('politically sensitive subject'[11]). The 1990s failures in this matter[12] are doomed to be repeated in an enlarged EU. The obstacle to the adoption of an EU energy and/or carbon dioxide regulatory tax represented by the unanimity rule in tax matters was not eased by the Treaty Establishing a Constitution for Europe.

2.2 Distribution of Competences in the Treaty Establishing a Constitution for Europe

Energy and environmental matters constitute policies complementing or flanking the single area competences and are shared by the EU and the Member States, and thus the EU must lay down the general rules. However, in all the cases of shared competences, the distribution of competences between the Member States and the Community is *de facto* determined as much by the treaty provisions as by the will of the Member States expressed in the way they involve themselves, via the Council, in the EU's decision-making procedures. Furthermore, paralysis caused by inter-governmental procedures and unanimous decision-making has resulted in competences theoretically conferred on the EU by the Treaties being retained, 'for no good reason', by the Member States.[13]

Competences not conferred upon the EU in the Constitution remain with the Member States (principle of conferral, art. I-11/2 of the Treaty Establishing a Constitution for Europe). The constitutional treaty (art. I-14/2) provides that the EU holds a shared competence with the Member States in the main areas related to energy taxation, that is, energy and environment. This means that the EU and the Member States shall have the power to legislate and adopt legally binding acts in these areas. Member States shall also exercise their competence to the extent that the EU has not exercised, or has decided to cease exercising, its competence (art. I-12/2). Though, on the environment the protective provisions adopted pursuant to art. III-234 shall not prevent any Member States from maintaining or introducing more stringent protective provisions, which shall be compatible with the Constitution and notified to the Commission (art. III-234/6).

As far as environmental policy is concerned, the EU is competent to enact European laws or framework laws establishing what action is to be taken in order to achieve the objectives referred to in art. III-233. Under normal circumstances, the EU legislative power is exercised in the following way. The European Parliament, jointly with the Council of Ministers, shall enact legislation (arts I-20/1 and I-23/1). Except where the Constitution provides otherwise, EU legislative acts can be adopted only on the basis of a Commission proposal (art. I-26/2). To fulfil the EU's objectives on environmental policy, legislative acts shall be adopted after consultation of the Committee of the Regions and the Economic and Social Committee (art. III-234/1). The same is required with regard to the EU policy on energy (art. III-256/2). However, art. III-234/2(c) introduces a derogation stating that the Council of Ministers shall unanimously adopt European laws or framework laws establishing measures primarily of a fiscal nature and measure significantly affecting a Member State's choice between different energy sources and the general structure of its energy supply.

The possibility of considering environmental taxes as instruments of environmental policy, and thus sheltered from the principle of unanimity, was already discussed by EU institutions. To clarify the issue, it was decided to expressly state that measures primarily of a fiscal nature within environmental policy shall be approved according to the principle of unanimity. This statement in art. III-234/2(a) of the Treaty empowers the Member States to block the adoption of new environmental tax measures on energy related aspects.[14] This is so despite the fact that the Community is legitimized to intervene by both the criteria of relevant scope and of synergy in environmental matters.

In spite of considering such measures environmental policy measures, otherwise they would not be mentioned in art. III-234, it was decided to value more strongly the arguments in favour of the maintenance of the bulk of fiscal and tax powers at national level than the motives supporting the sharing of competence in environmental policy. This option might require the clarification of what is a 'measure primarily of a fiscal nature' to avoid the election of the legal regime more according to the form than the substance of the intervention.

Furthermore, the step taken in the Treaty might indicate that Member States and the EU institutions are quite aware that many 'environmental taxes' are, in fact, aimed first at revenue collection and only secondarily at environmental improvements. This is an approach that can be especially common on the adoption of fiscal measures concerning energy matters. Therefore, it is not surprising the willingness to keep this realm sheltered from the qualified majority procedure.

Another aspect to consider is that the provision found in art. III-234/2 hinders not only the advancement of EU environmental policy but also of EU energy policy. This can be explained by the two following aspects. First, some of the main objectives pursued by EU energy policy are environmentally connected (art. III-256/1(c) – 'promote energy efficiency and saving and the development of new and renewable forms of energy'). Secondly, tax measures are among the most important instruments available to develop an EU energy policy. The decision rule within energy policy is 'qualified majority'. This is so by default, as the silence of art. III-256 concerning this aspect refers the legislator to art. I-23/3. However, the reference found in the last paragraph of art. III-256/2 to art. III-234/2 does not allow any debate on which should be the decision rule to apply when energy policy is developed via tax measures.

3. CRITICAL ANALYSIS OF DIRECTIVE 2003/96/EC

The rationale of the Energy Products Taxation Directive[15] is not the introduction of a new tax but rather a framework of rules to restructure and harmonize national tax systems within the context of the Single Market. Furthermore, it aims to widen the scope beyond mineral oils to include the competing sources of energy, such as coal, lignite and natural gas as well as electricity, so as to strengthen the incentives for a more efficient and less polluting use of energy.[16]

Building up an Internal Energy Market and promoting the reduction of environmental damage caused by energy production and consumption are two concerns which guide this directive. Keeping in mind these aspects we shall develop a critical analysis of some aspects of the regime laid down by the directive. These are points where we understand there is space for improvement either with gains for the internal market or the environment or for both.

3.1 Environmental Incentives Requiring Further Improvements

3.1.1 Price corrections

The Community framework allows Member States to exempt or reduce excise duties so as to promote biofuels. Although distortions of competition shall be limited and the incentive of a reduction in basic costs for producers and distributors of biofuels shall be maintained through periodic adjustments. These have to be carried out by Member States taking into account changes in raw materials prices (art. 16/3). This also means that the fiscal treatment of biofuels has to be corrected according to the evolution experienced in production costs.

The price evolution is, however, not taken into account in motor fuels, heating fuels and electricity taxation, since the minimum levels of taxation applicable (Annex I) are defined in euros by unit of energy source and are not indexed. A periodical renegotiation of the levels set in the Directive is to be expected, but this can end up being a very prolonged process, as the lengthy adoption process of the Directive itself shows. We have to recall that, in spite of the provision found in art. 10 of Directive 92/82/EEC of 19 October 1992 (which required the minimum rates set for the excise duties to be updated every 2 years), in 1998 no price correction had yet occurred. This regime will, therefore, tend to lead to an erosion of the tax incentive provided to renewable energy sources.

This flaw can be attenuated through an accurate application of the EU Community guidelines on State aid for environmental protection.[17] The Commission accepts payments to new plants producing renewable energy calculated on the basis of the external costs avoided. 'External costs avoided' are environmental costs that society would have to bear if the same quantity of energy were produced by a production plant operating with conventional forms of energy. These shall be calculated on the basis of the difference between, on the one hand, the external costs produced and not paid by renewable energy producers and, on the other hand, the external costs produced and not paid by non-renewable energy producers. Furthermore, this calculation shall be carried out based on internationally recognized methods.

In spite of the important corrective role the Community guidelines can play concerning the problem identified, these guidelines may be insufficient to overcome the flaw left in the Directive 2003/96/EC. The referred guidelines operate a positive discrimination which practical application depends on the amount of public financial resources available. Therefore, they can prove a delicate means of intervention in times of budget crisis. Moreover, the Polluter Pays Principle would recommend the internalization of external costs to reduce the traditional advantage provided by the *status quo* to polluting energy sources.

3.1.2 Energy intensive consumers

Some of the dispositions inserted in the Directive are not in agreement with environmental interests and some opportunities for improvement can be pinpointed, among these is the treatment provided to energy-intensive consumers. These consumers benefit from legal protection the more energy they use, being partially protected until their consumption for heating purposes reaches 50% of their costs per product and totally protected thereafter. In the latter case it is placed no restriction on use purposes.

This uneven distribution of emission control costs, also observable in most national energy tax systems already before 2003[18] is in breach of the Polluter Pays Principle. Different tax burdens lie on energy consumption depending on the identity of the consumer (households *versus* the industry, and different economic sectors) raise some problems. Special motivation is required to support different legal treatments among equally polluting subjects, namely equity concerns, competitiveness reasons, and feasibility arguments. Cost responsibility, ability to avoid the damage and efficiency are criteria which require a balanced distribution of the tax burden.

The Directive does not apply to electricity when it accounts for more than 50% of the cost of a product (art. 2/4(b)), being this cost calculated by the addition of effective costs at the level of the business and per unit on average. By failing to take as reference a benchmark, this provision does not pressure for energy efficiency. The criticism is all the more valid as we notice that, from a long-term perspective, competitiveness issues could be even better tackled if, instead of real costs, 'best possible performances' were taken into consideration. It is worth mentioning the Dutch case, where the referred exemption is only assigned when there is an environmental agreement (covenant) forcing the adoption of energy efficiency measures.[19]

3.1.3   Tax differentiation according to product characteristics and production inputs

Article 5 allows Member States to apply differentiated tax rates when these are directly linked to product quality. This enables tax differentiation according to environmental criteria based on the characteristics of the product. In those cases where the use of different energy sources to produce the final energy product does not affect its final characteristics, as occurs in the production of electricity with different kinds of energy sources, tax differentiation is not allowed by the Directive. Thus, the lack of equivalence between differentiation based on products' characteristics and processes' characteristics, common to the General Agreement on Tariffs and Trade (GATT)/World Trade Organization (WTO) regime, is adopted by this Directive.

The Directive also allows Member States to exempt energy products and electricity used to produce electricity. However, they may, for reasons of environmental policy, subject these products to taxation without having to respect minimum levels of taxation. In such case, the taxation of these products shall not be taken into account for the purpose of satisfying the minimum level of taxation on electricity laid down in art. 10 (art. 14/1(a)). Therefore, Member States that decide to tax the inputs used to produce electricity can do so but must also tax electricity according to the minimum taxation levels set in the Directive.

Taxation of inputs cannot be used to split the total tax burden levied on electricity, being only an added charge on the latter. Member States which act more effectively in environmental terms, promoting the use of renewable energy sources to produce electricity, are punishing the producers using traditional energy sources operating in their territory. Since they cannot differentiate the tax level applied to imported energy produced with renewable energy sources and traditional energy sources. The most rational behaviour from an economic perspective will then keep on being the one taken by the Finnish Government following the decision in the *Outokumpu Oy* case.

Attending to arts 5 and 14/1(a), the solution remains on being tax differentiation only at the output level, and not at the input level. Furthermore, art. 15/1(b) allows Member States to apply total or partial exemptions or reductions in the level of taxation to electricity obtained from renewable energy sources. This means that the Directive sets as the level playing field the taxation of electricity produced with traditional inputs, which is not the best expression of the Polluter Pays Principle, since such reference point assigns the right to the polluter while the referred principle gives it to the society. The combination of arts 5, 14/1(a) and 15/1(b) leaves Member States with a single option to internalize positive externalities. This regime hinders the possibility to internalize negative externalities associated to electricity production without harming the competitiveness of national industry in the short and medium term.

Another solution, that is, the differentiation of the tax level according to the inputs used, could only take place without harming the internal market if there was a uniform taxation of the inputs across the EU. However, even this solution would harm the competitiveness of the EU power industry as the GATT/WTO regime does not allow differentiation according to the characteristics of the process, where the characteristics of raw materials are included.

Tax differentiation at the output level is not the most effective way to promote cleaner electricity production. This difficulty is due to the characteristics of this product, ie, its homogeneity independent of the inputs used, and the moment when the tax becomes chargeable ('the time of supply by the distributor or re-distributor', art. 21/5), which is also a consequence of the characteristics of the product. However, the legal technique adopted is the solution most coherent with the destination principle that should guide consumption taxes. The problem does not occur in the case of production for self-consumption. Hence, Member States may exempt small producers of electricity for their own use provided that they tax energy products used for the production of that electricity (art. 21/5).

To facilitate trade in electricity produced from renewable energy sources and to increase transparency in consumers' choice between electricity produced from non-renewables and electricity produced from renewables, the guarantee of origin of such energy is necessary.[20] This guarantee allows the application of different tax regimes to electricity based on the energy sources used for its production, which previously was only possible at the domestic level. The issue of certificates can, thus, help to surpass the above-mentioned restriction, since it lays objective and non-discriminatory conditions to set such differentiated system, which promotes the use of cleaner energy sources to produce electricity. Hence, as far as electricity is concerned, conditions are created to overcome problems existent in the previously described legal framework.

Guarantee of origin can be a helpful instrument to promote renewable energy production at the national level without hindering the internal market. This is not, however, the only possible way to achieve such objective. The EU Directive allows Member States to use tax expenditure measures to promote biofuels. Notwithstanding, some Member States found it more useful to use direct subsidies rather than tax benefits. This approach is coherent both with the Energy Directive and the Community guidelines on state aid for environmental protection.[21] The direct subsidization approach was taken, for example, by the Netherlands and Finland.[22]

With subsidies to promote renewable energy sources instead of tax expenditure measures, the Dutch Government expected to be able to fine-tune its regulatory intervention.[23] It allowed it to take more carefully into consideration the differences among several energy sources and the precise kind of support needed by each type of producer. Furthermore, direct subsidization of the renewable energy sector avoided the leakage effect, which is inevitable when tax measures are used. Since, in the latter case, internal market regulations imposed the spread of the incentive provided to national producers to any foreign producer selling renewable energy to the Netherlands. The country was, thus, able to promote national production of renewable energy without breaching EU law and avoiding the problem Finland faced after the ECJ decision in the *Outokumpu Oy* case. Moreover, this was done without the necessity to use the guarantee of origin.

3.1.4 Taxing energy consumption

It is possible to raise the question whether, despite electricity production being a sector covered by the emission allowance trading scheme, there could still be a point in taxing energy consumption if this could bring an environmental effect in terms of energy efficiency. The caps put on energy industry, for instance in Germany, are considered so low that few price changes are expected from the application of the trading scheme.[24] Furthermore, electricity market liberalization might cause a production cost reduction. However, possible positive effects from taxing energy consumption might not be very relevant due to the low price elasticity of energy consumption.

Furthermore, there is a great downturn in taxing energy consumption and this is the regressive impact. The Netherlands considered the option and gave it up due to this motive and the lower efficiency such an approach entails when compared with the use of an emission allowance trading scheme covering the production sector, since taxation is a national measure while the emission allowance trading scheme has a European scope. Moreover, if energy consumption demand increases, the emission allowance trading scheme will respond via the market price increase of tradable allowances. In the Netherlands it was suggested to tax natural gas consumption instead since the EU allowance trading scheme does not cover its extraction.[25]

Article 15/1(g) of the 2003 Energy Directive allows Member States to apply total or partial exemptions or reductions in the level of taxation to natural gas if the market share of this product in final energy consumption was less than 15% in 2000 (infant industry argument). The tax benefit may apply for a maximum period of 10 years after the entry into force of the Directive, i.e., until 31 October 2013, or until the national share of natural gas in final energy consumption reaches 25%, when this occurs at an earlier date. Thus, complying with the conditions set, the Member State is allowed to exempt natural gas from energy taxation until this energy source reaches a maximum market share of 20% of final energy consumption.

However, even when the natural gas market share was below 15% in 2000, the total exemption can only apply until it reaches a 20% level. Between 20–25% of market share, the Member State has to apply a strictly positive level of taxation to this product. The level, which shall from then on, increase on a yearly basis in order to reach at least the minimum rate through which harmonization is attempted. This means that when the natural gas market share reaches one quarter of the market (25%) it is obligatory to tax the product according to the minimum levels of taxation set in the Directive.

3.2   Market Distortions not supported by Environmental Reasons

The 2003 Directive might allow some distortion between industries due to an asymmetric tax treatment of different energy sources neither supported by internal market reasons competitiveness arguments, nor environmental concerns. For instance, the aluminium industry mainly uses electricity, whereas the steel industry uses mainly coal. Both are energy intensive sectors but just the first one is entitled to benefit from tax reductions (art. 17), though the latter are explained by competitiveness reasons.

In line with art. 2/4(b) of the Directive, Member States can decide not to tax electricity used in metallurgical processes. Therefore, electricity consumed by the aluminium industry can have a tax treatment similar to the one used by the steel industry.[26] However, by not taking into consideration the use of different energy sources by these two European energy intensive industries, as no differentiation is done according to the inputs used to produce electricity (art. 14/1(a)), the Directive is allowing market distortions (aluminium and steel can be substitute goods in some cases), which might not entail any environmental benefit.[27] Thus, some more transparency is required in the design of energy tax exemptions.

Furthermore, energy products and electricity used to produce electricity shall be exempted from taxation (art. 14/1(a)). Member States may, for reasons of environmental policy, subject these products to taxation if they so wish. Such an option is, however, complex as already explained. Thus, as uranium is not subject to the Directive and the nuclear industry is not covered by the emission trading scheme, nuclear power plants are assigned an advantage over power plants using traditional energy sources and, more problematic, to further development of renewable energy sources as an alternative. This diverse tax treatment can explain some tax proposals on energy consumption.[28]

Some other special treatments allowed by the Directive might also undermine the development of renewable energy sources. Let us take, as an example, the possibility given to Portugal to apply levels of taxation on energy products and electricity consumed in the Autonomous Regions of Azores and Madeira lower than the minimum level of taxation laid down in the Directive in order to compensate for the transport costs incurred as a result of the insular and dispersed nature of these regions (art. 18/7). In spite of some caution introduced in the regime, expressed in the fact that the reduction may not be greater than the additional costs incurred in transporting the fuel oil to those regions (Annex II, point 12, third recital), incentives to the development of renewable energy sources within these regions are thereby eroded.

Isolated regions, such as islands and mountain regions, bear high costs to use traditional fuels. Since not only it is often difficult or even impossible, due to physical or economic reasons, to extend the electrical distribution networks to these areas, but also it can be quite costly to store electricity in the region. These obstacles create an opportunity for local development of renewable energy sources. By allowing the mentioned tax benefit, the Directive reduces the extra costs involved in the use of traditional energy sources and therefore the incentive to develop renewable energy sources in the Portuguese Autonomous Regions.

These aspects of energy taxation, and the huge amount of resources applied by the EU in the technological development of nuclear energy when compared with the amounts devoted to the development of renewable energy sources, seem to indicate that a silent but spread political choice was already made between nuclear energy and renewable energy sources as alternative to fossil fuels. This raises serious issues since for an important part of the European citizens nuclear energy is not a safe and desirable option for the future.

3.3 Total Energy Tax Burden on Road Transports

We will start the analysis of the energy tax burden on road transports set by Directive 2003/96/EC by introducing some of its most relevant provisions, namely the ones which are part of art. 7. In art. 7/4, it is possible to read the following:

> Notwithstanding paragraph 2, Member States which introduce a system of road user charges for motor vehicles or articulated vehicle combinations intended exclusively for the carriage of goods by road may apply a reduced rate on gas oil used by such vehicles, that goes below the national level of taxation in force on 1 January 2003, as long as the overall tax burden remains broadly equivalent, provided that the Community minimum levels are observed and that the national level of taxation in force on 1 January 2003 for gas oil used as propellant is at least twice as high as the minimum level of taxation applicable on 1 January 2004.

The above mentioned paragraph 2 (art. 7/2) has the following content:

> Member States may differentiate between commercial and non-commercial use of
> gas oil used as propellant, provided that the Community minimum levels are observed
> and the rate for commercial gas oil used as propellant does not fall bellow the national
> level of taxation in force on 1 January 2003, notwithstanding any derogations for this
> use laid down in this Directive.

Therefore, art. 7 of the Directive only exempts Member States from the second condition established in art. 7/2, this is 'the rate for commercial gas oil used as propellant does not fall below the national level of taxation in force on 1 January 2003'. The 'overall tax burden' mentioned means energy taxation plus road charges. Thus, it is required that the tax level on 1 January 2003 is twice the minimum level fixed for 1 January 2004 onwards. This is the minimum level of taxation by the Member States that is accepted by the Community.

If the level of taxation observed on 1 January 2003 can be lowered, as long as the new level of taxation on 1 January 2004 is not less than half of the one observed on 1 January 2003, the manoeuvring space for the Member States to compensate for a reduction in gas oil taxation with road charges is equivalent to half of the taxation level observed on 1 January 2003, but not more than that. The Member States can, thus, at most split the tax burden observed on 1 January 2003 on equal shares between fuel taxation and road charges. This condition, and the requirement that the Community minimum levels (art. 8/1, Annex I, Table B) are observed, prevents a strong reduction in the taxation level, which could harm the internal market and the tax base of neighbouring countries.

3.4 The Kerosene Issue

The possibility given to Member States to apply total or partial exemptions or reductions in the level of taxation to energy products and electricity used for the carriage of goods and passengers by rail, metro, tram and trolley bus (art. 15/1(e)) is especially relevant, considering the obligation to exempt from taxation energy products supplied for use as fuel for the purpose of air navigation other than private pleasure-flying (art. 14/1(b)). There is a legal constraint to tax air transports dating back to 7 December 1944, when the Convention on International Civil Aviation (also known as the Chicago Convention) was signed.

This Convention governs the treatment of fuel already on board an aircraft, and in its original version requires the contracting parties to exempt from excise duties such fuel (art. 24(a)). Since then, and in order to promote the development of this infant industry, several bilateral agreements have been signed exempting from tax fuels used both in international and domestic flights. Today, the treatment of fuel loaded on to an aircraft is governed by a large number (more than 120) of bilateral Air Service Agreements (ASAs), which exist between individual Member States and also between Member States and Third Countries. These usually contain a clause to the effect that both fuel in transit and fuel supplied in the territory of the contracting parties are exempt from fuel taxes.[29]

There are strong protests against the tax treatment of kerosene, due to the market distortion and environmental damage it induces. These protests have been voiced mainly by the rail industry (for example, the Deutschebahn[30]), which is losing market share to low-cost airlines in medium and short distance routes, and environmental NGOs have brought the issue to the political agenda. Following a study completed in 1998, the Commission suggested the taxation of all the fuel sold in EU airports as an alternative preferable to the taxation of intra-Community flights.[31] The arguments for the choice were mainly environmentally and revenue-related (employment promotion and the 'double dividend' argument).

The tax privilege given to this industry subsists, though. The obligatory exemption contained in Directive 92/81/EEC of 19 October 1992, art. 8/1(b), was maintained by Directive 2003/96/EC. To remove the exemption for fuel supplied on the territory of the contracting parties it is necessary to renegotiate the current legal regime established by those international treaties, which is extremely complex due to the myriad parties involved (the Chicago Convention had, in 2004, 188 contracting States), making it a lengthy and cumbersome process. Hence, there are high transaction costs involved (path dependencies), which create an institutional barrier to change. Furthermore, the air navigation industry is one of the EU strategic areas regarding its competitive position in the world economy vis-à-vis the USA (the Boeing/Airbus 'fight'). These aspects and the post 9/11 crisis have postponed the revision of the tax privilege benefiting air transport.

It is, however, worth noticing that the 2003 Directive (art. 14/2) allows Member States to limit the scope of the exemption to international and intra-Community transports, being able to tax fuel used in domestic flights. Furthermore, it does not create an obstacle for Member States to review their bilateral treaties which are constraining their tax behaviour concerning air navigation fuel. Where a Member State has entered into a bilateral agreement with another Member State, it may also waive the exemption. In such case, Member States may, however, apply a level of taxation below the minimum level set out in the Directive. The Community seems to accept that if Member States are already doing something in the right direction, voluntarily and at the expense of their own competitiveness, they should not be forced to go all the way through.

The International Civil Aviation Organization (ICAO), despite being a specialized agency of the UN linked to the Economic and Social Council, has a strong representation from civil aviation companies. This influence can be noticed in the lobby the ICAO was playing in 2004, to set a multilateral treaty to ensure the continuation of the kerosene exemption and to keep the aviation sector out of the emission allowance trading.[32] This could be proof of a blocking reaction to the discussion on the adoption of aviation and port charge directives taking place in the EU Commission and where such solutions were being considered as a way to compensate for fuel tax exemption[33], as the EU would 'lose its right to initiate policy' in the area.

Earlier on, in January 2001, European airports backed aviation carbon emission trading.[34] They had called for the aviation sector's carbon dioxide releases to be covered by the EU emissions trading scheme from 2008, with the trade body Airports Council International stating that trading should be embraced as preferable to charges or taxes, which were also being considered by the EU and labelled 'crude, blunt and unacceptable' alternatives.

Emissions from airport facilities were already covered by the emission trading scheme, therefore this position put the focus on the aviation industry. By then, a joint industry position was already forming in the UK, where the heads of two British airlines, British Airways and Virgin, had publicly declared their support for trading. Scandinavian, Dutch and French carriers were also thought to be in favour, though German and Italian airlines were less enthusiastic.

On 14 June 2005, the Commission launched the review of the EU Emissions Trading Scheme, whose results are scheduled to be fed into a Commission report to be submitted to Parliament and the Council by mid-2006. The survey will address the issue of international competitiveness that business associations have long been calling for. However, the expected news on major changes, namely the possible addition of new gases and sectors, including aviation, are to be brought in only for the second trading period in 2008–12, since even if proposals are brought forward, the length of the legislative procedure means they will not take effect before 2013. The second round of National Allocation Plans (NAPs) for the second trading period is due on 30 June 2006.

Therefore, by accepting to be submitted to the EU Emissions Trading Scheme as a condition to avoid taxation the aviation industry got a win-win result. With the manipulation of the regulatory process this industry played, it managed to award itself, at least for the moment, a market advantage expressed in the going on with its business-as-usual until 2013. In fact this is not an original strategy. The USA power industry also accepted a SO_2 tax, to which it had been opposed for a long time, when it realized that there was a high probability that the 1990 Clean Air Act would be approved.[35] Environmentally related taxes were attractive because, in a period of crisis for the industry (as the early 2000s have been), it was expected to be a successful political lobby concerning the tax level. However, for the European aviation industry the emission trading scheme offered multiple advantages. On one hand, this instrument could act as a barrier to new entrants in the market and, on the other, it allowed a pollution-for-free threshold. Moreover, it would be easier to postpone regulation if this took the form of emission tradable permits rather than taxes, due to the rigid timetable set for the legislative process involving the EU Emissions Trading Scheme.

3.5 Tax Treatments in Breach of EU Law

Another problem in the text of the Directive might arise from the fact that it allows most Member States (the exemptions are Denmark, Greece and Sweden) to provide tax benefits for waste oils which are reused as fuel, either directly after recovery or following a recycling process for waste oils, and where the reuse is subject to duty.[36] This might be considered in breach of EU law, Sweden having already being condemned at the ECJ in connection with this issue.

In Cases C-201/03, 30 March 2004, and C-424/02, 15 July 2004, the ECJ declared that by failing to take the necessary measures under art. 3/1 of Directive 75/439/EEC (16 June 1975) on the disposal of waste oils, as amended by Directive 87/101/EEC (22 December 1986), to ensure that priority is given to the processing of waste oils by regeneration where technical, economic and organizational constraints so allow, Sweden and the UK, respectively, have failed to fulfil their obligations under that directive. The ECJ also did not accept as a valid argument to not comply with the directive the existence of obstacles to waste oil regeneration, namely, the importance of the market for recovered waste oil for use as fuel and the weak market for regenerated oil.

Member States are required to promote waste management according to the following hierarchy: first reuse, second recycling and third recovery.[37] In the case of waste oils this requirement demands regeneration to take precedence over any other form of waste management, namely waste oils recovery as fuel. However, such hierarchy is not reflected in the energy taxation Directive, being indifferent for the concession of the tax benefit the precedence established in waste oils management.

4. CONCLUSION

On the way to an IEM in an enlarged EU, the need for an explicit authority for supranational treaty-making in the energy field builds up. This problem was addressed at the EU constitutional level in the Treaty Establishing a Constitution for Europe. However, an appropriate level of response is still missing, due both to the conservatory approach to the issue taken by the Treaty and the obstacles faced in its ratification process. The insertion of environmental concerns within energy policy still lays very much in the hands of the Member States.

Directive 2003/96/EC represents an important step towards a level playing field for renewable energy sources. However, it is still possible to find some space for improvement in the present legal framework. The promotion of renewable energy sources could be made more effective if corrections according to price evolution were imposed on the fiscal treatment conferred not only on bio-fuels but also to other fuels. Besides this aspect, concerning the stimulus affecting the choice over the kind of energy source, environmental interests could also be more actively pursued by the Directive in terms of energy efficiency. This would be the case if the treatment provided to energy-intensive consumers did not get more beneficial as increasing levels of consumption are reached.

Furthermore, some aspects of the legal framework set by Directive 2003/96/EC fall short of full compliance with the Polluter Pays Principle, while others might be considered in breach of the EU law. It is especially worth mentioning here the fact that the energy taxation Directive does not reflect the hierarchy set by the EU for waste management. The latter problem has already emerged in some decisions of the ECJ. It is therefore important to co-ordinate the legal provisions in order to preserve the systemic order of the *acqui communautaire*.

The EU Community guidelines on State aid for environmental protection can help to overcome the above mentioned uneven treatment, provided to biofuels and other fuels as far as price correction is concerned. However, the Polluter Pays Principle favours the internalization of external costs to reduce the traditional advantage provided by the *status quo* to polluting energy sources rather than the conferral of positive discrimination to cleaner ones.

Likewise, by allowing Member States to apply total or partial exemptions or reductions in the level of taxation to electricity obtained from renewable energy sources, the Directive sets as level playing field the taxation of electricity produced with traditional inputs. By doing this it conflicts with the Polluter Pays Principle, since such reference point assigns the right to the polluter, while the principle gives it to the society.

Moreover, the Directive allows differential tax treatments that simultaneously distort the market and harm the environment without being supported by solid competitiveness arguments. On one hand, the aluminium industry is favoured over the steel industry, despite not only the energy intensiveness common to both industries but also the intersubstitutability which characterizes some of their products. On the other hand, nuclear power plants are advantaged over power plants using either traditional energy sources or renewable energy sources.

The 2003 Directive opened the door for a more active role of national tax policies towards environmental protection. However, this was more an option given to Member States than an imposition upon them, the kerosene issue being a good example of this approach. In any case, the concern over the internal market worked as a constraint to the manoeuvring space allowed to Member States, as demonstrated by the restricted power they were assigned to compensate for a reduction in gas oil taxation with road charges.

NOTES

1. Lecturer in Tax Law, Energy Law and Environmental Taxation, Portuguese Catholic University, e-mail: casoares@porto.ucp.pt.
2. Wälde, 2003. On energy security consult the report prepared for DGTREN, 'Study on Energy Supply Security and Geopolitics', January 2004.
3. This critical approach was extended to the EU Community guidelines on State aid for environmental protection, with especial emphasis on support schemes for the development of renewable energy sources projects, in Claudia Dias Soares, 'The case of renewable energy sources in the EU: Between market liberalization and public intervention', Communication presented at the Sixth Annual Global Conference on Environmental Taxation: Issues, Experience and Potential, 22–24 September 2005, Leuven (Belgium), published in *Critical Issues in Environmental Taxation IV*, (2006) Richmond Law & Tax.
4. Directive 96/92/EC, 19 December 1996, OJ L 27, 30 January 1997, followed by Directive 98/30/EC, 22 June 1998, OJ L 204, 21 July 1998, for natural gas.
5. Further information can be found in the inventory study on subsidies in the energy sector published in December 2002 by the European Commission.
6. Judgment of the Court of Justice of 2 April 1998 in Case C-213/96 *Outokumpu* [1998] ECR 1-1777.
7. OJ L 283, 31 October 2003.
8. COM(2005) 265 final, 10 June 2005.
9. Hannes Farnleitner, Paper on the delimitation of competences between the EU and its Member States (CONV 58/02), 21.05.2002; Draft Report of the 'Complementary Competences' Working Group (WG V – WD 30), 02.10.2002; Contribution of Ponzano, Commission representative in the 'Complementary Competences' Working Group (WG V – WD 26), 24.09.2002; Contribution of Mr Farnleitner, Mr Einem and Mr Bösch (CONV 358/02), 22.10.2002.
10. The Community intervention is legitimized based only on the criterion of relevant scope, the criterion of synergy or the criterion of solidarity. According to the criterion of relevant scope, the Community is competent if the relevant scope of the proposed measure goes beyond the limits of a Member State and the measure might give rise to perverse effects (distortion or imbalance) for one or more Member State should it not be implemented at Community level. The criterion of synergy empowers the Community to act when the measure planned at Community level would generate, by comparison with similar measures implemented separately by individual Member States, substantial synergies in terms of effectiveness and economies of scale. The Community is also legitimized to intervene when the proposed measure meets a requirement for solidarity or cohesion which, in the light of disparities in development, cannot be met satisfactorily by the Member State acting alone (criterion of solidarity). Report of the Committee on Constitutional Affairs, A5-0133/2002, 24.04.2002, p. 7/44.
11. Newbery, 2005:1.
12. Boeshertz and Rosenstock, 2003: pp. 151–61.
13. Report of the Committee on Constitutional Affairs, A5-0133/2002, 24.04.2002, pp. 7–8/44.
14. The importance given by Member States to this issue has already motivated the following Declaration on art. 130r of the EEC Treaty: 'The Conference confirms that the Community's activities in the sphere of the environment may not interfere with national policies regarding the exploitation of energy resources' – European Union, 1999: p. 609.
15. Directive 2003/96/EC, 27 October 2003, Restructuring the Community framework for the taxation of energy products and electricity.
16. Jos Delbeke, DG Environment, European Commission, International Conference on the Chal-

lenge of Implementing New Regulatory Initiatives: State of affairs and critical issues of EU
Climate Change Policy, 18.09.2004, Leuven.

17.   OJ C 37, 3 February 2001, pp. 3–15.
18.   Hoerner and Bosquet, 2001.
19.   Personal communication by Bastiaan J. Gen, Dutch Ministry of Finance, 10.09.2004, Pavia.
20.   Cf. art. 5, Directive 2001/77/EC, OJ L 283, 27 October 2001.
21.   OJ C 37, 3 February 2001, pp. 3–15.
22.   Määttä, 1997: 305.
23.   Personal communication by Bastiaan J. Gen, Dutch Ministry of Finance, 09.09.2004, Pavia.
24.   Seiche, 2004.
25.   Personal communication by Coen Peel, Dutch Ministry of the Environment, 10.09.2004,
      Pavia.
26.   We are thankful to Manfred Rosenstock for drawing our attention to this aspect.
27.   Aluminium, highly electricity-intensive and internationally traded, will be specially affected
      by differential tax treatment of carbon across countries and less so by fuel taxes, as most alu-
      minium is hydro-based. Although, fuel taxes will affect the marginal cost of electricity in
      increasingly integrated EU electricity market – Newbery, 2005: 30.
28.   Seiche, 2004.
29.   Deketelaere, 2003.
30.   In 2004, the CEO of Deutschebahn decided to challenge the lack of taxation of aviation in
      Germany and the German Government has challenged the Commission for non-action with
      regard to the distortion of competition and eventually intends to take the issue to the ECJ; In-
      formation provided by Kai Schlegelmilch, Federal Ministry for the Environment, Nature
      Conservation and Nuclear Safety, Pavia, 11.09.2004.
31.   COM(2000) 110 final.
32.   *Environment Daily*, 1727, 16/09/04.
33.   Jos Delbeke, DG Environment, European Commission, 10.09.2004, Pavia. Further information
      on this issue is available at www.oecd.org/cem.
34.   *Environment Daily*, 1806, 21.01.05.
35.   Stavins, 1998: 75.
36.   I am thankful to Ludwig Krämer for having pointed out this aspect to me during the confer-
      ence taking place in Leuven, 17–18 September 2004.
37.   COM(96) 399 final, 30 July 1996.

REFERENCES

Boeshertz, Daniel and Rosenstock, Manfred (2003), 'Energy Taxes in the EU: a Case
   Study of the Implementation of Environmental Taxation at a Supranational Level', in
   Janet Milne, Kurt Deketelaere, Larry Kreiser and Hope Ashiabor (eds), *Critical Issues
   in International Environmental Taxation: International and Comparative Perspectives*,
   vol. I, Richmond, UK: Richmond Law & Tax, p. 175.
Commission of the European Communities (1999), *European Union, Selected Instru-
   ments taken from the Treaties*, Book I, vol. I, Luxembourg: Office for Official
   Publications of the European Communities.
Deketelaere, Kurt (2003), *EC Transport Policy and Environment and Energy Taxation*,
   Paper presented at the Fourth Annual Global Conference on Environmental Taxation:
   Issues, Experience and Potential, Sydney.
Fauchald, Ole Kristian (1998), *Environmental Taxes and Trade Discrimination*, London:
   Kluwer Law.
Hoerner, J. Andrew and Bosquet, Benoît (2001), *Environmental Tax Reform: The Euro-
   pean Experience*, Washington, DC: Center for a Sustainable Economy, February.

Määttä, Kalle (1997), *Environmental Taxes: From an Economic Idea to a Legal Institution*, Helsinki.

Newbery, David (2005), 'Why Tax Energy? Towards a More Rational Energy Policy', *CMI Working Paper 72*, Cambridge: CMI Working Paper Series.

Olivecrona, Christina (1995), 'Wind Energy in Denmark', in Robert Gale, Stephan Barg and Alexander Gillies (eds), *Green Budget Reform*, London: Earthscan Publications, p.55.

Seiche, Mathias (2004), 'The relation between emissions trading and ecological tax reform: for better harmony in the concert of instruments', Paper presented at the fifth Annual Global Conference on Environmental Taxation: Issues, Experience and Potential, 09–11 September 2004.

Stavins, Robert N. (1998), 'What Can We Learn from the Grand Policy Experiment? Lessons from SO_2 Allowance Trading', *Journal of Economic Perspective*, **12**(3), Summer 1998, p.69.

Wälde, Thomas W. (2003), *The International Dimension of EU Energy Law and Policy*, Advance Working Paper for forthcoming book on EU Energy Law, see www.cepmlp.org, 25 August 2005.

PART IV

Good governance for climate change:
Reflections and perspectives

15. Some reflections on the EU mix of instruments on climate change

Ludwig Krämer[1]

1. INTRODUCTION

For about 15 years, the struggle against man-made climate change has progressively become one of the top political priorities of the European Union (EU). In 1994, the EU concluded the United Nations Framework Convention on Climate Change (UNFCCC)[2] and subsequently deployed considerable efforts to reach a consensus on the Kyoto Protocol of 1997.[3] Both the Prodi Commission (1999–2004) and the new Barroso Commission (2004–09), supported in that from numerous declarations by the European Council, the EU Council and the European Parliament, declared the fight against climate change as a subject of political priority for the EU in the years to come. The importance of this position of the EU is not diminished by the fact that these political declarations, statements and resolutions also offered the EU an opportunity to gain worldwide credit and respect, in view of the well-known negative attitude of the United States (US) to take any serious measure to reduce the emission of greenhouse gases. Indeed, the EU has put its whole political and economic weight into the attempt to limit climate change that is caused by human activities and has accepted global leadership in the climate change discussion, though neither its internal structure nor the instruments and resources at its disposal enabled it to really assume such a leadership role.

This chapter neither tries to discuss the dispute between the US and the EU on climate change issues, nor other aspects of this global discussion. It rather concentrates on the EU's internal issues and considers more closely, from a point of view of the law,[4] the different instruments which the EU has put into operation or is trying to develop, in order to reach the targets which it has fixed for itself and/or which appear to be required. The contribution is based on the assumption – which hardly needs to be argued anymore, except perhaps by some people in the US – that there is a man-made climate change and that measures are necessary in order to contain that change within reasonable limits.

In a first section, the political and legal frame for an EU policy and measures will shortly be presented, followed by a description of the constitutional and

279

governance situation. Then, the different instruments which the EU has developed so far and which it intends to develop in order to fight climate change, will be reviewed. Some evaluations of the state of affairs within the EU and some suggestions for the years ahead will conclude this chapter.

2. THE EU'S POLITICAL AND LEGAL FRAME FOR CLIMATE CHANGE POLICY

2.1   The EU Constitution

The greatest part of all greenhouse gas emissions that are considered responsible for the warming up of the atmosphere and thus causing a change of the global climate, stem from the burning of fossil fuels, in particular petrol, coal, gas, lignite and wood. These materials are sources of energy. Measures that influence the production, supply and consumption of energy are normally bundled under the heading of 'energy policy'; transport of these fuels is normally classified under 'transport policy', while the effects on the global climate are grouped under 'environment policy'. Other political sectors that are affected by climate change measures are agricultural policy, taxation policy, and internal and external trade policy. Climate change is thus a very typical integrated subject, where several policy sectors need to be bundled under one heading.

The EC Treaty enables the EU to act within the limits of the powers conferred on the EU by it. The phrase 'climate change' or any similar notion does not exist in the EC Treaty. The EU also does not have a competence for energy policy, be it shared with the EU Member States: art. 3(1.u) only allows the EU to take 'measures' in the area of energy. This means that Member States alone are responsible for energy policy questions.[5]

This constitutional aspect had and has a number of consequences. Before going into them, it must be remembered that the EU is not a State; it does not have a government and its Parliament does not (alone) represent the sovereign will of the EU peoples. This makes the setting of political priorities on which 25 Member States have to be – and to remain during a certain time-span – in agreement, a rather difficult undertaking. There is great temptation to choose political objectives and make corresponding statements which are so general that every Member State can agree to them – even if within that Member State political majorities change. This is the reason why objectives such as 'sustainable development' or 'making the EU the most competitive economy by 2010', and even, perhaps, the objective to fight climate change are so attractive: they allow political leaders to demonstrate that they have visions and may easily forget that the devil is in the detail. Linguistic pollution adds much to the loss of credibility of environmental policy, globally, EU-wide or nationally.

2.2 EU Climate Change Programme

The lack of a general EU competence for energy policy has, as a consequence, meant that climate change issues were, politically, treated as particular environment protection issues. This means that Council resolutions and other statements were elaborated and adopted by the Environment Council,[6] but almost never by the Energy Council, the Transport Council or other Councils. In theory, of course, a Council resolution – or indeed, any other act – shall be politically binding[7] on all institutions and administrations at EU and national level. The administrative reality is, however, different. The rather vertical structure of administrations has meant that the energy administration – at the level of the Commission, the Council, the national, regional and international level – has their own agenda, priorities and objectives to reach. The same is true for the transport, agricultural or trade administrations.[8]

It follows from this that the different Environment Council resolutions on climate change have a political rather than legal relevance. Also, the extent to which these resolutions are politically relevant or important in the areas of energy, transport etc, depends on the political and administrative actors of that area.

In contrast to that, the decision to adopt the Sixth EU Environmental Action Programme[9] which contains long passages on climate change, was taken in a legally binding form (namely a decision) under art. 249 of the EC Treaty. However, again, it was the Environment Council which acted, with a very limited input from other political sectors; and any controversy between diverging interests in climate change was solved by a decision of the environmental ministers in the Council. This made Decision 1600/2002 not less binding; but reduces the political commitment of other administrative departments inside the Commission, the Council and Member States to make the climate change section of the Sixth Action Programme fully operational.

As a plan or programme – these two notions are indistinctively used in EU law – is 'an organized and co-ordinated system of objectives' that contains a timetable and is reviewed at regular intervals,[10] it can safely be argued that in any legal sense, there is, apart from the chapter on climate change in the Sixth Environmental Action Programme, no EU plan or programme on climate change. It is true that the Commission made, in 2000, a communication 'towards a European Climate Change programme (ECCP)'.[11] However, this Communication contained a list of possible measures that could be undertaken at EU level; and the corresponding Council Resolution[12] only indicated that the Council 'considered the list of priorities of action' and 'selected issues that have particular importance'. It then encouraged the Commission to make progress and propose concrete measures. Furthermore, the Council requested the Commission to 'study and prepare measures' in four areas of the transport sector, to 'develop policies and measures' in three areas of the energy sector and added a number of general requests for other sectors. It was thus correct that the European Parliament had regrets with regard to the Commission Communication – but this remained true also for the Council Resolution – that it 'contained more a list of wishes than a clear action plan with timetable'.[13] Furthermore, the Council resolution was not even officially published; a clear sign of the wish to further play down the political importance of the Resolution.

Of course, the legal nature of a Council act may not be that decisive, if only there is political consensus on how to proceed and which objectives to reach. The point made here, though, is that there was an interest in keeping the Resolution of 11 October 2000 as a resolution of the Environment Council, which allowed the Council in its configuration as Energy Council or Transport Council to largely ignore the Resolution of 11 October 2000 – and to allow also the Commission administrations of energy, transport and others to ignore that Resolution.

This policy aspect becomes obvious with a comparison between the Council Resolution of 11 October 2000 and the Commission's implementation report one year later.[14]

From this comparison it appears that the Commission itself did not consider the Council resolution in any way politically binding on it. Rather, the Commission selected those items from its previous list and of the Council Resolution for which it considered an EU measure appropriate.

A rather similar result is reached, when one compares the different detailed actions in the energy, transport, industry and other sectors (which the European Parliament and the Council required under the Sixth Environmental Action Programme) with the actual proposals made by the Commission some 30 months later.[18] The requests[19] include the following aspects:

- an inventory and review of subsidies that counteract an efficient and sustainable use of energy with a view to gradually phasing them out;
- encourage renewable and lower carbon fossil fuels for power generation;
- prevent and reduce methane emissions from energy production and distribution;
- identifying and undertaking specific action to reduce greenhouse gas emissions from aviation;
- identifying and undertaking specific actions to reduce greenhouse gas emissions from marine shipping;
- inviting the Commission to submit, by the end of 2002, a communication on quantified environmental objectives for a sustainable transport system;
- identifying and undertaking further specific action, including any appro-

Table 15.1

Council request in Resolution of 11 October 2000	Commission Action by 23 October 2001, Communication COM(2001) 580
Measures to reduce CO_2 emissions from light utility vehicles; the Council is ready, if necessary, to consider future measures for the reduction of CO_2 emissions from new passenger cars in order to reach the Community's established commitments.	The Communication does not mention this action at all.
Effective limitation of the increase of Greenhouse gas emissions in air transportation, bearing in mind the importance of achieving progress at the international level.	The Communication does not mention this action at all.
Limitation/reduction of CO_2 emissions from road transport by promoting the use of rail freight and passenger transportation, as well as intermodal and combined transport.	A package of actions should ensure that the growths of the road sector is curbed and the modal split of 1998 is again recovered in 2010. No concrete measure is announced.
Promote the increased use of combined heat and power (CHP), aiming at achieving the indicative target of 18% share of CHP by 2010.	The Commission intends to present a proposal in 2002.[15]
Promoted the increased use and improved competitiveness of renewable energy sources aiming at achieving the indicative target of 12% of energy consumptionby 2010.	The Communication does not mention this action at all.
Enhance the energy efficiency in buildings; of industrial processes; of equipment; and of appliances, by strengthening energy efficiency through appropriate measures, including voluntary measures.	(a) elaboration of a horizontal Best Reference Document (BREF) document on generic energy efficiency techniques will be asked;[16] (b) framework directive for minimum efficiency requirements for end-use equipment.[17]

    priate legislation, to reduce greenhouse gas emissions from motor vehicles
    including N_2O;
- promoting the development and use of alternative fuel and of low-fuel-
  consuming vehicles, with the aim of substantially and continually
  increasing their share;
- promoting measures to reflect the full environmental costs in the price of
  transport;
- developing means to assist small and medium sized enterprises to adapt,
  innovate and improve performance;
- promoting energy efficiency – notably for heating, cooling and hot water
  in the design of buildings;
- promoting the use of fiscal measures.

From all these facts, one can only conclude that the different Commission administrations and the Council (in its different formations) had and are pursuing
– as regards climate change – their own agenda.[20] An EU climate change action
programme in the legal sense – which would, in this author's opinion, have to
be elaborated in the form of art. 175(3) of the EC Treaty, i.e. with the legislative
participation of the European Parliament – does not exist, though the notion
'climate change programme' is frequently used.

2.3   Commission Climate Change Structure

The Commission itself contributed to the continuity of the situation of vertical
structures, as it never seriously tried to integrate the different administrations
that were affected by issues of climate change. Neither did it formally set up
any internal horizontal structure – such as a working group of three or four
members of the Commission or a task force – to discuss and decide climate
change issues.[21] Nor did it ever consider charging one member of the Commission with the portfolio of 'climate change'. Instead, it has allowed the different
administrations – environment, energy, transport, agriculture etc. – to pursue
their own objectives, priorities and legislative action. As in all sectors mentioned, there was enough work to do, new initiatives could be announced which
also contributed somehow to improving energy supply, energy efficiency or favoured modal shift in transport – an objective that is publicly favoured for more
than 40 years – but masked the fact that a systematic approach on reducing
greenhouse gas emissions existed more in public announcements than in
reality.

    The best illustration for this might be the Commissions White Paper on the
future of Transport Policy.[22] Though this 124 page Paper indicates that greenhouse gas emissions from transport might increase by about 50% by 2010
(compared with 1998 levels), it did not even spend one section on climate

change, the car and truck contributions to climate change, or ways and means to reduce this impact. Instead, undistorted and unhampered transport within the EU remained the overriding priority.

Over the last 10 years, there were a further three amendments of the EU Treaty which were discussed; the Amsterdam Treaty, the Nice Treaty, and the Constitution for Europe. At no time during the European or intergovernmental discussions has the Commission launched the discussion about whether the outdated Treaties on Coal and Steel (since 2002 this Treaty has expired) and Euratom (following a major blocking from France (President de Gaulle) in the 1960s, this Treaty, in very large parts, never became operational at European level) should be replaced by a more modern and up-to-date European Treaty on alternative energies, renewable energy or other forms of addressing the future needs of the EU economy. Any such step would not have been more audacious than the setting up of an Euratom Treaty in the beginning. The political and economic importance of addressing the relationship between energy questions and climate change in a Treaty for the twenty-first century need not be further elaborated here.

In conclusion, it must be stated that neither the Council, the Commission, nor Member States showed any vision to confront climate change issues for the EU as a whole. Member States continued and are continuing their national strategies, thereby maintaining the power positions and power sharing structures which national administrations have established together with national energy producers, distributors and consumers.

In view of the above-mentioned competence sharing in the EU Treaty, a change of the status quo would not have been and is not possible with legal instruments alone, but requires first political will, as well as strategic and visionary capacity. It might well be that environmental administrations in the Member States, the Environment Council and the environmental European administration had some political will to change the status quo. However, they do not have the political weight to impose their opinions and convictions on the other actors at national or EU level.

3.   THE EU MIX OF INSTRUMENTS TO PREVENT CLIMATE CHANGE

In order to reach the political objectives which it fixes for itself, the EU may use the binding and non-binding instruments which are foreseen in art. 249 of the EC Treaty. A short overview of the instruments used to reduce the emissions of greenhouse gases[23] provides a good understanding.

3.1 Burden Sharing

The EU-15 Member States[24] and the EU itself are committed by virtue of the Kyoto Protocol to reduce their greenhouse gas emissions by 2012, at the latest, by 8%, compared with 1990 levels. In order to take into account the differences in economic development, the EU agreed a burden sharing: some Member States had to achieve higher, other less high reductions or were even allowed to increase their emissions. The political agreement of 1998 was made legally binding by a Council Decision in 2003.[25] The burden sharing brings on board all EU Member States, some of which might otherwise have been reluctant to see their economic development hampered by the obligations under the Kyoto Protocol.

3.2 Emission Trading

The Kyoto Protocol, which came into force in February 2005, provides for the introduction of a global system for emission trading by 2008. The EU anticipated this date and adopted a directive on emission trading, which it completed later on with further details concerning the Kyoto provisions.[26] The trading system does not reduce the quantity of emissions per se, but allows the trading of greenhouse gas emission rights within the EU and the implementation of Member States' commitments, in order to ensure that investments to reduce greenhouse gas emissions are made at economic costs that are as low as possible. Transport and private households are not affected by the Directives, which cover about 45% of all greenhouse gas generators.

3.3 Taxes

There is no EU tax for greenhouse gas emissions. A proposal to adopt a combined CO_2-energy tax, did not reach the necessary unanimity. Some Member States introduced such taxes at national level, but saw the limits of this instrument, as a further increase would bring competitive disadvantage to their own industry. While there is almost no controversy over the idea that greenhouse gas taxes would accelerate the introduction of alternative energies, no concrete proposals exist for EU-wide taxes.

In its White Paper on transport policy,[27] the Commission developed plans for EU charges for the use of roads and other infrastructure and, furthermore, harmonized taxes for petrol and diesel[28] and taxes for air transport. These taxes and charges were also to cover the environmental costs of transport. In 1999, a directive was adopted on the charging of heavy goods vehicles.[29] Until now, however, such plans were not followed by concrete actions.

3.4 Ban of Substances

There are no EU provisions to prohibit products – such as certain fuels – that emit greenhouse gases. A proposal for a regulation on the restriction of use of the three greenhouse gases HFCs, PFCs and SF6 in a number of products and industrial processes was made in 2003.[30]

3.5 Energy Saving and Alternative Energies

Since the mid-1970s, there have been a number of Council recommendations on the rational use of energy which seem to have had almost no impact. Measures to improve energy saving and alternative energies in view of climate change were first adopted in 1993, and later pursued in the form of decisions which provided for financial support of specific projects. At present, a Decision of 2003 provides for financial support for 2003–06 and for four programmes on (1) energy saving (SAVE); (2) new and renewable energy sources (ALTENER); (3) energy aspects of transport (STEER); and (4) renewable energy sources and energy efficiency in developing countries (COOPENER).[31] The total sum made available under this Decision is €200 million for the 4 years.

A directive on the energy performance of buildings was adopted in 2002.[32] It provides for minimum energy performance requirements for new buildings which are to be fixed by Member States. Existing buildings which undergo major renovation shall comply with such requirements 'in so far as this is technically, functionally and economically feasible'. As buildings have a lifetime of between 50–100 years, the energy saving effect of this Directive is limited.

3.6 Voluntary Agreements

There are commitments by the car manufacturers' association to reduce the CO_2 emissions from cars by 2008, which were followed by corresponding Commission recommendations.[33] The commitments refer to the emission per car and thus did not recur to the Kyoto commitment to reduce the overall emissions from the car fleet by 8%. An increase of the number of cars or an increase of the kilometres driven per car might thus offset advantages reached in emission reductions, without the commitment being breached.

Television and video cassette recorder manufacturers concluded an agreement under art. 81 of the EC Treaty which the Commission, having negotiated this agreement with industry, approved and which provided for the reduction of stand-by energy losses of this electronic equipment.[34] The Commission announced its intention to favour the making of further negotiated agreements, in particular to voluntarily reduce greenhouse gas emissions. Progress though is rather slow; also, because (1) the EC Treaty does not provide for negotiated

agreements to be formally concluded by EC institutions and (2) economic operators are mainly interested in obtaining legal guarantees that any agreement
would not be, at a later stage, offset by legislation – a commitment which the
legislature cannot make in law.[35]

3.7   Other Measures

There are numerous other measures which directly or indirectly influence the
emission of greenhouse gases, such as a directive on landfills[36] which tries,
through better monitoring of landfills and the reduction of land filling of biode-
gradable waste, to reduce methane gas generation; a directive on the promotion
of electricity from renewable energy sources;[37] a directive on the promotion of
the use of bio-fuels or other renewable fuels for transport;[38] a decision to monitor
the emission of greenhouse gases;[39] and other measures, in particular in the areas
of transport, energy and agriculture policy.

4.   ASSESSING THE INSTRUMENT MIX

If one looks at the different organizational and structural measures which the
EU has taken from the climate change discussions at end of the 1980s until now
(early 2005), in order to reduce greenhouse gas emissions, the following observations can be made:

- Global warming was, for the first time, addressed at EU level by a Commission Communication of 1989[40] to which the Council reacted with a
  resolution in which the commitment was made to stabilize CO_2 emissions
  by 2000 (to the level of 1990).

Subsequently, the challenge of climate change and the need to find adequate
answers to this challenge induced cooperation of EU Member States within the
EU to an extent which exceeded expectancies, and which progressively led to
Member States orienting their policies and measures to the decisions which
were taken by the EU. The environmental administrations in almost all Member
States were the driving forces for addressing climate change issues and progressively aligning national measures to the EU orientation.

- The attitude of the US, since the beginning of the discussions on climate
  change, not to take up the challenge of leading the world into the battle
  against climate change, but to largely ignore the issue, brought the EU
  into a role of global leadership which it did not seek, but which it accepted
  and tried to fill, be it at the price of many compromises.

Again, this international discussion was led by the environmental administrations rather than by foreign policy, trade or energy administrations.

- The EU managed to largely overcome the constitutional problems which were linked to the fact that it is not a State and that it did not have a general competence for an EU energy policy. It based most of its measures on the environmental provision of art. 175(1), though climate change issues also affect many other policy sectors; in particular agriculture, transport, energy, trade, development, competition, regional policy and foreign policy.
- A climate change programme was not elaborated. Instead, the EU listed measures which might be taken at EU level and indicated the potential for reducing greenhouse gas emissions of the different measures. This approach went in parallel with the discussions at international level, where also the climate change issues were progressively addressed: the Climate Change Convention 1992[41] requested the contracting parties to stabilize their greenhouse gas emissions by 2000 (the level of 1990). The Kyoto Protocol also looked for greenhouse gas emissions from industrialized countries to be reduced by some 5% by 2012 at the latest (compared with 1990).
- At present, the EU measures are concentrated to reach the reduction that the EU had committed itself to in the Kyoto Protocol, i.e. a reduction of 8% by 2012 at the latest. Though many measures will have effects that go beyond 2012 – an example is the Directive on energy performance of buildings which aims at reducing the need of heating – the measures undertaken so far are almost entirely targeted to reach the 2012 objectives, but do not really aim at reaching reductions beyond that date.

Whether the measures taken, and those which will still be adopted before 2012, will be sufficient to reach the obligation under the Kyoto Protocol – the famous 8% reduction of emissions with regard to 1990 is, as yet, uncertain. All depends on the following question whether the measures which were agreed upon will actually be applied; whether Member States will set up and implement greenhouse gas emission reduction programmes; whether economic operators will look for alternative substances, alternative fuels, and other measures to reduce their emissions.

- The instrument mix which the EU has set up until now, appears to be rather full. The EU used law and order instruments, economic incentives, subventions and state aids. It went to the negotiations on the Kyoto Protocol with the intention to resist attempts by the US to introduce emission trading systems. Being confronted by the US with the alternative to either not have a Kyoto Protocol or to accept a trading system, the EU accepted the flexible instruments in the Kyoto Protocol, in order to have a global agreement. It then tried to make the best of it, anticipating the entry into effect of a global emission trading system and also adopting legislation to introduce the other flexible mechanisms under the Kyoto Protocol.

In view of the constitutional and institutional barriers for setting up an EU climate change policy which were mentioned in section 1 above, and in view of the EU having almost no institutional capacity to act as a global player, the achievements of the EU climate change policy are very considerable. There does not seem to be any developed State which has, with the same political determination, invented a climate change policy and put it into operation. It is against the background of these general observations with regard to the EU climate change policy that the following critical remarks are made, though it is one thing to conceive and make operational a policy and another to criticize it. Nevertheless, it is not desirable to declare everything that has been done as perfect, because in such a case the many different and diverging interests which led to the present state of affairs might be encouraged to increase their efforts in order not to improve climate change policies, but rather to promote their own interests.

5.   THE NEED OF AN EU CLIMATE CHANGE POLICY

The first aspect to criticize appears to be the absence of a scientific advisory body. In the framework of the UN, an 'Intergovernmental Panel on Climate Change' (IPCC) was set up by the World Meteorological Organization and the Environmental Programme UNEP, in order to give scientific, economic and political advice on climate change issues. This Panel has gained general recognition. Its findings and statements are not seriously contested, though, of course, its advice is not always, and not even regularly followed.

There is no scientific advisory body on climate change issues at EU level. Such a body would not be superfluous in the view of the UN panel, as the assessment of the European evolution and the assessment of the global situation from a European viewpoint is very necessary in many situations. The EU, however, has not even made the attempt to set up such an independent, scientific body which could give advice away from considerations of national political and vested interests.[42] The EU has scientific committees for food and animal feedingstuff, for animal welfare, plants, cosmetics and pharmaceuticals,[43] but obviously is of the opinion that to have enough scientific know-how in climate change issues is not necessary. This is typical for administrative over-estimation: the administration knows best. At the same time, in the absence of expert advice, the discussion on climate change is more strongly exposed to political and economic lobbying which influences its outcome.

The second aspect has already been mentioned. In view of the long-term importance and evolution of climate change issues, it appears necessary to have, inside the European Commission, a political/administrative structure which is politically and administratively in charge of climate change issues. A Commissioner for climate change who would keep this topic in the public discussion, stimulate research and innovation, and who would accumulate responsibilities from the energy, transport, agricultural and environmental sector in the area of climate change – could give a political dimension to this discussion – which, for the 50 and more years to come, is likely to impinge more and more on economic and private activities. At the same time, it would affirm the EU's leading role in the world in climate change issues – there is not much else – where the EU is, at present, capable of taking a leading role.

As regards the EU funding which is made available to stimulate energy saving, increase energy efficiency or develop alternative energies – solar and wind, water and soil energy etc. – the amounts which the EU is making available are ridiculously low. As mentioned above, there are some €200 million made available for a period of 4 years. One should compare this figure with the information given by the Commission[44] that it promoted nuclear energy between 1988–98 with about €310 million per year, and that between 1986–95 Member States promoted nuclear energy on average with €1470 million per year. In the period between 1974–98, aid granted by EU Member States for nuclear technology amounted to $55 billion, an average of $2.2 billion per year for the whole EU. Also, Member States alone granted, in 2001, €6.3 billion as State aid to the coal industry.

These figures show that the conversion of energy in Europe to a low carbon (renewable) energy is not yet taken seriously. There is a lot of discussion on energy saving, energy efficiency and renewable and alternative energies, but the conservative structures, including coal and oil support in infrastructure, public aid, distribution of energy, tax promotion and other aspects, remain intact, continue to receive their traditional public aid and are difficult to change.

This leads to the most relevant aspect. The Kyoto Protocol provided for a reduction of the greenhouse gases emitted by industrialized countries by about 5% by 2012.[45] Scientists, though, are convinced that neither the date nor the amount of emission reductions agreed in Kyoto is appropriate to prevent serious global warming during the next few decades. In its proposal for the Sixth Environmental Action Programme, the Commission formulated:[46]

> Objective: In line with the aim of the United Nations Framework Convention on Climate Change, to stabilise the atmospheric concentration of greenhouse gases at a level that will not cause unnatural variations of the earth's climate. Scientists estimate

that to achieve this objective global emissions of greenhouse gases need to be reduced by approximately 70% over 1990 levels in the longer term. Given the long-term objective, a global reduction in the order of 20–40% (depending on actual rates of economic growth and thus greenhouse gas emissions as well as the success of measures taken to combat climate change) over 1990 by 2020 will need to be aimed at, by means of an effective international agreement.

The European Parliament and the Council declared:[47]

> a long term objective of a maximum global temperature increase of 2° Celsius over pre-industrial levels and a CO_2 concentration below 550 ppm shall guide the Programme. In the longer term this is likely to require a global reduction in emissions of greenhouse gases by 70% as compared to the Intergovernmental Panel on Climate Change (IPCC).

Global discussions on the post-2012 climate change measures have just started. In view of the opposition of the US to agree any reduction of emissions and of the reticence of developing countries – including China, India, Brazil, South Korea, Mexico and other threshold countries – as well of other industrialized States such as Russia, Japan and Australia, to agree any further reduction, the international discussions are everything but easy. The outcome is as yet more than uncertain. Under these circumstances, the EU can only hope to be able to lead the way towards a global policy, if it makes climate change – or the change towards low carbon energy – a political issue, in other terms, if it makes the shift to low carbon energy its political priority. This would suppose that the issue of climate change obtains another political value, with the EU institutions in Brussels as well as in all 25 capitals of the Member States. In order to get there, it would be necessary for the European Commission to begin to treat climate change issues differently than it has until now, and make it the first priority of its foreign and transport, energy and environment, development and research policy. Climate change policy would have to become a peace policy of the importance of the Near East Policy – with the difference that it would require, first, the cooperation of all 25 Member States and, secondly, a gradual extension to the global scene. Climate change cannot be prevented by some countries alone, but a policy to promote climate change needs driving forces. The divisions which are available within the United Nations are not sufficient to act as such a driving force. As the US is unable to assume its political function as global leader and prefers to base its foreign policy on arms and trade, there is no other driving force visible than the EU.

If this assessment is correct, then questions on instrument mix at the level of the EU become rather trivial. The instrument mix is not wrong, but the measures taken to date are not even a sufficient beginning for the necessary changes which are needed in the next 50 years. With regard to the setting up of this EU, the 'ever-closing union of the peoples', the founders of the EU had sufficient vision to establish structures for decades ahead. The perspectives of success were at that time not particularly bright. The big question mark is, whether present political leaders in the EU, in capitals as well as in the European institutions, are capable of developing visions for a change in energy policy. In public, there are already sufficient speeches by Member States' leaders which identify climate change as the biggest challenge this planet is confronted with. However, these speeches, for the time being, are not more than lip-service; there is no political roadmap to put the necessities into political and economic reality. As the EU institutions, in particular the European Commission, are the only bodies within the EU which have to look after the general European interest, the responsibility of launching this policy lies most of all with the European Commission.

I am well aware of the proverb that a journey of thousand miles starts with the first step. The problem is that the present discussion gives the impression that neither the first step is being considered, let alone taken, nor that the way to go these thousand miles is examined, planned or taken. One may act by committing a wrong; but one may also commit a wrong by not acting. That is the situation where the EU climate change policy stands at the beginning of 2005.

This, then, would be the list of *desiderata* for an EU climate change policy:

(1) Conceive and put into operation an EU climate change policy, which is more than the co-ordination of five or six diverging national climate change policies.

(2) Establish clear EU responsibilities for climate change issues. Charge one member of the Commission with the responsibility for such a policy.

(3) Promote research in order to better understand the origin and extension of climate change and its consequences in Europe and worldwide.

(4) Intensify research and investment in alternative energies and clean technologies.

(5) Take up the role of global leadership in climate change issues, which would prevail over interests in export and trade.

(6) Take up a leading role in environmental questions that are linked to climate change issues, and link development policy, economic policy and trade policy to climate change issues.

(7) Ensure that climate change issues will progressively gain an importance that is similar to human rights issues.

(8) Fully comply with international commitments under the Climate Change Convention and the Kyoto Protocol, including the financial commitments.

(9) Integrate the climate change policy into the EU policy on sustainable development, which is based on the concepts of an open political society, absence of wars and military-economic interventions, respect for human

rights, economic growth, social justice and adequate environmental protection, in order to demonstrate that this can be the societal model of the future and that neither the US-model nor the Chinese-Indian model – economic growth at the expense of social and environmental concerns – are capable of reaching (in the long term), comparable results.

(10) Discuss and fix objectives for the reduction of greenhouse gas emissions by 2020, even against internal arguments that this would lead to competitive disadvantages.

NOTES

1. Professor Dr Ludwig Krämer, University of Bremen.
2. Decision 94/69 (1994) OJ L 33, p. 11; the Convention is published at p. 13.
3. Kyoto Protocol to the UN Framework Convention on Climate Change, UN Doc.FCCC/ CP1997/Add.1. The EU concluded this Protocol by Decision 2002/358 (2002) OJ L 130, p. 1; the Protocol is published at p. 4.
4. A discussion on EU law will almost never be able to be contained to purely legal questions. In particular, on a topic such as climate change, where 'there are many uncertainties in predictions of climate change, particularly with regard to the timing, magnitude and regional patterns thereof' (UN Convention, n. 2 above, 5th recital) questions of legal policy, policy, economics, social issues and others come into play. While law is thus only one aspect of the total (global) problem, it might contribute, by making the issues more transparent, to put the right questions, look for the right answers and assist political leaders in taking the right decisions.
5. The Draft Treaty establishing a Constitution for Europe, which still has to be ratified by the EU Member States in order to come into effect, provides in art. I-13 that, in future, there shall be shared competence in the area of 'energy'.
6. See, for example, Council Resolutions on climate change of 16–17 June 1998, 10 October 2000 and of 12 December 2001.
7. A Council resolution is not legally binding, as it is not enumerated in art. 249 of the EC Treaty as one of the instruments with binding effect. Few people know that following a non-written consensus among Member States, Council resolutions are adopted unanimously.
8. Article 3 of the Treaty on European Union – see also the similar provision in art. III-1 of the Draft Treaty establishing a Constitution for Europe – provides that the 'Union shall be served by a single institutional framework which shall ensure the consistency and the continuity of the activities carried out'. Unfortunately, this provision is all too often ignored in practice. See also the integration requirement in art. 6 of the EC Treaty.
9. Decision 1600/2002 of the European Parliament and of the Council laying down the Sixth Community Environmental Action Programme (2002) OJ L 242, p. 1.
10. Court of Justice, Case C-347/97 *Commission* v. *Belgium* (1999) ECR I-309 (concerning a programme for batteries).
11. European Commission, *Communication on EU Policies and Measures to Reduce Greenhouse Gas Emissions: Towards a European Climate Change Programme* (ECCP), COM/2000/88 Final of 8 March 2000.
12. Resolution of 11 October 2000, Doc.12240/00.
13. European Parliament, Resolution of 26 October 2000, (2001) OJ C 197, p. 397, n. 6; note that this resolution was adopted later than the Council resolution.
14. European Commission: Communication on the implementation of the first phase of the European Climate Change Programme, COM/2001/580 of 23 October 2001.
15. The proposal was made and in the meantime adopted: Directive 2004/8/EC on the promotion of cogeneration based on a useful heat demand in the internal energy market, (2004) OJ L 52,

p. 50. See also the related Directive on the eco-design of energy-using products, 2005/32/EC of 6 July 2005, PbL 191, 22 July 2005, pp. 29–58.
16. The website of the European Integrated Pollution and Prevention Control Bureau indicated, in early 2005, that no trace of work having started on this issue.
17. The proposal was made in the meantime: proposal for a directive on energy end-use efficiency and energy services, COM/2003/739 of 10 December 2003.
18. See Decision 1600/2002 (n. 9 above), art. 5.
19. See also art. 208 of the EC Treaty: 'The Council may request the Commission to submit to it any appropriate proposals'. A similar provision exists in art. 192(2) of the EC Treaty for requests by the European Parliament. The majority of legal authors is of the opinion that the Commission is bound to follow such a request.
20. See also Commission, Green Paper: Towards a European strategy for the security of energy supply COM/2000/769; White Paper: European transport policy for 2010: time to decide, COM/2001/370 of 12 September 2001 'which contains some 60 measures to be taken at Community level, several of which will contribute to the reduction of greenhouse gas emissions' (Commission, COM/2001/580 (n. 14 above), para. 2.3.
21. It is true, though, that the Commission, following its Communication of March 2000 (n. 12 above) instituted a steering committee and seven working groups which discussed possible measures at EU level. Some 40 measures were identified and included by the Commission in its Communication of October 2001 (n. 11 above).
22. Commission, White Paper (n. 20 above).
23. The following comments refer to the six greenhouse gases which are covered by the Kyoto Protocol, ie Carbon dioxide (CO_2), Methane (CH_4), Nitrous oxide (N_2O), Hydrofluorocarbons (HFCs), Perfluorocarbons (PFCs) and Sulphur hexafluoride (SF_6). The greenhouse gases which also deplete the ozone layer and are covered by the Montreal Protocol on substances that deplete the ozone layer are not discussed, though they also have a greenhouse effect.
24. The 10 Member States which joined the EU in 2004 are committed to reach the following reductions in percentages, with regard to 1990): Czechia 8; Estonia 8; Hungary 6; Latvia 8; Lithuania 8; Poland 6; Slovakia 8 and Slovenia 8. Malta and Cyprus have no commitments.
25. Decision 2003/358 (n. 3 above) The following percentage figures were agreed (compared to 1990): Austria –13; Belgium –7,5; Denmark –21; Finland no change; France no change; Germany –21; Greece +25; Ireland +13; Italy –6,5; Luxembourg –28; Netherlands –6; Portugal +27; Spain +15; Sweden +4; United Kingdom –12.5.
26. Directive 2003/87/EC establishing a scheme for greenhouse gas emission allowance trading within the Community and amending Directive 96/61 (2003) OJ L 275, p. 32; Directive 2004/101/EC amending Directive 2003/87/EC in respect of the Kyoto Protocol's project mechanisms, (2004) OJ L 338 p. 18.
27. Commission, White paper (n. 20 above).
28. See in this regard Directive 2003/96/EC restructuring the Community framework for the taxation of energy products and electricity, (2003) OJ L 283 p. 51.
29. Directive 1999/62/EC on the charging of heavy goods vehicles for the use of certain infrastructures (1999) OJ L 187 p. 42.
30. Commission proposal for a regulation on certain fluorinated gases, COM/2003/492 of 11 August 2003. The explanatory memorandum indicates that the emissions of these gases in 1995 were about 2% of all six greenhouse gases.
31. Decision 1230/2003 of the European Parliament and the Council adopting a multi-annual programme of action in the field of energy: 'Intelligent Energy-Europe' (2003) OJ L 176, p. 29.
32. Directive 2002/91/EC on the energy performance of buildings, (2003) OJ L 1, p. 65.
33. Commission, COM/1998/495 of 29 July 1998 and Recommendations 1999/125 (1999) OJ L 40, p. 49, 2000/303 (2000) OJ L 100, p. 55 and p. 57.
34. For details see (1998) OJ C 12 p. 3 and COM/1999/120 of 15 March 1999.
35. For the general problems see Krämer, L. (2003), *EC Environmental Law*, 5th edn, London: Street & Maxwell, p. 56 and p. 283. See also Commission, Environmental Agreements at Community level, COM/2002/412 of 17 July 2002, p. 4: 'environmental agreements are not negotiated with the Commission'; this contradicts COM/1999/120 (n. 34 above).

36. Directive 1999/31/EC on landfills (1999) OJ L 182, p. 1.
37. Directive 2001/77/EC on the promotion of electricity from renewable energy sources in the internal market (2001) OJ L 283 p. 33.
38. Directive 2003/30/EC on the promotion of bio-fuels or other renewable fuels for transport (2003) OL L 123 p. 42.
39. Decision 280/2004/EC concerning a mechanism for monitoring Community greenhouse gas emissions and for implementing the Kyoto Protocol (2004) OJ L 49 p. 1.
40. Commission, The greenhouse effect and the Community, COM/1988/656 of 16 January 1989.
41. UN Framework Convention on Climate Change of 9 May 1992. The EU adhered to the Convention by Decision 94/69 (1994) OJ L 33 p. 11. The United States is of the opinion that the Convention does not contain a binding commitment. The EU is of a different opinion.
42. In 1997, the Commission had set up a Consultative Forum for environment and sustainable development, but dissolved this Forum some years later.
43. Decision 97/579, (1997) OJ L 237 p. 18.
44. Commission, Written Question E-4133/97 (Telkämper) (1998) OJ C 310 p. 13.
45. Kyoto Protocol (n. 3 above).
46. Commission, Environment 2010, our future, our choice; COM/2001/31 of 24 January 2001, para. 3.2.
47. Decision 1600/2002 (n. 9 above), art. 2(2).

16. Good governance and climate change: Recommendations from a North-South perspective

Joyeeta Gupta[1]

1. INTRODUCTION

Much has already been said about the challenges facing the European Union (EU) with respect to the current regulatory regime on the climate change problem. Pallemaerts and Williams (Chapter 2) have described the legal evolution of the global climate change agreements and have highlighted the evolving nature of EU policy. The specific technical and sometimes political aspects of the various instruments being developed within the multilateral regime and the European framework have been discussed extensively by the other authors. This chapter continues where Krämer (Chapter 15) ends with his critical evaluation of EU policy, to place the discussion of climate change in the context of the key North-South challenges in the climate change regime.

If we accept the fact that the climate change problem is a global commons problem, then there is always the risk of free-riders. Where there is a risk of free-riders, it becomes essential to develop a global regulatory regime to address the problem successfully. If we accept the fact that the climate change problem is a long-term problem, then there is a risk that short-term solutions that are seen as politically feasible are not durable. Where there is such a risk, we must seek to design a long-term durable regulatory regime. If we accept that climate change may have severe impacts that are both immediately visible (such as the melting of the glaciers and the bleaching of coral reefs) as well as being visible for centuries to come (such as the rising of the sea-level which will continue long after concentrations of greenhouse gases have stabilized), we must seek to address both the immediate impacts as well as the long-term impacts. In other words, we need a durable, global regime that addresses both current and future impacts.

Others may argue that a global regime is not possible, and since we cannot avoid current impacts or prove that they are caused by climate change, perhaps we should focus on limiting the damage in the future through the development of bilateral, regional, sectoral, public private and voluntary agreements.[2] This approach focuses on designing the regime to suit the politics of international negotiations. The motivation is clear: can one seek to define solutions that have a high likelihood of being accepted by the major actors in the regime in the short-term?

However, I have set a different task for myself in this chapter. I am going to look at what is necessary to solve the climate change problem. In doing so, I am clearly not going to focus as much on the transatlantic dimension (except in passing) but much more on the North-South dimension. My justification for doing so is that if the problem is to be solved, we need the support not only of the United States (US) but also the developing countries. We cannot engage the developing countries by simply forcing them to sign on to an agreement they cannot implement. We need to understand more clearly the key North-South issues that affect the climate change discussions. In attempting to answer this question, I will identify 10 key challenges, and then perhaps seek to examine possible solutions to these challenges.

2.   THE CHALLENGES

2.1   Problem Definition

The first challenge is that the international regime to deal with climate change was based on a problem definition that emissions of greenhouse gases are a global problem which need a global solution, while impacts are a local problem which need local solutions.[3] This definition, that split the problem into two sub-problems, justified the initial design of a regime in terms of a 'leadership' regime as opposed to a polluter pays regime with a strong element of liability and compensation built into it.[4] Furthermore, climate change was seen as an economic and technological problem and the solution was, hence, cast in terms of economic instruments that would help promote technology transfer.[5] The Intergovernmental Panel on Climate Change helped support this definition by itself taking an apolitical stance – none of the reports thus far has had a chapter devoted to the political science aspects of climate change.

The implication of the definition for the developed countries was that they could be seen as generous leaders instead of careless polluters, and since solutions could be crafted in terms of new technologies, the climate regime gave an opportunity for developing new markets in the developing countries. The implication of this for the developing countries was that they did not have to do anything quantitatively measurable, nor potentially take responsibility for future pollution, and they waited naively for the technology transfer to come without strings attached.

However, for a leadership regime to work, there has to be a leader. In the absence of the US in the regime, the EU does figure as leader, but its leadership may be too little too late and too reluctant for many developing countries (see also section 2.2)![6] The problem needs to be defined much more clearly in terms of liability and politics, before solutions will become evident.

2.2 Changing Political Paradigms

The climate change regime began, thus, as a leadership regime. The word leadership is to be found in several Political Declarations[7] preceding the negotiations and is also included in the United Nations Framework Convention on Climate Change (UNFCCC).[8] In 1992, there was confidence that the North could lead, and the South would follow. By 1997, the US senate was beginning to redefine the problem in terms of free-riders. It started defining the developing countries as free-riders, and used that as an argument to legitimize the lack of US action.[9] This free-rider discussion was possible because the climate change problem was not seen as a liability problem, since the emissions and impacts issue had been explicitly disconnected in the regime behind the implicit argument of scientific uncertainty and the notion that we are all responsible (common but differentiated responsibilities). Hence, when the Kyoto Protocol (the Protocol) was adopted in 1997, the Netherlands made clear that it would not ratify the Protocol until and unless Japan and the US also did so. The US said that it would wait until the key developing countries undertook meaningful action and the developing countries were still waiting for the North to take action. This was the conditional leadership paradigm.

This was clearly also the strategy of the EU. In 2001, when the Bush administration clarified that it would not ratify the Kyoto Protocol it, in effect, called the bluff of the EU. Surprising many political scientists and game-theorists, the EU then went ahead and ratified the Protocol and persuaded several other countries to do so, including recalcitrant Russia. The Protocol entered into force in 2005. However, with the US gone, the regime took a bad hit.

The implication of the changing paradigms for the developing countries is that they no longer have the bargaining position that many thought they had in the early 1990s. If the developing countries decide to walk out, the regime will collapse. In other words, they have to now reflect – is the climate change problem a serious threat to their countries and what can they do to avoid it, rather than fall back on the rhetoric: the North is responsible and the North must take action. Unless the climate change regime is redefined in terms of liability, the South cannot force the North to take action. In the meanwhile, they can only hope for a series of technological breakthroughs.

2.3 The Deepening Transatlantic Divide

It is clear that the South is facing a deepening transatlantic divide. How do you deal with a North that is clearly splitting up along new lines. In pre-1990 days, the North was split into the capitalist West and the communist East. In those days, developing countries could play one against the other. Both sides needed followers and the developing countries were in demand. However, since 1990, the world has developed along unipolar lines and the Eastern bloc now, instead of being a potential patron, is a competitor for the meagre funds from the developed countries. For example, the Global Environment Facility provides funds to both developing countries and countries with economies in transition. Against this background, the much weakened developing countries now have to face the new split between the EU and the US. This divide, less apparent in the days of the cold war, has now surfaced as a dominant topic for discussion.

It is, therefore, not surprising that scholars from the US (and elsewhere) argue that public international law is not really law and that international law is the subject of intense bargaining between unequal states. Kagan explains that the US is from Mars and Europe is from Venus and that American political and legal thought is inspired by Hobbes – that the world is an anarchy and the most powerful will win. Europeans, on the other hand, are much more inclined to pursue peace and rule-based international order.[10] This implies that legal and economic thought within the US is coloured much more by its political and economic vision of the world and this is less compatible with a proactive, creative push to re-design the global order in order to make it truly sustainable. On the climate change issue there are marked differences in how the EU perceives climate change[11] and how the US perceives the necessary developments in climate change governance.[12] While the US may tend to argue that potential future increases in emissions in developing countries might undermine any efforts of the developed countries to reduce their own emissions,[13] Wallström argues that 'If the United States rightly calls for involving the developing countries more in the global effort to combat climate change, it can do so credibly only if it reduces its own emissions first'.[14] While EU climate scientists (including social scientists) focus on the need to broaden (seek ways to increase the engagement of developing countries) and deepen (further develop mechanisms under the UNFCCC) the climate change regime, American scholars focus more on bilateral, sectoral and technology oriented proposals. In the meanwhile, the US has developed a number of unilateral initiatives at the international level. The US has launched several initiatives including a Methane-to-Markets Partnership, an International Partnership for a Hydrogen Economy and a Carbon Sequestration Leadership Forum, a Generation IV International Forum to promote proliferation resistant nuclear energy systems, a Renewable Energy and Efficiency Partnership, several bilateral and regional agreements on climate change, a

Tropical Forest Conservation Act and an initiative against illegal logging.[15] In July 2005, it also promoted an agreement called the Asia Pacific Partnership on Clean Development and Climate.

However, while Europeans and Americans argue on how to develop the international legal order and address climate change, the developing countries become increasingly vulnerable. For, both the EU and the US are able to protect their own interests. Both parties are using the existing science to defend a range of policy options. In the process, developing country negotiators are getting increasingly confused about whether climate change is a serious problem, and about what policy measure would be most relevant given the current geo-politics. In a recent discussion with a senior developing country negotiator on the current melting of glaciers and the impacts on infrastructure, roads, water flow etc, I was informed by the negotiator that top US officials do not see any connection between glaciers melting and climate change. A few weeks later in a *Time Magazine* 'Verbatim' section, reference was made to a sentence on the relationship between global warming and mountain glaciers being deleted by staff from the White House Council on Environmental Quality in a report on global warming.[16] This is just one of many anecdotes showing how confused negotiators are becoming by the way in which science is being presented. Furthermore, most developing countries already find it very difficult to cope with the multiple negotiating sessions within the UNFCCC, let alone having to deal with a number of unrelated initiatives at the international level on climate change. The transatlantic divide will further disenfranchise the developing countries and the competing international negotiations are becoming increasingly confused through this divide (see below).

2.4 Processes of Negotiation

One element that also needs further discussion is the nature of international negotiations and treaty making. International negotiations increasingly take a standard North-South approach to simplify matters. However in doing so, the average differences between the two groups become marginalized by the similarities between countries at the periphery of both groups. Thus, it is not in the interest of the richer developing countries to impose responsibilities on the North, because they may be next in line to being subjected to such responsibilities. The poor in the North see themselves as having to pay too high a price for belonging to the rich group. A more rational way to divide countries may serve global interests better. The challenge, however, is that many countries bring their geo-political alliances to the negotiating table and it becomes difficult to convince them to think otherwise.

The other problem is a legal one. This focuses on the rules of procedure at international negotiations. The UN ostensibly calls for all plenary sessions to be held in all six UN languages, and for formal rules of procedure to be obeyed. The law of treaties assumes that only those with an understanding of the issues are sent to negotiate. International law tends to give the impression to countries that only agreements ratified by domestic parliaments are legally binding. These three basic assumptions are all easy to falsify at the international level. The climate change negotiations are so complex that informal meetings are regularly set up, working groups and friends of the chair are established and a multitude of meetings take place simultaneously (often in only one language) completely marginalizing one or two delegate countries. Decisions are then developed so far in these groups that it is practically very difficult to change them in plenary session. Since climate change is so complex, it is virtually impossible to have an omnipotent negotiator who single-handedly can effectively negotiate for his country. Finally, taking shelter under the idea that only the ratified treaty is legally binding will prove to be no shelter at all, because if legal scholars are right, we may be witnessing a dramatic shift in the way regimes are negotiated as the law making process takes place continuously and not at discrete intervals, and the annual decisions of the Conference of the Parties (COPs) are possibly binding, and at any rate irreversible. Besides, new science will lead to new law, mostly untried and untested; and negotiators will have to think on their feet to be able to project and predict how these will impact on their countries. The veil of ignorance hangs heavily over the developing countries, less so over the developed countries and this will skew the process even further.[17]

2.5   Long-term Target – Conspicuous by its Absence

A serious challenge facing the climate change regime is the articulation of its long-term goal. Article 2 of the UNFCCC states that:

> The ultimate objective of this Convention and any related legal instruments that the Conference of the Parties may adopt is to achieve, in accordance with the relevant provisions of the Convention, the stabilization of greenhouse gas concentrations in the atmosphere at a level that would prevent dangerous anthropogenic interference with the climate system. Such a level should be achieved within a time-frame sufficient to allow ecosystems to adapt naturally to climate change, to ensure that food production is not threatened and to enable economic development to proceed in a sustainable manner.

However, Art. 2 has yet to be further elaborated in terms of concentration levels to be achieved within specified time-frames. To the uninitiated this may appear to be a relatively simple problem. However, since the impacts of climate change are likely to be different from region to region, and some countries may be essentially losers, while others may be more of a winner than a loser, the more powerful countries are hedging their bets. In the meantime, in the international negotiations very little progress has been made towards defining this goal.

However, such a goal needs to be defined because without such a definition, it is impossible to see if the Convention process is on track to addressing the most important impacts.[18] Since the first range of impacts, such as the melting of glaciers and bleaching of corral reefs are already visible (whether directly attributable to climate change or not), this is a burning question. Scientists have thus far been reluctant to address this issue, arguing that danger is a matter of perception.[19] Politicians have been nervous to push the discussion further, arguing that this is politically very contentious. However, a number of recent projects[20] have started to examine this issue closely and in the Netherlands we have developed a participative integrated methodology to deal with this problem.[21] This method begins from the presumption that concentration levels and average temperatures are abstract issues and most people do not understand these or relate to them. However, people can relate to specific impacts. In the Netherlands we have tried to identify the key indicators of climate change that may become visible in the domestic context or may affect the perceptions of people and have then developed threshold levels of acceptable and unacceptable change. Unacceptable threshold levels have been translated, to the extent possible, back to concentration levels and temperature rise. The approach tentatively concluded that a 2 degree rise in temperature over a hundred year period should be the absolute upper limit for a rich country like the Netherlands. This does not imply that such an upper limit is acceptable for other more vulnerable countries.[22] An examination of the national reports submitted by 126 developing countries suggests that the anticipated impacts may be severe. Developing countries need to seriously consider what sort of long-term target would be consistent with protecting their own economies.

2.6 Short-term Targets Based on Horse-trading and Poor Precedent

Short-term targets are one way of ensuring whether a regime is set to meet its ultimate objective. The Kyoto Protocol specified targets for the developed countries, however, these targets were more the result of horse-trading than any clear and systematic application of principles. As a result some very rich and highly polluting countries such as Australia, Iceland and Norway were given the right to increase their emissions. This was possible because within the EU, some EU Member States were allowed to increase their emissions by as much as 27%![23] However, if the global community needs to drastically reduce emissions, and the initial agreement was that the developed countries would reduce their emissions to make room for the growth of developing country emissions, then targets to increase emissions for individual developed countries send the wrong message; both in terms of substance (rich countries can increase their emissions) but also in terms of procedure (if you are powerful you can get concessions in your target).

There are a large number of proposals in the market that are very complex and the implications for the developing countries are not transparent. These include the method on contraction and convergence, the carbon intensity approach, the Centre for Science and Environment (CSE) approach and the Tryptich approach.[24] Sets of approaches have been evaluated by several authors.[25] My own feeling is that although all these proposals are based on specific choices of principles, they are very complicated and not transparent for developing countries. It becomes difficult for many countries to see at a glance what could be expected from them in the future.

2.7   Best Available Practices to Modernize the Globe

In the climate change regime, scientists are vying with each other to develop more and more sophisticated policy measures to address the problem. They are trying to find the best possible solutions to addressing global problems. They find highly modern tools that have worked in the richest countries of the world and suggest them as a model for the poorer developing countries. This leads to suggestions such as emissions trading, which ultimately was adopted in the Kyoto Protocol 1997. However, tools such as emissions trading require very complex and sophisticated accounting systems in countries, good monitoring and enforcement. The possibility of such a system working at the global level is mind boggling for those of us who work on developing country issues and have explored these aspects further.

At the same time, Tariq Banuri argues that the world is a third world country. He makes the point that all the characteristics of a third world country (lack of good governance, lack of rule of law, corruption, power politics, huge differences in incomes between the rich and poor) are also visible at the international level.[26] This implies that one has to find policy solutions that work in developing countries and then try to internationalize them, instead of the other way around.

What one could argue is that what we are witnessing in the global arena is a competition of ideas. Each country is trying to sell policies and practices that work best within the domestic context as a way in which to ensure that implementation is less expensive for it. However, since this inevitably implies that many complex developed country systems are being adopted at the international level, these are then subsequently being downloaded into domestic systems in the developing world, where they are, by definition, unlikely to work effectively. Subsequently many legal scholars then talk about the implementation deficit in developing countries, when in my view the problem is more that the design of the instrument is inappropriate for developing countries.[27]

2.8 Finding Mechanisms, Technology Transfer and Capacity Building

Developing countries have, throughout recent history, asked for funding from the developed countries to support their development process and to support the adoption of environment friendly technology. Since the climate change regime was not developed as a polluter pays regime, it relies on voluntary contributions made by the developed countries. Over the last few years, we have seen several funds being established including the Global Environment Facility, established in 1990, which is used as the operating entity of the funding mechanism under the Convention and the Special Climate Change Fund established in 2000 to finance some adaptation, transfer of technologies in the fields of energy, transport, industry, agriculture, forestry and waste management and to help developing countries diversify their economies. The projects need to be country-driven, based on national priorities and aim at sustainable development. There is also a Least Developed Country (LDC) Fund also initiated in 2000, which will support LDCs to prepare and implement National Adaptation Programmes of Action. The adaptation fund, established in the 1997 Kyoto Protocol is to be financed by a share of the proceeds on the Clean Development Mechanism (CDM) in the order of 2% of Certified Emissions Reduction (CERs) – and by other sources of funding.

However, despite the proliferation of funds, the amount of resources available in these funds is quite limited and the conditions for accessing these funds very complex. Most developed countries have been unable and unwilling to reach the 0.7% of Gross National Product (GNP) target for development funding as recommended by the Conference on Financing the Environment, and have also not really been able to generate new and additional funding for these environmental funds.

It is argued that if developing countries had accelerated access to modern technologies, they could avoid polluting as much as the developed countries did when they had poorer technologies. While this argument was accepted as far back as in 1990 at the Second World Climate Conference, the idea that developing countries might need help to develop the capacity to deal with this problem, only achieved formal recognition within the climate change regime in 2000.

With respect to technology transfer, there are some key problems. Most modern technologies are in the hands of the private sector and are new and, hence, costly. This puts them out of the reach of developing countries, unless they are helped financially to do so. Since the funds are limited, there is only a limited amount of technology transfer taking place under the auspices of the climate change convention. In the meantime normal technology transfer is occurring selling affordable and, hence, older technologies to the developing countries, thus effectively reducing the effectiveness of technology transfer as an instrument for sustainable development.

With respect to capacity building, the key argument is that this helps with 'soft' knowledge and will assist both developing countries and countries with economies in transition to promote sustainable development while meeting the objectives of the Convention. The capacity-building provisions for developing countries, adopted in the Marrakesh Accords of 2000 emphasize, inter alia, that there is no 'one size fits all' formula for capacity building. It also emphasizes that capacity building is a continuous, progressive and iterative process and should be integrated, programmatic and aim at maximizing synergies with other conventions.

The problem is that domestic and international scientific and legal knowledge affects the determination of national interests and negotiating positions at international negotiations, as well as the domestic implementation of (inter) national policy. It is assumed that developing countries lack both knowledge and capacity to determine their own interests and to implement internationally negotiated agreements. As a result, capacity building aims to transfer Western 'sustainability' knowledge to developing countries. This is an appropriate response if the problem is that developing countries do not have the knowledge and the developed countries do have the knowledge. However, if we accept the argument that the developed countries, themselves do not have a good idea how they will reduce emissions by more than 60%, then what kind of knowledge will they be transferring to the South? If we accept the argument that what we need is a fundamental overhaul of production and consumption processes in the North to deal with climate change and sustainability issues, then such knowledge needs to be first created before it can be transferred. Furthermore, will it be context relevant and affordable? In the meanwhile, transfer of such knowledge to developing countries might lead to the very same technological and ideological lock-in that we possibly seek to avoid.[28]

2.9   Multiple Messages from Different Regimes

Closely related to the last point is the fact that developing countries are at the receiving end of different messages from the global order. On the one hand, key UN agencies focus on meeting basic needs (e.g. the Millennium Development Goals) and dealing with environmental issues (e.g. the UN environmental treaties). On the other hand, the multiple processes of globalization and the solutions marketed by the World Trade Organization and the World Bank promote the concept that free trade and private sector participation in environmental resource management will address key challenges for all countries. The jury is still out as to whether the competing goals and ideologies of the two processes can be reconciled. The UN Secretary General's Global Compact is an experiment in this direction.

Bhagwati makes the point that globalization is a powerful tool for economic growth and countries have to seek to identify ways and means to unleash its potential.[29] Das goes on to argue that liberalization will ensure that these economies will be released from the constraints that they have imposed themselves on development.[30] At a practical level, the argument is that the private sector has the money to invest and is efficient and, hence, one should invite it to participate in the management of resources.

However, the alternative discourse argues that globalization may empower some at the cost of others and provides story lines and cites global statistics to prove its point. Khor[31] points out that globalization will benefit the top layers of society but the benefits will not trickle down to the poor, and De Rivero[32] submits that the development model and the liberalization theory that it is encapsulated in will export models that are not viable for most countries in the world. At a practical level, the question raised is – the private sector aims at profits, and selling water or energy to the poor is neither a lucrative activity nor an objective with these bodies. So why should they be better at providing water, energy, transport or health services to the poor? This is not to deny that the public sector needs to become much more efficient, but possibly it should not seek efficiency at the cost of service to the people.

2.10　Key Domestic Challenges

Consequently, at the ideological level, most developing countries face the key challenge of trying to understand which ideological commitment will help them achieve sustainability within their own country. At a political level, they face different ideological approaches in different arenas and they have to respond very much in an ad hoc manner, not having settled the philosophical question. At a policy level, they are left with the difficult choices of how to choose between large hydro, fossil fuels and nuclear energy, all three with substantial negative environmental and social impacts, while the renewables are both perceived as intermittent and expensive. At an economic level, they are often not in a bargaining position to be able to withstand being the live experimental fields for the development policies of the developed countries and the Banks (e.g. structural adjustment policies; privatization of water – a policy which is now being reversed). At a practical level, they are often left to cope with highly sophisticated and complex mechanisms designed in the brains of a sophisticated scientist in the West (e.g. CDM and possibly emission trading in the future).

While the definition of the climate change problem is fairly straightforward for the small island states for whom the potential impacts of climate change are likely to be overwhelming, for the larger countries, there is inadequate (domestic) science to show whether the potential local impacts will be devastating for the economy or not. In the meantime, there are pressing current problems that make climate change appear to be an esoteric, abstract issue.

3.   DESIGNING CONSIDERATIONS THAT CONSIDER DEVELOPING COUNTRY ISSUES

3.1   Introduction

The above section identified 10 North-South challenges that ostensibly create problems for developing country participation and implementation of policy measures. If we accept the underlying arguments, even if not in their entirety, but partially, this would still imply that we need to keep some design considerations in mind in developing a regime on climate change. I would argue that four criteria and two behavioural changes need to be explored. The four criteria are the need to focus on multilateralism, redefining the problem and the paradigm to make it viable, simplicity as the key element in broadening and deepening the agenda, and the effective use of issue-linkages to widen the regime. The two behavioural elements that will be discussed below – are retrospection for the North, and proactivity for the South.

3.2   Multilateralism: From 'divide and rule' to 'divided we fall'

The idea that we live in a global anarchy and, hence, we must use our power to divide and rule is being explored currently in a project on Multilateralism in a transatlantic United Nations University project. My own feeling is that the more we assume that we live in an anarchy, the more we create conditions in which anarchy can flourish, and the more we provide opportunities for those who wish to use their power to assert themselves for good or evil.

The more we see the world in Huntingdom's terms, the more will this come to haunt us within our own multicultural societies, a product of colonial relations, international transport and modern day globalization. The more we see the world through Fukuyama's eyes, the more we will close our eyes to environmental and social issues. Since Schumacher's world, however desirable, does not immediately help to solve global problems, we may have no choice but to pursue the (revised) Brundtland dream of trying to solve problems together. Divide and rule, and the rational actor model may just lead to a situation in which 'divided we fall'.

3.3   Problem Definition Revisited

I would submit that when a problem is defined in terms of technology and economics, developed countries will need a continuous source of funding to generate the new technologies and to adapt and transfer them to countries that can ill-afford them. This calls for a continuous source of funding from the State to both push for technology development and support technology transfer. This is an unsustainable position for any State to take, and as such, a leadership paradigm based on this idea is likely to run out of steam as soon as it runs out of money.

I would further submit that even though by dividing the problem into one of emissions and one of adaptation, and by avoiding the discussion on liability, the developed countries avoid taking responsibility for their pollution and paying accordingly, by doing so they also give a licence to the future powers in the South – China, India, South Africa, Brazil – the licence to do the same in the future. How will anyone ever force China to take action? By defining the problem in terms of polluter pays and ability to pay, there is potential to generate funds from all polluters, including the developing countries, and one sends a clear and uncompromising long-term signal to all, based on a well-known principle of international law – the no-harm principle. Having to pay now (for those who can afford it) and later (for those who are likely to be soon able to afford it) puts all producers on notice and provides the necessary economic signals to all in society that they need an ingenious and creative solution.

I am hoping that the recent rise in litigation within the US and outside on climate change will provide the added impetus to push this issue further. Although it is still early days to see how these cases will fare in court, there can be no doubt that these cases will add to the political pressure on the US Government to possibly reconsider its position. As Kilaparti Ramakrishna of the Woods Hole Research Centre puts it: 'The most important impact of this case internationally will be to send another signal to the rest of the world that they do not have to give up on the US as a potential future partner in combating greenhouse gas emissions'.[33] There are at present court cases in Australia as well. These cases may shape the way governments think in the coming decade. The rise in PhDs exploring this issue shows that the legal community is once more becoming active to try and find a solution to these problems.

3.4 Simplicity and Equity in Design for Broadening and Deepening

I would make the plea that as the regime is broadened and deepened, the key criteria for accepting the design should be: is it simple and equitable?

This means that we need to negotiate an ultimate objective in terms of targets and timetables and develop milestones to achieve this target. Some scientists as mentioned earlier are already engaging in this. The argument of mainstream scientists that this cannot be undertaken in a scientifically responsible manner, I see as scientific abdication of responsibility. Scholars from different disciplines need to sit and see if they can find a long-term objective that is consistent with protecting the weakest and most vulnerable rather than the strongest. This should be undertaken irrespective of scientific arguments as to whether it is feasible economically, politically or scientifically, which should be seen as the second step. This will make the proposed system transparent, predictable and effective; and we will be able to measure whether proposed policy steps are in line with achieving these goals.

Simplicity in designing short-term targets should be such that the system should be predictable and treat like countries alike. So countries can already see that if they reach a certain welfare level and emit emissions above a certain level they will be on notice to take on specific responsibilities. This too sends clear and unambiguous signals to all countries. This would immediately result in heavier responsibilities for the rich countries hiding among the poor (e.g. Singapore) and would provide some relief to the poorer countries in the club of the rich (e.g. Turkey). The KISS model (Keep it simple, stupid; see Table 16.1) attempts to do just that. Any country, the weakest and the poorest, can see what its CO_2 emissions are and what its GNP is, since information on these two criteria is readily available and generally not controversial. It can then see which package of limitation and adaptation measures it will be subject to under the system. Countries may choose from a list of criteria to argue why this simple system may work against their interests and this is then decided by an adjudication body that develops an internally consistent set of precedents.

Table 16.1   The packages of responsibilities[34]

	Emissions below minimum per capita	Emissions in the middle category	Emissions above maximum per capita
Very high income per capita	Package 1	Package 2	Package 3
High income per capita	Package 4	Package 5	Package 2
Medium income per capita	Package 6	Package 7	Package 5
Low income per capita	Package 8	Package 9	Package 7

Source: Gupta 2003.

The concept of Best Available Practices needs to be replaced with Regionally Appropriate Practices to ensure that suggestions are tested for their applicability and implementability in the countries in which they are meant to be undertaken.

The literature recommends a number of follow-up Protocols to the UNFCCC, although many explicitly argue that these can be pursued independently of the UNFCCC. These include Protocols Promoting Technology Standards, Research and Development and, among others Protocols for Promoting Cooperation in Specific Sectors. In designing such Protocols, care must be taken to ensure that they are effective, give clear long-term signals, but also do not necessarily prejudice the interests of developing countries.

3.5   Issue-linkages for Widening to Sustainable Development

If we accept the argument that climate change is both closely linked to most sectors of society and that it also to some extent questions our production and consumption patterns, then it becomes imperative to identify the key links between climate change policies and a host of other policies at international and national level.[35] An exploration of issue-linkages yields a whole set of policy options which can be included in Memoranda of Understanding with other UN Agencies and Agreements; or can be tabled in Joint Working Groups. Such arrangements/fora may, in the long-term, allow for discussion between the different ideological dogmas driving different regimes. Policies may be mainstreamed in other UN Agencies or included in specific sections of their policy-plans and legal arrangements. This could help to align the goals of the competing, conflicting or synergetic regimes in a manner that the synergies are highlighted and conflicts minimized.

3.6   Behavioural Change: Retrospection for the North and Proactive Behaviour for the South

In the ultimate analysis, what appears very important is the need for capacity building in the North to understand how the South functions and what works in Southern contexts and what does not. This would allow for a much more open discussion of what is possible to achieve within these countries, rather than mechanisms focused on removing barriers to policies in developing countries. An exploratory and explanatory approach, rather than an instrumental approach, may lead to different results. At the same time, there is a need for retrospection in the North to examine key theoretical frameworks and to explore what ideological perspectives are more consistent with achieving sustainable development.

For the South, I think it becomes increasingly more imperative that it devotes greater attention to foreign policy and international relations. Domestic politics are important, but not to the exclusion of understanding and playing a key role in shaping global politics. This implies taking a much more proactive approach to international decision making. Lack of personnel and facilities can be compensated with creative solutions. There are students who can function as interns and do research for government officials; there are NGOs waiting to serve; there are possibilities for offering sabbaticals to senior researchers within government to help develop policies; the internet offers considerable scope for cheap communication; and global media offers a forum to voice one's opinions and to generate platforms for discussion. The system will only help those who help themselves.

4.   CONCLUSION

This chapter has argued first that climate change – being a long-term problem affecting the global commons and having both immediate and cumulative impacts – must be addressed within the context of a durable global regime. It has further highlighted 10 critical areas of concern from a North-South perspective and these critical areas affect developing countries negatively. These include the way the problem was defined, the changing political paradigms in the past, the deepening transatlantic divide, disempowerment through the practical implications of negotiation, the absence of a long-term target and guiding principles for short-term targets, scientific and policy competition, aid fatigue despite the proliferation of funding mechanisms, the multiple ideological messages at the global level and the key domestic challenges.

If these North-South issues need to be taken into account, then the design of a durable long-term regime must meet some additional criteria; namely it must be based on a strong multilateral framework and all bilateral, sectoral, regional and other initiatives must be undertaken in line with the agreements reached within the multilateral framework. Secondly, in deepening and widening the regime, the goal must be to avoid complexity and develop simple and predictable rules which makes it easy for even the poorest countries to know that their interests will be taken into account. Thirdly, the regime must extensively develop issue-linkages to ensure consistency between different but related policy fields. Fourthly, the issue of liability needs to be reconsidered seriously within the context of the regime. In other words, there needs to be a team that looks at the long-term design aspects every time short-term design elements are proposed.

ACKNOWLEDGEMENTS

This paper has been written in the context of the project on inter-governmental and private environmental regimes and compatibility with good governance, rule of law and sustainable development – financed by the Netherlands Organization for Scientific Research – Contact: 452-02-031.

NOTES

1. Joyeeta Gupta is Professor on Climate Change Law and Policy at the Vrije Universiteit in Amsterdam and Professor on Policy and Law on Water Resources and the Environment at UNESCO-IHE Institute for Water Education in Delft. Her email address is joyeeta.gupta@ivm.falw.vu.nl.
2. Cf. Bodansky, Daniel (ed.) (2003), *Climate Commitments: Assessing the Options. Beyond Kyoto, Advancing the International Effort Against Climate Change*, Arlington: Pew Center on Global Climate Change.
3. Bodansky, Daniel (1993), 'The United Nations Framework Convention on Climate Change: A Commentary', *Yale Journal of International Law*, **18**, 451–588.
4. Gupta, Joyeeta (1998), 'Leadership in the climate regime: Inspiring the commitment of developing countries in the post-Kyoto phase', *Review of European Community and International Environmental Law*, **7**(2), pp. 178–88; Gupta, Joyeeta (2001), *Our Simmering Planet: What to do About Global Warming*, London: Zed Publishers.
5. Gupta, Joyeeta (1997), *The Climate Change Convention and Developing Countries – From Conflict to Consensus?*, Dordrecht: Kluwer Academic Publishers.
6. Cf. Gupta, Joyeeta and Grubb, Michael (eds) (2000), *Climate Change and European Leadership: A Sustainable Role for Europe*, Dordrecht: Kluwer Academic Publishers; Gupta, Joyeeta and Ringius, Lasse (2001), 'The EU's climate leadership: Between ambition and reality', *International Environmental Agreements: Politics, Law and Economics*, **1**(2), 281–99.
7. These include the Declaration of The Hague adopted by the Meeting of Heads of State on 11 March 1989; the Noordwijk Declaration on Climate Change, adopted by environmental ministers in November 1989 in the Netherlands, and the 1990 Ministerial Declaration of the Second World Climate Conference, Geneva.
8. Gupta, Joyeeta 1998, n. 4 above.
9. See, for example, Byrd-Hagel Resolution (1997), Congressional Record: 3 October 1997 (senate) page S10308-S10311; available at http://www.microtech.com.au/daly/hagel.htm.
10. Kagan, Robert (2004), *Of Paradise and Power: America and Europe in the New World Order*, New York: Vintage Books.
11. Eg Wallström, Margot (2003), 'Meeting the long-term, challenge of global warming: A European perspective', in Michel, David (ed.), *Climate Policy for the 21st Century: Meeting the Long-Term Challenge of Global Warming*, Washington DC: Centre for Transatlantic Relations, Johns Hopkins University, pp. vii–ix; and the EUs support for a multilateral framework).
12. Hagel, Chuck (2003), 'Meeting the Long-Term Challenge of Global Warming: An American Perspective', in, Michel, David (ed.), *Climate Policy for the 21st Century: Meeting the Long-Term Challenge of Global Warming*, Washington DC: Centre for Transatlantic Relations, Johns Hopkins University, pp. x–xi; Victor, David (2004), *Climate Change: Debating America's Policy Options*, Council of Foreign Relations, New York: Brookings Institution Press.
13. Cf. Victor 2004, n. 12 above, at p. 6.
14. Wallström 2003, n. 11 above, p. ix.
15. *US Climate Change Policy: The Bush Administration's Actions on Global Climate Change*, Fact Sheet released by the White House, Office of the Press Secretary, Washington DC, 19 November 2004; available at http://www.state.gov/g/oes/rls/fs/2004/38641.
16. Anon (2005), Verbatim, *Time Magazine*, 20 June 2005: p. 13.
17. Gupta 1997, see n. 5 above; Gupta, Joyeeta (2000), 'North-South aspects of the climate change issue: Towards a negotiating theory and strategy for developing countries', *International Journal of Sustainable Development*, **3**(2), 115–35; Chasek, Pamela and Lavanya Rajamani (2003), 'Steps toward enhanced parity: Negotiating capacity and strategies of developing countries', in Inge Kaul, Pedro Conceicão, Katell Le Goulven and Ronald U. Mendoza (eds), *Providing Global Public Goods: Managing Globalization*, Oxford: Oxford University Press, pp. 245–61.
18. Ott, Konrad, Gernot Klepper, Stefan Lingner, Achim Schafer, Juergen Scheffran, Detlef Sprinz (2004), *Reasoning Goals of Climate Protection. Specification of Art. 2 UNFCCC*, Berlin: Umweltbundesamt. p. 27; O'Neill, Brian C. and Oppenheimer, Michael (2002), 'Dangerous

climate impacts and the Kyoto Protocol', *Science* **296**(5575), 1971–72; Oppenheimer, Michael and Petsonk, Annie (2003). 'Global Warming: The intersection of long-term goals and near-term policy, in Michel, David (ed.), *Climate Policy for the 21st Century: Meeting the Long-Term Challenge of Global Warming*, Washington: Center for Transatlantic Relations, Johns Hopkins University, pp. 79–112; Gupta, Joyeeta and van Asselt, Harro (eds) (2004), *Re-evaluation of the Netherlands' Long Term Climate Targets*, Report E-04/07, Amsterdam: Institute for Environmental Studies, Vrije Universiteit; Hare, W. (2003), *Assessment of Knowledge on Impacts of Climate Change – Contribution to the Specification of Art. 2 of the UNFCC*, Berlin: WBGU.

19. It was agreed only in 2003 to include key vulnerabilities (including references to what is dangerous climate change and hence how long-term objectives could be defined) as one of the cross-cutting themes for AR4. Patwardhan, Anand, Schneider, Stephen H., Serguie M. Semenov (2003), 'Assessing the Science to Address UNFCCC Article 2', IPCC Concept Paper available at http://www.ipcc.ch/activity/cct3.pdf, 3 January 2005; IPCC (2003), *Report of the Twentieth Session of the Intergovernmental Panel on Climate Change*, IPCC: Paris, 19–21 February 2003, available at http://www.ipcc.ch/meet/session20/finalreport20.pdf, 3 January 2005.

20. Brooks, Nick; Gash, John; Hulme, Mike; Huntingford, Chris; Kjellen, Bo; Köhler, Jonathan; Starkey, Richard; Warren, Rachel (2005), *Climate stabilisation and 'dangerous' climate change: A review of the relevant issues*, London: Tyndall Centre Working Paper, in press, available at http://www.cru.uea.ac.uk/~e118/publications/stabil-WP.pdf, 3 January 2005; Dessai, Suraje, Adger, Neil, Hulme, Mike, Koehler, Jonathan M., Turnpenny, John, and Warren, Rachel (2003), 'Defining and experiencing dangerous climate change' *Climatic Change*, **64**(1), 11–25; Gupta and van Asselt (eds) (2004), n. 18 above; Hare (2003), n. 18 above; IPCC (2004), IPCC Expert meeting on The Science to Address UNFCCC Article 2 including Key Vulnerabilities, short report available at http://www.ipcc.ch/wg2sr.pdf, 3 January 2005; O'Neill and Oppenheimer (2002) n. 18 above; Oppenheimer and Petsonk (2003); n. 18 above; Ott et al. (2004) n. 18 above.

21. Gupta and van Asselt 2004 (eds), n. 18 above.

22. Parry, M., Arnell, N., McMichael, T. et al. (2001), 'Millions at risk: defining critical climate threats and targets', *Global Environmental Change*, **11**(3), 181–83; Hare (2003), n. 18 above; Baer, Paul and Athanasiou, Tom (2004), Honesty about dangerous climate change, available at http://www.ecoequity.org/ceo/ceo_8_2.htm, 3 January 2005.

23. Gupta (2001), n. 4 above.

24. Meyer, Audrey (2000), *Contraction and Convergence: The solution to Climate Change*, Devon: Green Books Ltd; Groenenberg, Heleen, Phylipsen, Dian and Blok, Kornelis (2001), 'Differentiating the burden worldwide – Global burden differentiation of GHG emissions reductions based on the Triptych approach: a preliminary assessment', *Energy Policy*, **29**, 1007–30; Agarwal, Anil (2001), 'Making the Kyoto Protocol work: Ecological and economic effectiveness and equity in the climate regime', New Delhi: CSE Statement; Berk, Marcel, Gupta, Joyeeta and Jansen, Jaap (2003), 'Comprehensive Approaches to Differentiation of Future Climate Commitments – Some Examples', in Ierland, v. Ekko; Gupta, Joyeeta and Kok, Marcel (eds), *Issues in International Climate Policy: Theory and Policy*, Cheltenham: Edward Elgar, pp. 171–200; Claussen, Eileen and McNeilly, Lisa (1998), *Equity and Global Climate Change, The Complex Elements of Global Fairness*, Arlington: PEW Centre on Global Climate Change; Sijm, Jos, Jansen, Jaap and Torvanger, Asbjorn (2001), 'Differentiation of mitigation commitments: the Multi-Sector Convergence approach', *Climate Policy*, **1**(4), 481–97; Torvanger, Asbjorn and Godal, Odd (1999), *A Survey of Differentiation Methods for National Greenhouse Gas Reduction Targets*, Oslo: Center for International Climate and Environmental Research Report 1999: 5.

25. Bodanksy (2003), n. 2 above; Hoehne, Niklas (2005), *What is Next after the Kyoto Protocol? Assessment of Options for International Climate Policy – Post 2012*, PhD Dissertation, Utrecht: Utrecht University.

26. Banuri, Tariq (1996), 'The South and the Governability of the Planet : A Question of Justice', in Theys, Jaques (ed.), *The Environment in the 21st century: The issues*, vol. I, Paris: Themis, pp. 405–14.

27. Gupta, Joyeeta (2006), 'Regulatory Competition and Developing Countries and the Challenge for Compliance-Push and Pull Measures, in Winter, Gerd (ed.), *Multilateral Governance of Global Environmental Change*, Cambridge University Press, UK, pp. 455–69.
28. For more details on this argument see Gupta, Joyeeta (2004), 'Capacity building and sustainability knowledge in the climate regime', Berlin: Proceedings of the 2002 Berlin Conference on the Human Dimensions of Global Environmental Change, pp. 157–64.
29. Bhagwati, Jagdish (2004), *In Defense of Globalization*, Oxford: Oxford University Press.
30. Das, Gurcharan (2002), *India Unbound, The Social and Economic Revolution From Independence to the Global Information Age*, New York: Anchor Books.
31. Khor, Martin (2001), *Rethinking Globalization: Critical Issues and Policy Choices*, London and New York.: Zed Books.
32. De Rivero, Oswaldo (2001), *The Myth of Development*, London and New York: Zed Books.
33. Interviews with experts in the article entitled 'States File Suit Against major US Utilities', The Climate Group, http://www.theclimategroup.org/index.pjhp?pid=490.
34. Gupta, Joyeeta (2003), 'Engaging Developing Countries in Climate Change: (KISS and make-up!)', in David Michel (ed.), *Climate Policy for the 21st Century: Meeting the Long-term Challenge of Global Warming*, Washington DC: Center for Transatlantic Relations, Johns Hopkins University, pp. 233–64.
35. Asselt, Harro van; Biermann, Frank and Gupta, Joyeeta (2004), 'Institutional Interlinkages of Global Climate Governance', in Kok, Marcel T.J. and de Coninck, Heleen C (eds), *Beyond Climate: Options for Broadening Climate Policy*, Bilthoven: Netherlands Research Programme on Climate Change: Scientific Assessment and Policy Analysis, Report No. 500036 001, pp. 221–46.

Index